Transforming Cities by Controlling Pollution and Traffic Congestion

Md. Shafiqul Islam
https://orcid.org/0000-0002-5547-046X

International University of Business Agriculture and Technology, Bangladesh

IGI Global
Scientific Publishing
Publishing Tomorrow's Research Today

Vice President of Editorial	Melissa Wagner
Director of Acquisitions	Mikaela Felty
Director of Book Development	Jocelynn Hessler
Production Manager	Mike Brehm
Cover Design	Jose Rosado

Published in the United States of America by
IGI Global Scientific Publishing
701 East Chocolate Avenue
Hershey, PA, 17033, USA
Tel: 717-533-8845 | Fax: 717-533-7115
Website: https://www.igi-global.com E-mail: cust@igi-global.com

Library of Congress Cataloging-in-Publication Data

LCCN: 2025027389 (CIP Data Pending)
ISBN13: 9798337344225
Isbn13Softcover: 9798337344232
EISBN13: 9798337344249
British Cataloguing in Publication Data
A Cataloguing in Publication record for this book is available from the British Library.

Editorial Advisory Board

Table of Contents

Detailed Table of Contents

Rapid urbanization, population growth, and motor vehicle use in developing countries have intensified urban issues such as traffic congestion and pollution. These challenges, particularly acute in megacities like Dhaka, reduce quality of life, cause economic losses, and harm the environment. This study highlights structural and systemic failures in urban planning and emphasizes ecological and human-centered solutions. It proposes integrated strategies including pedestrian-priority zones, green infrastructure, alternative transit, and greening efforts like roadside landscaping and green islands to mitigate air and noise pollution. Using both qualitative (literature review, stakeholder interviews) and quantitative (traffic density, air quality, road length, green space per capita) data, the study finds that monocentric urban models worsen traffic, suggesting a shift toward polycentric planning. Ultimately, traffic issues are not just technical but social and environmental, requiring interdisciplinary, sustainable approaches to improve urban life.

Noise and air pollution significantly affect human health, the environment, and reduce the quality of life. In Dhaka city, air and noise pollution rank as the third most severe form of environmental pollution. This study seeks to evaluate the role of plant species in reducing temperature and mitigating air and noise pollution. Data were collected on multiple air and noise parameters in both high-traffic areas during peak and off-peak hours, weekend and free days, using a Class 1 Sound Level Meter and a portable air quality meter. It was recorded an average highest noise level of 87.7 dB(A) was recorded in Bijoy Shoroni, and a minimum of 65.8

dB(A) was recorded in Ramna Park. Air quality (including PM10, PM2.5, CO, CO_2, O_3, and temperature) showed maximum values of 864 μg/m³, 321 μg/m³, 7.0 ppm, 630 ppm, 0.0 μg/m³, and 29°Cat Bijoy Shoroni, while minimum values of 62 μg/m³, 34 μg/m³, 1.0 ppm, 429 ppm, 0.0 μg/m³, and 23.3°C were recorded at Ramna Park. The presence of plant species has shown lower noise levels, lower air pollution, and lower temperatures in Ramna Park. On the other hand, Bijoy Shoroni experienced higher levels of noise and air pollution due to a scarcity of vegetation. Therefore, enhancing urban greenery through plantation and implementing policies that promote eco-friendly transportation are vital steps toward mitigating noise and air pollution in Dhaka city.

R. Kathaian, Vel Tech Rangarajan Dr. Sagunthala R&D Institute of Science and Technology, India
Rajani H. Pillai, Mount Carmel College, India
Roopa Adarsh, Mount Carmel College, India
M. Thenkovan, Guru Nanak College, Chennai, India
P. Kumarasamy, Dhanalakshmi Srinivasan University, India
U. S. Syama, Kristu Jayanti University, India

Cities in India are making a pivotal contribution to economic growth and substantially driving the gross domestic product (GDP). The top cities, namely Delhi, Mumbai, Kolkata, Bangalore, Chennai, Hyderabad, Pune, Ahmedabad, Surat, and Jaipur, contribute to industrial growth, employment generation, and economic development. Subsequently, urbanisation also creates numerous issues and challenges, including environmental pollution and sustainability concerns. This chapter aims to assess the various positive impacts of pollution control practices and policies of the Government of India. The government has taken many efforts in the form of policy to arrest the pollution, such as industrial pollution, vehicle pollution, and traffic congestion. The government of India's policies, such as the National Clean Air Programme, measures for control of vehicular emissions, measures for control of emissions from vehicles. Based on the research results and conclusions, an appropriate suggestion will be given for the sustainable transformation of cities in India.

Air pollution is a global concern as it influences the health of humans and the environment because of increasing industrialization, urbanization, as well as other anthropogenic activities, several air pollutants like carbon monoxide (CO), nitrogen oxides (NOx), particulate matter (PM), sulfur dioxide (SO2), volatile organic compounds (VOCs), etc. are released into the atmosphere. These pollutants are associated with serious health problems such as cancers, premature deaths, neurological disorders, respiratory problems, as well as cardiovascular illnesses. These effects of air pollution are observed in susceptible inhabitants which including children, the elderly, and low-income groups. Sustainable air quality administration is needed to mitigate these influences and preserve public health. This chapter highlights the outline, sources, as well as impact of air pollution on the health of humans. Different air pollutants, sustainable management of air quality, recommendations, and future directions, as well as innovations, are also focused on in this chapter.

Preface

The city is the living place where one can find almost all facilities to make life easier than rural life. With the pass of time and modernization, the city is becoming heart of development. Urbanization denotes the movement of populations from rural areas to urban centers. In 1950, merely 30 percent global population resided in urban areas, the percent that increased to 55 percent by 2018. It is predicted that by the end of 2050, 80 percent people will inhabit urban areas. People are moving or migrating to the urban and peri-urban areas from remote rural areas for better income and opportunities. As a result, rapid urbanization is giving birth to urban pollution (air, water, soil, noise and light), huge traffic congestion, and lack of accommodation (shelter) and non-availability of green space in the cities.

This book explores the transformation process of the polluted cities to green cities highlighting the present situation of modern cities with the problems, remedial measures and future direction for transition, planning, development, engineering control and simple initiatives.

The city is composed of human life, infrastructures (buildings, roads, culverts, drainage), transportation system, green space (parks, play ground, lakes), facilities (electricity, water connectivity, internet), animals including insects, birds, and microorganisms are closely linked.

Destruction of urban environment, air pollution, water pollution, noise pollution and plastic pollution are common issues everywhere in the mega city. The sadistic cycle of urbanization and tertiary luxurious demand is dominating our society, making people toxic in their minds, culture, practices, behaviors and consumption patterns. Green cities and practices are new challenges for urban residents that make a critical challenge to urban ecosystems as recognized as a global priority.

The creation and dissemination of green city concept is important for students, educators, practitioners and researchers. Textbooks give practical resources for gaining new knowledge with simplified information. These are essential teaching supports within any educational framework. Active learning, united with liberal education, not only simplifies degree accomplishment but also prepares individuals with up-to-date, world-class knowledge. The global population is confronted with the associated risks of rapid urbanization, traffic congestion, air pollution and the environment. It needs pace to make a city green and livable. Diverse greening options can reduce pollution (air, water, and soil). This means that there is an urgent need to transform the cities to green by managing green space, play ground, park, lake, pedestrian ways and urban transport sustainably. It is aligned with the huge demand of a growing population, especially to fulfill the crying need for facilities and other necessities.

Innovative research and dynamic research on green cities are important in the world. This book builds the capacity of the global community to face rapid urbanization related challenges in the real world. The book contains the following objectives: i) to explore the objectives and strategies of green city; ii) to connect the cultural, social, environmental, and economic dimensions of urbanization; iii) to identify the problems and challenges (related to pollution, traffic and green space) faced by the urban community, along with the available options for sustainable city and urban development; and iv) to guide the students, researchers, and practitioners to ensure collective efforts for innovations and solutions.

This book organized nine chapters with the innovative knowledge and practices. The objective is to present sustainable knowledge for students, serving as a textbook for both undergraduate and graduate programs in Urban Ecosystems, Urban Forestry, Urban Planning, Landscape restoration, Sustainable Transportation Management, Climate Change, Urban Agriculture, Environmental Science, Environment and Development Studies, and Sustainability across various universities.

Urbanization has transformed quickly the way people live, often creating to environmental challenges including pollution, reduced green space, and increasing global temperatures. **Chapter One** explores sustainable response to these issues, emphasizing ecological balance, energy efficiency, and enhancing quality of urban life. It discusses the ideas behind Green Cities and why they are critical to the future urban development. It explores the ways to integrating green infrastructure, renewable energy, and eco friendly design to make the cities resilient, livable and environmental sustainability.

Chapter Two discusses how noise and air pollution affect human health, the environment, and reduces the quality of life especially in Dhaka city. The study shows the situation of Dhaka city considering air and noise pollution and environmental pollution. This study seeks to evaluate the role of plant species in reducing temperature and mitigating air and noise pollution. It shows that the average highest noise level in Bijoy Shoroni and a minimum in Ramna Park. Air quality (including PM10, PM2.5, CO, CO2, O3, and temperature) shows the worst situation at Bijoy Shoroni. The presence of plant species has shown lower noise levels, low air pollution and temperature in Ramna Park.

Chapter three explores the leading cities, such as Delhi, Mumbai, Kolkata, Bangalore, Chennai, Hyderabad, Pune, Ahmedabad, Surat, and Jaipur, that play a significant role in fostering industrial growth, generating employment, and promoting economic development. The research identifies various challenges, including issues related to environmental pollution and sustainability. Its objective is to evaluate the diverse effects of pollution control measures and the policies implemented by the Government of India. The chapter investigates governmental initiatives aimed at mitigating pollution, addressing concerns such as industrial emissions, vehicular pollution, and traffic congestion.

Air pollution represents a worldwide issue that affects both human health and the environment, driven by escalating industrialization, urban expansion, and other human activities. **This chapter (Four)** details various air pollutants, including carbon monoxide (CO), nitrogen oxides (NOx), particulate matter (PM), sulfur dioxide (SO2), and volatile organic compounds (VOCs), which are emitted into the atmosphere. The repercussions of these pollutants are linked to severe health issues, including cancers, premature mortality, neurological disorders, and respiratory ailments, as well as cardiovascular diseases. This chapter delves into the origins and effects of air pollution on human health.

Ecosystems, human health, and long-term sustainability are all linked with groundwater. **Chapter five** discusses the opportunity of ML and AI that offer promising new approaches to manage critical problems. This study uses AI and ML to make decision-support frameworks for policymaking and resource allocation, predictive modeling for the spread of contaminants, and real-time monitoring of water quality. This chapter delves into the topic of how to enhance data collection and get valuable insights by combining AI-driven solutions with IoT devices. It explores how AI and ML have the ability to revolutionize methods for controlling groundwater pollution and fostering sustainable management of water resources.

This paper explores **(Chapter six)** drainage problems surrounding the cities, from past inequalities to present planning gaps, and how impact the communities. Based on evidence from different cities such as Delhi, Mumbai, and Bengaluru to compare it with similar problems in Lagos, Nigeria the research demonstrates how open drains spread disease, cause flooding, and lead to environmental degradation. Sponge parking and permeable pavements facilitate better deal with water sustainably. Inadequate coordination across departments and financial constraints remain as obstacles. Planning, innovation, community participation, and efficient resources utilization, can ensure safer, cleaner, and stronger urban areas in India.

Chapter seven discusses the urban mobility can be changed with adaptive routing and intelligent signal systems (ISS), as it is an efficient way to improve traffic flow and emissions and optimize the journeys of people commuting within smart cities. The study explores technologies, including AI, IoT, GIS, and V2X communication to deliver real-time services and data-based traffic control. The chapter analyzes the global trends of implementation, assesses the constraints of legacy systems, and explores the issue of socio-spatial equity, cybersecurity threats, and forms of governance. It combines interdisciplinary strategies such as sociotechnical systems, multi-level governance, and innovation diffusion to postulate a strategic model of inclusive efficiency and futuristic urban mobility.

Chapter eight examines the evolution of urban traffic management within the framework of Transforming Cities by Controlling Pollution and Traffic Congestion, showcasing how AI-driven signal control and alternative routing strategies can reshape city mobility. It opens with a review of adaptive signal systems that harness real-time IoT and camera data to optimize phase timings and coordinate "green waves," then delves into eco-routing algorithms that factor emissions, road grades, and congestion for cleaner path selection. It emphasizes the multimodal integration supporting pedestrians, cyclists, and shared micro-mobility and the use of digital twins and simulation platforms for policy testing and risk reduction.

Chapter nine discovers youth perspectives on traffic governance, focusing satisfaction levels and leveraging both regression and ML applications for predictive insights. It discloses the gap between enforcement mechanisms and public perception; especially the younger road users and their behaviors significantly impact road safety outcomes. The chapter identifies that safety risk perceptions including dangers of speeding and mobile phone use during driving are strong positive predictors, while operational governance factors like prompt violation response, road maintenance, and clear signage enhance satisfaction. AI-based monitoring system showed a negative association with the technology use.

Finally the book provides conclusion, recommendations and implication process of solutions transforming the cities to green and sustainable urban development.

It considers that the book surely contribute in understanding and advancement of research in transforming cities to green, practices and managing healthy urban life and environment. Welcome suggestions from readers to improve the content and structure of this book in future endeavor.

Md. Shafiqul Islam
International University of Business Agriculture and Technology

Introduction

City is becoming the first home of the people, for the nation and the first generation with modernity. At the earlier time, it was mentioned that more than eighty percent of people lived in the rural areas. Mass people, small cities, huge traffic, pollution altogether along with ineffective urban planning are making the city worse in the world especially in the developing countries. This book is distinctive in its nature, addressing various dimensions of environmental pollution, traffic congestion, and the necessity for urban planning through alternative routing systems.

The aim of the book is to explore effective urban development strategies to reduce car dependency, minimize expensive land occupancy and easy accessibility and mobility through eco-friendly and public transportation. Other intermediate objectives are to: i) find out higher occupancy modes of public travel; ii) a reduction of traffic congestion; iii) diverting traffic that arises from livelihood and demographic changes; iv) ensure better traffic management generated by land use changes; v) find more efficient land use arrangements; vi) explore how best to use eco-friendly transport modes; vii) develop a low cost technology to the movement of people and goods; viii) improve safety and maximizing benefits to the users.

Major three parts of the book have been rationalized to cover different dimensions including pollution control, traffic congestion and routing system, and pinning next to prediction and youth engagement transforming the city. The book provides details on pollution control through electrical vehicles (EV), rooftop gardening is the means of pollution control, open sewers and veiled menace, transfoming urban environment, reshaping the city, plant contribution in pollution control, policy governance and practices and public health concern of air pollution.

The focus of this book is on the pressing issues faced by urban populations in developing nations, particularly in the context of Dhaka, Bangladesh. It offers sustainable strategies such as avenue planting, integrated green pedestrian planning, alternative routing, and eco-friendly mass transport systems to control traffic congestion. The daily struggles of individuals due to traffic jams and inadequate traffic signal systems are highlighted, along with the significant costs associated with these issues, including lost productivity, emissions, and fuel consumption, which collectively impose a substantial financial burden on urban residents.

The book seeks answers to the so-called orthodox sustainable transport vision, by considering immense issues; growth, environmental pollutions and spheres of social justice. On the basis of past and present cases from developing countries with a strong commuting culture and a traditional tendency towards multimodality, the book shows sustainability discourse into policy actions. On the one hand, it discusses a number of factors where additional travel is often frequent and unintentionally stimulated, with an unforeseen contribution to climate change effects. Further, these efforts refer to cases where subsidies for sustainable transport modes seem to be geared towards wealthier groups, while poorer groups are exposed to additional expenses. It is essential to integrate into the policy process particularly the problem identification, potential solutions, and the design of the chosen plan or programme, and illustrate the mechanisms underlying goal-displacements. The urban form of many global cities is not adapted to current demands of travel, and therefore very much saturated, urging for redevelopment within the next decades. Aiming the objectives, stakeholders (the inhabitants, students, professors, policy makers who live and work in Brussels Metropolitan areas) consultation is important to develop new knowledge and solutions for the urban transportation planning process. The book attempts to integrate land use and transport development with the view to build new towns that would provide public housing for an increasing population, and physical transformation of the urban area from a monocentric to a polycentric spatial structure as to decentralize commercial and employment activities from the core city. The expected outcome of the book is to make a linkage with the role of polycentric urban development on car dependency and transport-land use interaction and development of a polycentric urban model, though the useful concentric urban development as the innovative ways to urban transportation planning and land use to explain growth and negative consequences of monocentric urban development. Urban dynamics is the main focus in polycentric urban development. Numerous factors influence urban dynamics, while transportation is one of the high significance factors. Urban transport investments are expected to improve the mobility of passengers and freight, which in turn improve the economic performance of a country. The book is also concentrate on the importance of polycentric urban development in reducing car dependency, land use interaction through several changing policies (very slow changes, slow changes, fast changes and very fast changes) to overcome congestions by rerouting vehicles.

As for example from developed countries, Brussels has been called the 'European traffic jam capital' (Hubert et al., 2013). Thus the statement depicts the scenario of mobility in Brussels. Brussels has taken rank among the most congested cities in the world (Inrix, 2014; Tomtom, 2012 cited in Lebeau and Macharis, 2014). Mobility in Brussels is also delayed due to traffic congestion. Traffic is the biggest problem in Brussels (Lebeau and Macharis, 2014). Mobility is synonomus with travel or movement in physical space. Spatial or geographic mobility is broken down into different temporal scales (daily, weekly, monthly, seasonal or and over life time). Hubert et al., (2013) mentioned that Brussels faces three major challenges which includes population growth, expansion of activities and accessibility of the city. The current urban transportation system is not efficient in accommodating the increased population, industries and monocentric urban development. Long back ago in 1925, Burges developed an urban land-use model that subdivided cities into several concentric circles to relocate from downtown to the suburbs.This model provided empirical evidence and a relationship between the income of households and the distance. This monocentric model described six concentric zones as the process ofexpansion and reconversion of land use. Urban transportation planning is an essential need for urban development and planning processes. Efficient decisions on how to build urban facilities are the basic elements of travel forecasting. It should be the key notion to identify the causes of car dependency, inaccessibility, less mobility, lengthy trips, traffic congestion, urban air pollution and inappropriate land use pattern in the urban. Eventually urban land use classifications should be reviewed considering population density, residence, commerce, industry, institution, recreation and urban agriculture.The emergence of polycentricity originally aims at resolving a series of urban problems arousing from monocentric development, such as uneconomical energy use, long-distance commutes, traffic congestion, and environmental deterioration (Wang et al., 2019). The more neighborhood density and diverse land use enhance more activities (i.e. walking and cycling), non-work trips by inspiring urban residents for shopping and recreations (Naess et al., 2019 cited in Hou, 2019). It was reported that the transformation from monocentric urban to a polycentric form makes travel more efficient and trips shorter (Gordon et al., 1989).The low socioeconomic level, the lack of room in dwellings in Brussels and cultural obstacles, explain the low rate of bicycle owner-ship in Brussels: only 39.9% of Brussels households have one compared to 74.8% of households on the outskirts. And the situation is not progressing. Furthermore, three times more households in Brussels (29.7%) than households on the outskirts (8.5%) claim that they do not have enough room in their home for a bicycle (Hubert et al., 2013). The measurement of morphological polycentricity is based on various datasets such as remote sensing image, land use maps, and a population census dataset (Cai et al., 2017). The decrease not only in the use but also in the ownership of cars is a crucial challenge in the development of public spaces. According to Hubert et. al., (2013), the estimated 360,000 to 380,000 cars (2010) used by the inhabitants of Brussels, parked (mainly at home) for approximately 95% of their lifetime, represent a true burden for the city and its public spaces, in a context of population growth. This takes place at a time when the decreasing size of dwellings and the demand for a better quality of life in the city lead to the emergence of a need for car-free, high-quality and multifunctional public spaces.Boussauw et al. (2010) examined the relationship between commuting trip lengths and number of spatial characteristics. Boussauw and Vanoutrive (2017) reported that sustainable development schemes are being required at three stages; i) defining the problems and selection of indicators; ii) solution preferences (e.g building of rail line); and iii) bureaucratic system (execution, design of chosen plan). Mismatch observed between orthodox sustainable transport discourse and aspects of limits to growth and social justice in the transport policy and practices (Boussauw and Vanoutrive, 2017). It was mentioned the best practices and good design of a research is essential in urban transportation planning (Boussauw and Vanin, 2018). Several strategies and sustainable urban land use & transport policies are required for polycentric urban development. The book looks the followings:

- How can strategies and policies contribute to polycentric urban development and more sustainable land-use travel interaction in the city areas?
- To what extent can car dependency be minimized by utilizing urban pollution and routing efficiently in relation to travel interaction?
- Could the polycentric urban development concept reduce traffic congestion in city's transportat system?

The book aims to explore effective polycentric urban development strategies to reduce car dependency, minimize expensive land occupancy and easy accesibility and mobility through eco-friendly and public transportation. Other intermediate objectives are to: i) find out higher occupancy modes of public travel; ii) a reduction of traffic congestion; iii) diverting traffic that arises from livelihood and demographic changes; iv) ensure better traffic management generated by land use changes; v) find more efficient land use arrangements; vi) explore how best to use eco-friendly transport modes; vii) develop a low cost technology to the movement of people and goods; viii) improve safety and maximizing benefits to the users. The book creates room for new opportunities for travel time, delays in mobility and search for alternatives for car dependency. It also looks for eco-friendly transport, new technology for pollution management, road zoning and minimum uses of valueable land. The book links social and economic drivers. Transportation and land use have a reciprocal relationship between transport supply and demand, a relationship that can be expressed by the mobility of passengers and freight. The maximization of mobility of people and goods can add value of fresh air for breathing. Acessibility and mobility is possible through minimizing travel time, costs and effective land use planning and polycentric urban development concepts. The most significant relationship between land use include commuting, professional movements, personal movements and concern of distribution. The interactions between transportation and land use are a complex phenomenon that includes economic, political, demographic and technological changes. Splitting of working hours may reduce trafic congestion but it depends on human behavior and social acceptance. Many people use shorter route that also cause traffic congestion. Road zoning is the key concern to make sure the spatial quantification of land use and demography of the urban people. The changes in urban settlement created more motorized traffic. Thus it encourages utmost adverse effects including i) more energy uses and urban air pollution; ii) negative impacts on environment and society; iii) reduce accessibility; iv) construction of infrastuctures have negative effects rather than solutions; v) limited mobility considering mode of transport. Shifting from car dependency to other modes is the key solution to mitigate the negative effects of transportation. Going through all consequences, human travel interaction is essentially important. The interconnected urban transport planning is required to create effective urban transportation focusing on polycentric urban development. The aspect of social and economy should be considered including housing patterns, culture, behvior and income status of the urban residents. Integrated planning approaches should be considered from land use and availability of public transport services. Furthermore, urban area should be divided into several travel zones-pedestrian for the people, public transport and car zones. The zoning provides the key benefit and explanation on the impacts of transportation on land use planning, accessibility, mobility, society, environment and urban development process.

To synthesis, the book deployed Remote sensing data, geographic information systems, and landscape ecology in the morphological study of urban development. It embraces several phases including assessment and analysis, design and planning phases and regression and relationships development with the explanatory variables. The explanatory variables are i) accessibility; ii) residential density; iii) land use diversity; iv) work density; v) minimum commuting distance; vi) freight flaws (in tonnage)per mode of transport; vi) land value; vii) proximity of facilities; viii) population density; ix) modes of transport; x) fuel consumption/km; xi) mobility of passengers and freight. Land economy influences the nature, pattern, distribution and land use for polycentric urban development. The assessment phase includes i) the selection of geographical locations of the study area; ii) existing road connections and facilities; iii) population density; iv) existing transport options (modes of transport, number of public buses, cars, and mototorized and non motorized vehicles, travel time, average travel time); v) road neworks (number and length, internal vs. external); vi) railway (number, line and length); vii) water way (number, river/channel, length); viii) airways connections (number of flights, landing, departure); ix) land occupied by the road networks (pedestrian, road, rail, water, air, structures, housing, green space, agriculture etc); x) travel costs (trips, per person, kilometers);

xi) fuel types, consumption and costs (car, public transport, automovile, rail, water transport etc); xii) traffic congestion (man hour, emission, fuel consumption, delayed hour); xiii) pollution from transport (air, water); xiv) existing pollution mitigation measures (soil, air, water, noise). The analysis has been done using appropriate software to get information and output on demography of the population (density, resident types, income, occupation, gender, age, occupation, household size and education), existing transport facilities . Pollution hazards (air quality, water quality, noise pollution); availability of of green space and land (road length, pedestrian road, infrastructures, other road network, % of green space, occupancy of green space, plant types, aesthetic view, parks, lakes and residents satisfaction level etc.); planning or solutions for polycentric urban development (public transport, no car day, modal shift, time based tax policy, eco friendly transport mainly electric or natural gas or ethanol fuelled, improved mobility, accessibility, road zonings, pollution mitigation measures, choice of plant species for green space).

The book also looks at opportunities of renewable energy (solar, bio-gas or bio-diesel etc.) to explore the usability in transport planning. Land use travel associated with density, design, diversity and distance transit and destination accessibility. It provides focus on the reality of sustainable transport, by considering the two challenging issues of growth, and social justice. A paradigm shift is therefore possible, which consists in considering ourselves to be not only pedestrians, but 'enhanced' pedestrians due to the use of public transport, bicycles, taxis or cars. Considering all the issues, primary data on mobility, socioeconomics, population changes, land uses, modes of transport, accessibility, freight, fuel type and consumptions have been collected using several techniques (survey, satellite data, remote sensing, focus group discussions and key informant's interviews). Secondary data have been generated from project reports, journal articles and from reliable data sources. Analysis has been done through developing relationships with the dependent variables and explanatory variables for this book. OLS models has been used to determine the level of significance of numerous factors that influence polycentric urban development and land use system of the city. The outcomes of the book are: i) relationships with commuting (home to work), professional movements (work to work), personal movements (shopping, social interactions and recreation), distribution (procurement, commercial deliveries, home deliveries) and relationships of core activities, central activities and peripheral activities; ii) build relationships with land zoning and transportation planning (functional zoning, form based zoning, intensity zoning and incentive zoning; iii) formulate different policies for urban transportation, land use planning and tax incentive policy.

Md. Shafiqul Islam

International University of Business Agriculture and Technology, Bangladesh

REFERENCES

Boussauw, K., & Vanoutrive, T. (2017). Transport policy in Belgium: Translating sustainability discourses into unsustainable outcomes. *Transport Policy*, *53*, 11–19. DOI: 10.1016/j.tranpol.2016.08.009

Cai, J., Huang, B., & Song, Y. (2017). Using multi-source geospatial big data to identify the structure of polycentric cities. *Remote Sensing of Environment*, *202*, 210–221.

Gordon, P., Kumar, A., & Richardson, H. W. (1989). The influence of metropolitan spatial structure on commuting time. *Journal of Urban Economics*, *26*(2), 138–151. DOI: 10.1016/0094-1190(89)90013-2

Hou, Y. (2019). Polycentric urban form and non-work travel in Singapore: A focus on seniors. *Transportation Research Part D, Transport and Environment*, *73*, 245–275. DOI: 10.1016/j.trd.2019.07.003

Hubert, M., Lebrun, K., Huynen, P., & Dobruszkes, F. (2013). BSI synopsis. Daily mobility in Brussels: Challenges, tools and priority undertakings. *Brussels Studies*, *71*, 1–27.

Lebeau, P., & Macharis, C. (2014). Freight transport in Brussels and its impact on road traffic?. Brussels Studies. La revue scientifique électronique pour les recherches sur Bruxelles/Het elektronisch wetenschappelijk tijdschrift voor onderzoek over Brussel/The e-journal for academic research on Brussels.

Næss, P., Strand, A., Wolday, F., & Stefansdottir, H. (2019). Residential location, commuting and non-work travel in two urban areas of different size and with different center structures. *Progress in Planning*, *128*, 1–36. DOI: 10.1016/j.progress.2017.10.002

Wang, M., Derudder, B., & Liu, X. (2019). Polycentric urban development and economic productivity in China: A multiscalar analysis. Environment and Planning A. *Economy and Space*, *51*(8), 1622–1643.

Chapter 1
Urban Transformation:
Addressing Pollution and Traffic Congestion to Reshape Cities

Mücella Ates
http://orcid.org/0000-0003-1449-0605
Necmettin Erbakan University, Turkey

ABSTRACT

Rapid urbanization, population growth, and motor vehicle use in developing countries have intensified urban issues such as traffic congestion and pollution. These challenges, particularly acute in megacities like Dhaka, reduce quality of life, cause economic losses, and harm the environment. This study highlights structural and systemic failures in urban planning and emphasizes ecological and human-centered solutions. It proposes integrated strategies including pedestrian-priority zones, green infrastructure, alternative transit, and greening efforts like roadside landscaping and green islands to mitigate air and noise pollution. Using both qualitative (literature review, stakeholder interviews) and quantitative (traffic density, air quality, road length, green space per capita) data, the study finds that monocentric urban models worsen traffic, suggesting a shift toward polycentric planning. Ultimately, traffic issues are not just technical but social and environmental, requiring interdisciplinary, sustainable approaches to improve urban life.

INTRODUCTION

Rapid urbanization, population growth, and motorization have become defining trends in developing countries during the 21st century. Cities are expanding rapidly, absorbing more people than ever before, especially in Asia and Africa, where urban populations continue to increase exponentially (World Bank, 2020). This accelerated

DOI: 10.4018/979-8-3373-4422-5.ch001

urban growth has brought numerous benefits, such as economic opportunities and social development, but it has also intensified critical urban challenges, notably traffic congestion and environmental pollution (Newman & Kenworthy, 2015). These issues are particularly severe in large metropolitan areas of developing countries, where infrastructure and governance mechanisms often struggle to keep pace with growth (Satterthwaite, 2016).

Traffic congestion causes significant economic losses, including wasted time, fuel inefficiency, and increased operational costs for businesses and commuters (Downs, 2004). In addition, environmental pollution-air, noise, and water pollution-poses severe public health risks and degrades urban ecosystems (World Health Organization, 2017). According to the World Health Organization (2021), air pollution is among the leading environmental risk factors for premature mortality worldwide, disproportionately affecting residents in densely populated urban areas with poor air quality management. Moreover, environmental degradation undermines the sustainability and livability of cities, limiting their long-term resilience in the face of climate change (Dodman et al., 2022).

Dhaka, the capital city of Bangladesh, exemplifies the challenges faced by rapidly growing urban centers in developing countries. With over 21 million inhabitants, Dhaka is among the world's fastest-growing megacities, grappling with severe traffic congestion and critical levels of pollution (Ahsan et al., 2016). The city's urban infrastructure, designed for far smaller populations, is overburdened, leading to economic inefficiencies, heightened accident risks, and deteriorating living conditions (Hossain et al., 2020). As such, Dhaka presents a critical case for exploring how urban transformation efforts can address these intertwined challenges.

Despite considerable efforts by governments and international organizations to mitigate urban traffic and pollution problems, many cities in developing countries continue to experience worsening conditions. The structural and systemic failures in urban planning, such as reliance on monocentric city models, insufficient green infrastructure, and fragmented transport networks, contribute to persistent congestion and pollution (Dodson & Sipe, 2008). The monocentric urban form-in which a single central business district attracts the majority of economic and social activities-results in lengthy, high-volume commuter flows that overload transportation networks and increase emissions (Batty, 2013; Gössling et al., 2016).

In Dhaka and similar cities, these challenges are compounded by rapid population influx, informal settlements, limited public transport options, and weak governance structures (Karim & Hasan, 2013). The lack of integrated, sustainable urban planning approaches has led to inefficient land use, poor access to green spaces, and increased environmental and social inequalities (Islam & Hasan, 2021). Moreover, residents' daily routines, mobility preferences, and environmental awareness are often overlooked in policy-making, which hinders the adoption of sustainable transport and ecological initiatives (Zhao et al., 2018). Consequently, the problem extends beyond transportation capacity; it encompasses the need to redesign urban life around human and environmental needs.

The existing literature on urban traffic congestion and pollution includes numerous studies focused on engineering solutions, such as road widening, traffic signal optimization, and emission control technologies (Litman, 2021; Papageorgiou et al., 2003). However, these approaches alone have proved insufficient in addressing the root causes of urban transport and environmental issues, especially in the context of developing countries (Gwilliam, 2011). Research exploring holistic and integrated urban transformation strategies remains limited, particularly those that combine urban design, social behavior, environmental sustainability, and governance.

In response to these limitations, greater analytical emphasis must be placed on governance structures, public participation mechanisms, and socioeconomic inequalities that shape urban mobility outcomes. In many developing cities, including Dhaka, fragmented institutional arrangements, weak regulatory enforcement, and limited inter-agency coordination significantly constrain the effectiveness of technical transport interventions. Moreover, unequal access to transport services disproportionately affects low-income groups, informal workers, women, and peripheral communities, reinforcing broader patterns of social exclusion. Public participation remains underutilized in transport and environmental planning, despite growing evidence that inclusive decision-making enhances policy legitimacy and long-term sustainability. Beyond identifying solutions, clearer prioritization of strategies is also required, distinguishing between short-term, high-impact interventions and longer-term structural reforms. Such prioritization should be guided by socioeconomic vulnerability, environmental risk exposure, and institutional feasibility. By foregrounding governance capacity, participatory planning, and equity considerations, the chapter can move beyond problem description toward a more actionable and socially grounded framework for urban transformation.

Studies on ecological urbanism and sustainable transport increasingly highlight the need for green infrastructure, pedestrian-friendly environments, and polycentric urban forms as means to alleviate congestion and pollution (Beatley, 2011; Cervero, 2013). Yet, empirical investigations focusing on cities like Dhaka, which face unique socio-economic and infrastructural constraints, are sparse (Rahman & Hossain, 2020).

Furthermore, there is a lack of research that integrates qualitative insights-such as stakeholder perspectives and behavioral patterns-with quantitative metrics like air quality indices, traffic density, and green space distribution to develop context-sensitive and actionable urban transformation frameworks (Alves & Ferreira, 2021). This study addresses this gap by providing a comprehensive, interdisciplinary analysis of traffic and pollution problems within the urban transformation paradigm, emphasizing human-centered ecological planning, social equity, and environmental health.

This chapter focuses on urban transformation strategies aimed at reducing traffic congestion and environmental pollution in rapidly urbanizing developing countries, with Dhaka as a focal case study

The research is guided by the following questions:

1. What are the key structural and systemic factors contributing to traffic congestion and environmental pollution in rapidly growing cities like Dhaka?
2. How can polycentric urban models and integrated land use planning reduce transportation demand and pollution in densely populated urban areas?
3. What role do green infrastructure and ecological public transport systems play in mitigating air and noise pollution?
4. How can urban design and social behavior be harmonized to foster sustainable mobility and environmental awareness among urban residents?
5. What policy and planning recommendations can support sustainable urban transformation in developing countries?

Research Aim

The primary aim of this study is to explore and propose sustainable and practical solutions to the intertwined problems of traffic congestion and environmental pollution in rapidly growing urban centers of developing countries. By focusing on Dhaka, the study seeks to develop an integrated urban transformation framework that addresses economic, social, and environmental dimensions, improving both urban functionality and residents' quality of life.

Specifically, the study aims to:

- Identify the structural and systemic urban planning failures leading to congestion and pollution.
- Advocate for the adoption of polycentric urban development and balanced land use to reduce commuting distances and transportation demand.
- Highlight the importance of green infrastructure, pedestrian-friendly environments, and ecological transport systems in mitigating pollution.
- Incorporate qualitative insights and quantitative data to develop contextually relevant urban transformation strategies.
- Provide actionable policy recommendations to guide future urban planning and research efforts in developing countries.

Significance and Contributions of the Study

This chapter makes several important contributions to the literature and practice of urban transformation in developing countries.

Holistic Approach Integrating Multiple Disciplines

Unlike many studies focusing narrowly on engineering or environmental interventions, this research integrates urban design, social behavior, environmental science, and policy analysis to provide a comprehensive understanding of urban traffic and pollution challenges. It emphasizes that technical solutions must be complemented by ecological and human-centered approaches for lasting impact (Newman & Kenworthy, 2015; Beatley, 2011).

Empirical Focus on a Rapidly Growing Megacity in a Developing Country

The detailed case study of Dhaka offers critical empirical evidence on how rapid urban growth in developing contexts exacerbates traffic and pollution problems, compounded by governance and infrastructural weaknesses (Hossain et al., 2020). By grounding theoretical discussions in this context, the study addresses a gap where most research is skewed toward developed world examples (Cervero, 2013).

Integration of Qualitative and Quantitative Data

The research uses a mixed-methods approach, combining literature review, stakeholder interviews, field observations, and analysis of urban indicators such as air quality, road networks, and green space distribution. This methodological pluralism allows for a nuanced understanding of complex urban phenomena and supports the development of evidence-based and practical recommendations (Alves & Ferreira, 2021).

Advocacy for Polycentric Urbanism and Balanced Land Use

The study's promotion of polycentric urban forms and land use integration responds to global critiques of monocentric city planning, which exacerbate congestion and pollution (Batty, 2013). By encouraging spatial restructuring and decentralization of urban functions, it aligns with emerging urban resilience and sustainability paradigms (Dodman et al., 2022).

Emphasis on Community Participation and Social Equity

Recognizing the role of citizens' daily routines, mobility preferences, and environmental awareness, the study underscores the importance of social equity and community engagement in shaping sustainable urban futures. This approach challenges technocratic planning models and advocates for inclusive, participatory urban governance (Satterthwaite, 2016; Zhao et al., 2018).

Practical Policy Guidance and Future Research Directions

Finally, the chapter offers actionable policy recommendations tailored to the realities of developing countries and provides a foundation for further academic inquiry into integrated urban transformation strategies.

BACKGROUND

In developing nations, urbanization and fast population growth have made problems with traffic jams, environmental damage, and deteriorating urban quality of life worse. According to international assessments, unplanned urban growth and inadequate transportation infrastructure greatly raise emissions, travel time losses, and environmental risk exposure (World Bank, 2020; UN-Habitat, 2022).

The economic feasibility of these urban interventions is increasingly scrutinized, as high-density developments require significant upfront capital but offer long-term reductions in per-capita infrastructure maintenance costs (OECD, 2020). These difficulties are now more widely recognized as the results of deeper structural problems with urban spatial form, land-use distribution, and governance capacity rather than just being the result of growing car ownership.

Specifically, governance challenges such as institutional fragmentation and lack of inter-agency coordination are identified as primary barriers to the successful implementation of integrated land-use policies (UN-Habitat, 2022).

Contemporary urban theory suggests that the "15-minute city" model- where essential services are accessible within a short walk or cycle- is a critical framework for reducing vehicle kilometers traveled (VKT) and enhancing urban livability. While originally popularized in Europe, global case studies such as Bogotá's "Barrios Vitales" demonstrate how decentralized, mixed-use planning can be adapted to high-density Global South contexts to reclaim street space for pedestrians (Moreno et al., 2021).

The role of smart city technologies is becoming central to this transition, where IoT-based traffic management and real-time data analytics are utilized to optimize the existing street capacity and reduce congestion without physical expansion.

Furthermore, recent meta-analyses confirm that Transit-Oriented Development (TOD) strategies, when integrated with high-density residential zones, can reduce household carbon footprints significantly compared to sprawled developments (Stevens, 2023).

Numerous studies show that one of the main factors influencing mobility patterns and environmental performance is the interplay between land use and transportation systems. According to basic empirical research, compact, mixed-use, and polycentric urban structures improve accessibility while lowering vehicle dependence and vehicle kilometers traveled (Cervero & Kockelman, 1997; Ewing & Cervero, 2010). Higher urban density and coordinated land-use planning are strongly linked to lower transportation energy consumption and more sustainable mobility outcomes, according to comparative international studies (Newman & Kenworthy, 2015).

Further validating these claims, a recent meta-analysis of Transit-Oriented Development (TOD) across 50 global cities confirms that integrated spatial planning can reduce household transport emissions by up to 25% (Stevens, 2023).

According to more recent studies on urban modeling, traffic congestion itself may be the primary cause of the formation of several subcenters, which would reduce commute times and redistribute travel demand in growing metropolitan areas (Louf & Barthélemy, 2013).

It is commonly acknowledged that traffic congestion is a complex urban issue with substantial negative effects on the economy, society, and environment. Global transportation studies consistently identify three major consequences: declining air quality, increased fuel consumption, and lost productivity (Santos et al., 2010). Recent studies also show that transportation-related emissions are increasingly contributing to the effects of urban climate change, especially in Asiatic and African cities that are motorizing quickly (Ma et al., 2021).

In many cities in the Global South, fossil fuel-based mobility systems continue to be closely associated with environmental pollution, particularly air and noise pollution (Gurjar et al., 2010). As a result, the idea of green infrastructure has become more well-known as an integrated planning approach. Urban vegetation can improve urban microclimates, lessen the effects of urban heat islands, and lower particulate matter concentrations, according to empirical data (Gill et al., 2007; Nowak et al., 2014).

To ensure these environmental benefits are equitably distributed, inclusive urban planning frameworks are adopted to prevent the marginalization of low-income residents during the implementation of green corridors (Harris & De Leeuw, 2022).

A concrete global example is Medellín's "Green Corridors" (Pasillos Verdes) project, which created interconnected nature routes that lowered city temperatures by 3°C and significantly improved air quality in dense urban canyons (World Economic Forum, 2022).

Green infrastructure also supports ecosystem services, improves biodiversity, and promotes public health and well-being, according to reviews of ecosystem-based urban planning (Roy et al., 2012; World Health Organization, 2017). Green infrastructure is currently positioned in international policy frameworks as an affordable addition to traditional gray infrastructure for enhancing urban resilience (OECD, 2020; UN-Habitat, 2022).

In the context of flood-prone megacities, China's "Sponge City" initiative illustrates how permeable green infrastructure serves as a scalable model for managing water hazards while increasing urban green space (Li et al., 2023).

The role of institutional and behavioral factors in influencing sustainable mobility transitions is becoming more and more prominent in the literature, in addition to physical infrastructure. Researchers contend that without adjustments to travel habits, social norms, and governance structures, technological or spatial interventions alone are inadequate (Banister, 2008; Cohen et al., 2014).

Recent studies on Mobility as a Service (MaaS) emphasize that digital behavioral shifts must be inclusive to prevent "digital exclusion" in developing economies (Pangbourne et al., 2020).

Therefore, it is emphasized that inclusive governance and participatory planning are essential tools for enhancing the legitimacy of policies and their successful implementation (Healey, 1997; Arnstein, 1969).

Rapidly expanding South Asian megacities like Dhaka represent particularly important but underrepresented contexts within this larger global discourse. Congestion, emissions, and infrastructure deficiencies are exacerbated in these cities by severe density pressures, limited transportation capacity, and institutional fragmentation (Revi et al., 2014; UN-Habitat, 2022). With extremely low average traffic speeds and significant economic losses due to mobility inefficiencies, Dhaka is commonly listed as one of the most congested urban areas in the world.

Recent research highlights the pressing need for integrated planning frameworks that concurrently address environmental management, land-use structure, transportation systems, and governance reform in light of these circumstances. In rapidly urbanizing metropolitan areas, such strategies are increasingly seen as crucial for enhancing livability and lowering long-term sustainability risks.

Similarities and Distinctions with Existing Literature

This part of the paper covers a wide range of urban studies literature capturing various aspects of the relationships between urban spatial configuration, transport networks, environmental conditions, and social structures. There is an increasing agreement in the literature regarding the urban structure and transport demand relationship, as well as the relevance of planning theories to explain the varied and often contradictory patterns of the extent of mobility and environmental consequences of planning. The first cross-national studies of urban structure and transport systems by Newman and Kenworthy identified a range of urban densities, dispersed land-use patterns and flexible transport systems that achieved reduced levels of car dependence and facilitated the emergence of sustainable travel patterns (Newman & Kenworthy, 1991; 1999). Their studies have demonstrated that a concentrated or polycentric urban structure can enhance accessibility while reducing travel distances and total energy expenditure.

Further studies validate the trend of shifting from monocentric to multicentric urban structures as cities expand. For example, some empirical studies and models of urban growth show that traffic congestion can stimulate the formation of multiple subcenters, which reduce average commuting distances and redistribute the demand for mobility (Louf & Barthélemy, 2013). Many current planning models, including the "15-minute city" proposal, illustrate the growing need for decentralised and more accessible, mixed-use land-use planning which improves pedestrian access to services (Graells-Garrido et al., 2021).

This section substantiates the claims found within the literature regarding the environmental and health impacts of urban green infrastructures. Interlinked green environments like parks, green roofs, and ecological corridors are proven to improve \ air quality, decrease urban heat islands, increase green spaces, and improve mental and physical health (Hanna & Comín, 2021; OECD, 2020; Zhang et al., 2022). Spanning all domains, the multiple benefits of green infrastructures are increasingly recognized as environmental strategies and economically viable alternatives to integrated traditional gray infrastructures, promoting urban resilience and reducing long-term infrastructure costs (OECD, 2020). Furthermore, evidence suggests that the rapid growth of urban areas leads to insufficient green spaces, exacerbating heat, emissions, and health concerns (Hanna & Comín, 2021; Li et al., 2023).

The rest of the sustainable urbanization discussion backed by global institutions can be reviewed here. UN-Habitat (2022) articulates global policy frameworks which describe the role of integrated urban planning, inclusive governance, and the just distribution of services, in the sustainable improvement of urban life and the mitigation of urban eco-logical footprints. With specific reference to the Global South, the unled, accelerated urbanization process is a growing contributor to the emissions and climate-related risks. This justifies the call for the fusion of frameworks guiding spatial planning, the environment, and transport (Revi et al., 2014; Dodman et al., 2022).

Alongside the ecological and the spatial dimensions, the chapter also builds upon the participatory and user-oriented planning approaches that underscore the significance of civic engagement and social practices in the sustainable urban development process. The emerging studies on urban resilience and green infrastructure have started to incorporate, to varying degrees, health equity, social inclusion, and place-based governance in their planning frameworks (Harris & De Leeuw, 2022). Although there is considerable convergence in the approaches, this chapter uniquely attempts to synthesize the perspectives of traffic engineering, ecological planning, spatial morphology, and behavioral science. Most of the studies that examine traffic congestion or the deterioration of the environment focus on them in isolation; however, this study argues that they affect one another in the context of urban everyday practices and people's mobility choices. The integrated ecological-human interface developed in this study conceptualizes traffic as more than just a problem of capacity. In fact, it is a socio-spatial phenomenon, which is determined by the configuration of land-use, lifestyle patterns, and the level of environmental consciousness.

Another distinctive merit is the emphasis placed on Dhaka as a prototypical megacity of South Asia. Research on urban sustainability has been geographically imbalanced, with the majority of studies located in the Western or East Asian context and the fast growing cities of the Global South being largely overlooked. With the most severe urban growth and infrastructure deficits in the world, Asia and Africa supply critical case studies that help understand the governance bottlenecks, density, and environmental challenges (UN-Habitat, 2022; Kund).

Specifically, the work prioritizes:

1. **Integrated Solutions:** Advocating for coordinated urban policies that merge transport planning, land use management, environmental protection, and social equity to address traffic and pollution holistically.
2. **Ecological and Human-Centered Urban Planning:** Emphasizing the role of green spaces, appropriate plant species, and green corridors in improving air quality and noise reduction, alongside pedestrian-priority zones and alternative transport routes that foster active mobility and social well-being.
3. **Polycentric Urban Forms:** Proposing spatial restructuring towards polycentric models to reduce commuting distances, balance land use, and alleviate traffic loads.
4. **Behavioral and Participatory Dimensions:** Highlighting the importance of citizen participation, public awareness, and behavior change as indispensable for sustainable urban mobility solutions.

5. **Policy-Relevant Recommendations:** Providing actionable guidance for policy-makers and planners to develop interdisciplinary urban transformation strategies suitable for the socio-economic and environmental context of Dhaka and similar cities.

Relation to Existing Literature and Contribution to Knowledge

It is such a very remarkable chapter that contributes highly to already available scientific literature on urban spatial structure, sustainable mobility and environmental performance, but broadens this with global case studies and interdisciplinary knowledge. Studies on the relation between land use and transportatıon one of the best studied area in which linkages between urban for travel and environmental consequences can be observed is the relationship between urban form and activity patterns. The pioneering work of Cervero and Kockelman (1997) and complemented by Ewing and Cervero (2010) offer strong empirical support to the idea that compact, mixed-use, polycentric urban forms reduce the amount of vehicle kilometers traveled and automobile dependence. In addition, the extensive cross-national research of Newman and Kenworthy highlights the fundamental relationship between density, urban form, and sustainable travel outcomes in rapidly growing conurbations (Newman & Kenworthy, 2015).

In addition to enhancing transport efficiency, this also adds to a growing body of research highlighting the environmental and health benefits of urban greening. The key studies by Gill et al. (2007) offer strong evidence on how green infrastructure (GI) effectively reduces the Urban Heat Island (UHI) effect and boosts climate adaptation efforts. Supporting research shows that urban vegetation contributes to air purification, noise reduction, increased ecosystem services, and especially public health improvements, even in densely populated cities (Roy et al., 2012; Nowak et al., 2014; World Health Organization, 2017). These findings are further supported by international comparisons and strategic policy frameworks that emphasize the need to integrate green infrastructure as a crucial component.

This chapter reflects the interdisciplinary effort that combines behavioral and social aspects into urban sustainability assessments. It is clear that infrastructure alone cannot ensure sustainable long-term outcomes unless social norms, daily practices, and institutional frameworks are also transformed (Cohen et al., 2014; Banister, 2008). This aligns with current planning theories advocating for participatory governance, social equity, and human-centered design to address urban challenges (Healey, 1997).

By raising Dhaka's urban issues under this broadened theoretical empirical scope, the chapter contributes to a critical gap in literature. A great deal of the existing literature on sustainable mobility, green infrastructure and urban form has been largely oriented to cities in Europe, North America or East Asia (UN-DESA, 2018) which are while South Asia's rapidly expanding megacity leviathans have remained stubbornly underrepresented as they recoil from extreme congestion and pollution along with infrastructure scarcity (Revi et al., 2014; UN-Habitat, 2022). The Dhaka case study adds tangible empirical content to the global discourse and demonstrates that traditional sustainability frames need adaptation when taken into settings of extreme density, informalization, governance frailty and poverty.

The chapter differs from other processes by using an all-round and integrated analysis of technical, ecological, and social measures. It dismisses outright the idea of treating traffic, pollution and urban form as separate problems. Instead, it treats them as interrelated components of a complex urban system that evolves through positive and negative feedback among infrastructure deployment, land-use development and human individual attitudes. This system-oriented approach also responds to criticisms of sectoral planning and is a reflection of appeals for more interdisciplinary approaches for urban change (Banister, 2008; UN-Habitat, 2022).

Lastly, this chapter rounds off the contribution to literature by stressing relevance for application and policy. Unlike pure theory and descriptive studies, it is grounded in quantifiable indicators and evidence from Dhaka and thus offers a tangible model for how research can be operationalized. This chapter serves as an important reference for policy makers, planners and researchers interested in replicating successful context- specific solutions in other developing cities experiencing such mobility and environmental problems.

In summary, the theoretical background underpinning the chapter is deeply rooted in well-established urban theories concerning spatial structure, transportation demand, environmental sustainability, and social participation. The chapter parallels and extends these literatures by:

- Emphasizing integrated, interdisciplinary solutions.
- Highlighting the ecological-human interface in urban transformation.
- Providing a focused, empirical case study from Dhaka.
- Addressing a critical gap in the representation of developing world megacities.
- Advocating for policy-oriented, actionable strategies grounded in local data.

This nuanced synthesis of existing knowledge combined with original contextual insights positions the chapter as a meaningful contribution to the scholarship on sustainable urban transformation in the Global South.

MAIN FOCUS OF THE CHAPTER

This chapter primarily focuses on addressing the intertwined challenges of traffic congestion and environmental pollution in Dhaka, one of the fastest-growing megacities in the Global South, through the lens of sustainable urban transformation. The complexity of Dhaka's urban dynamics-with its high population density, unplanned urban sprawl, inadequate infrastructure, and institutional challenges-requires a comprehensive and multidimensional approach that integrates spatial planning, ecological interventions, transportation management, and social behavior change. This chapter aims to contribute to both the theoretical understanding and practical solutions by developing a holistic framework tailored to Dhaka's unique socio-environmental context, which can also be extrapolated to similar rapidly urbanizing cities in developing countries.

1. Integrated Understanding of Traffic Congestion and Pollution

Traffic congestion and environmental pollution are frequently studied separately, but this chapter foregrounds their inextricable linkages and mutual reinforcement. Dhaka's notorious traffic jams not only cause economic losses through wasted time and fuel but also significantly worsen air and noise pollution, directly impacting public health and urban livability. Existing studies have documented these effects, yet the chapter emphasizes the importance of treating traffic congestion and pollution as two sides of the same coin requiring joint intervention strategies.

Unlike fragmented approaches that target either vehicular emissions or traffic flow independently, the chapter advocates for integrated urban transformation strategies that simultaneously reduce congestion and mitigate pollution through coordinated land use planning, transport infrastructure, and green urban design. This holistic approach recognizes that improving one aspect (e.g., reducing traffic) without addressing the other (e.g., air pollution sources) can lead to suboptimal or temporary outcomes.

2. Emphasis on Spatial Restructuring Toward Polycentric Urban Form

A central argument of the chapter is the critical role of urban spatial form in shaping transport demand and pollution patterns. Dhaka currently exhibits a highly monocentric urban structure, with a dense and congested central business district attracting daily commutes from sprawling residential peripheries. This configuration exacerbates peak-hour traffic and concentrates pollution hotspots.

The chapter focuses on exploring polycentric urban development as a viable pathway to decongest Dhaka's core and reduce travel distances. By promoting multiple decentralized activity centers distributed across the metropolitan region, polycentricity can encourage local employment, diversify land use, and foster sustainable mobility patterns such as walking, cycling, and public transit use. This strategic spatial restructuring is not merely a theoretical exercise but is linked to specific policy proposals and urban design interventions contextualized for Dhaka.

3. Integration of Green Infrastructure and Ecological Urbanism

Another distinct focus is the incorporation of ecological urbanism principles-particularly green infrastructure and urban greening-as an essential pillar of sustainable urban transformation. Dhaka's urban heat island effect, poor air quality, and noise pollution are partly attributable to limited green cover and high impervious surface areas.

This chapter highlights the multifunctional benefits of green infrastructure, including urban parks, green corridors, roadside vegetation, and street trees, as critical tools to filter air pollutants, reduce ambient temperatures, and improve mental health and well-being of residents. It also discusses the selection of plant species best suited for pollution absorption and climatic resilience in Dhaka's subtropical environment.

Importantly, green infrastructure is integrated with transport and spatial planning interventions-for example, green corridors along pedestrian and cycling paths-which enhances both ecological function and active mobility. This integrated ecological-human framework aligns with emerging global best practices but is adapted and fine-tuned for Dhaka's local conditions.

4. Human-Centered Urban Design and Mobility

The chapter stresses that sustainable urban transformation is not purely a technical or ecological challenge but fundamentally a human-centered one. Citizen behavior, perceptions, and social dynamics strongly influence the success or failure of any urban policy or intervention.

Therefore, the chapter focuses on designing urban environments that prioritize pedestrian comfort, safety, and accessibility, fostering a shift away from private vehicle dependency. This includes advocating for pedestrian zones, traffic calming measures, and accessible public transport linked with green and public spaces to create inviting urban environments that encourage walking, cycling, and community interaction.

Additionally, the chapter underscores the importance of public awareness campaigns and participatory planning processes that involve residents in decision-making and create a sense of ownership and responsibility toward urban sustainability goals. This approach draws from social behavior theories and urban governance literature emphasizing co-production and citizen engagement.

5. Context-Specific Policy and Planning Recommendations

While grounded in global sustainability and urban planning theory, the chapter's main focus is on generating actionable, context-specific recommendations tailored to Dhaka's particular institutional, socio-economic, and environmental realities. The city's governance fragmentation, informal settlements, and resource constraints require pragmatic solutions that can be realistically implemented in the short and medium term.

The chapter therefore bridges theory and practice by identifying policy levers and planning interventions at multiple scales, including:

- Revising land use policies to encourage mixed-use and polycentric development;
- Strengthening institutional coordination between transport, environment, and urban planning agencies;
- Implementing pilot projects on green corridors and pedestrian zones in strategic neighborhoods;
- Enhancing data collection and monitoring to inform evidence-based decision-making;
- Promoting multi-stakeholder partnerships including local communities, NGOs, and private sector actors.

This pragmatic orientation distinguishes the chapter from more abstract or idealistic studies and provides a valuable roadmap for urban planners, policymakers, and civil society actors engaged in Dhaka's urban transformation.

6. Contribution to Theory and Practice in the Global South Context

A key dimension of the chapter's focus is its contribution to filling the empirical and theoretical gap in sustainable urban transformation scholarship regarding megacities in the Global South. Much of the existing literature and policy discourse has historically centered on Western or East Asian cities with different developmental trajectories and institutional capacities.

By concentrating on Dhaka, the chapter brings forth a nuanced understanding of how the intertwined issues of traffic congestion, pollution, and spatial planning manifest in a densely populated South Asian metropolis with severe resource constraints. It integrates interdisciplinary perspectives spanning urban geography, environmental science, transportation engineering, and social behavior to propose comprehensive solutions.

This synthesis offers valuable insights that extend beyond Dhaka to other similarly situated cities in Asia, Africa, and Latin America, making the chapter relevant for a broad audience of scholars and practitioners concerned with sustainable urban futures in rapidly urbanizing contexts.

The chapter advocates for an incremental and adaptive planning process that integrates short-term achievable projects with long-term strategic spatial restructuring and ecological enhancements. It also promotes resilience thinking to prepare Dhaka for future urban pressures, including climate change and demographic shifts.

This forward-looking perspective encourages stakeholders to envision a transformed Dhaka that is more livable, equitable, and environmentally sustainable, setting a model for other rapidly growing megacities in the Global South.

In summary, the main focus of this chapter is to provide a comprehensive, contextually grounded, and actionable framework to tackle the complex challenges of traffic congestion and environmental pollution through sustainable urban transformation in Dhaka. By integrating spatial restructuring, green infrastructure, human-centered mobility design, participatory governance, and data-driven policy analysis, the chapter aims to contribute significantly to theory and practice in urban sustainability, especially within the underrepresented Global South context.

SOLUTIONS AND RECOMMENDATIONS

The complexities of traffic congestion and environmental pollution in Dhaka demand a multidimensional response that moves beyond short-term fixes to encompass systemic urban transformation. This chapter presents comprehensive solutions and targeted recommendations grounded in principles of sustainability, spatial planning, and participatory governance. These proposals are developed specifically for Dhaka's socio-political and infrastructural context, while also being framed within widely adopted global urban transformation agendas, including SDG 11, the New Urban Agenda, and international low-carbon mobility strategies. New findings from UN-Habitat (2024), emphasises that achieving integration will be essential; it is anticipated that integrated spatial planning will decrease urban carbon emissions by as much as 25% in a high-density Global South urban context (Stevens, 2023). The goal is to mitigate existing urban pressures while steering the city toward a more resilient, equitable, and environmentally sound future.

Effective urban transformation in Dhaka necessitates not only spatial and environmental considerations but also attention to long-term urban trends observed in rapidly growing cities of the Global South. According to World Bank (2023) reports, the epicenter of urban growth is shifting toward South Asian corridors, necessitating a transition toward "Sponge City" resilience and nature-based solutions to manage climate-induced risks (Li et al., 2023).

Comparative evidence from other megacities facing similar demographic pressure and mobility challenges highlights that spatial and transportation strategies remain ineffective without inclusive planning frameworks, governance capacity, economic viability, and technological readiness. Successful implementations, such as Bogotá's "Barrios Vitales" and Medellín's "Green Corridors," further demonstrate that operationalizing sustainable mobility depends on decentralizing urban functions to reduce vehicle dependency (Moreno et al., 2021; World Economic Forum, 2022).

In this context, well-designed interventions risk remaining aspirational rather than operational in the absence of financially feasible implementation models and coordinated institutional structures.

1. Spatial and Land Use Restructuring: Toward a Polycentric Dhaka

One of the root causes of Dhaka's congestion and pollution is its monocentric urban form, which concentrates economic activity, services, and administrative functions in a narrow central corridor. This spatial imbalance mirrors patterns observed in other high-density developing cities and generates long commuting distances, peak-hour pressure on infrastructure, and elevated emissions from private vehicles.

Recommendation:

Adopt a polycentric spatial development strategy by creating multiple urban subcenters or growth nodes in peripheral areas such as Uttara, Mirpur, Purbachal, and Keraniganj. These nodes should integrate residential, commercial, and institutional land uses to reduce the necessity for cross-city travel.

Action Steps:

- Update and implement Dhaka's Structure Plan and Detail Area Plans (DAPs) to prioritize mixed-use zoning and transit-oriented development (TOD).

- Decentralize key government offices, universities, and business hubs to secondary centers.
- Provide economic incentives for businesses relocating to peripheral nodes.

Expected impact:

This restructuring would reduce cross-city commuting, ease central traffic congestion, and distribute environmental burdens more evenly across the metropolitan area.

2. Sustainable Public Transport Reform

Dhaka's transport system is dominated by inefficient buses, informal modes, and rapidly increasing private vehicle ownership- trends commonly documented in fast-urbanizing cities where public transport capacity has not kept pace with growth.

Recommendation:

- Develop an integrated, reliable, and affordable mass transit system centered on Metro Rail (MRT), Bus Rapid Transit (BRT), and coordinated feeder services.

Action Steps:

- Prioritize completion and expansion of MRT and BRT corridors with full network integration.
- Establish multimodal hubs connecting buses, metro, rickshaws, and bicycles.
- Provide incentives for low-emission or electric buses on high-demand routes.
- Regulate informal transport modes through digital fare and route management.

Expected impact:

Reduced private vehicle dependence, improved air quality, shorter travel times, and enhanced mobility equity.

3. Promotion of Non-Motorized and Active Transport

Despite congestion, a significant share of Dhaka's population relies on walking and rickshaws. However, the lack of safe infrastructure discourages active transport, reflecting a broader trend in developing cities where motorized mobility dominates planning priorities.

4. Urban Greening and Environmental Buffer Zones

Green space per capita in Dhaka is far below World Health Organization (WHO) recommendations, limiting the city's natural ability to absorb pollutants and regulate urban temperatures.

Recommendation:

Expand urban green infrastructure to enhance ecological resilience and improve air quality.

Action Steps:

- Develop linear parks and green belts along major roadways and waterways.
- Enforce tree-planting mandates along new transport corridors and sidewalks.
- Rehabilitate and protect wetlands and open spaces from encroachment through zoning enforcement.
- Utilize green roofs and vertical gardens on public buildings to maximize space use.

Expected impact:

Reduction in particulate matter and urban heat island effects, increased biodiversity, and improved mental well-being.

5. Air Pollution Monitoring and Regulation

Addressing environmental pollution requires real-time data and stringent regulation of pollution sources, including vehicles, industries, and construction activities.

Recommendation:

Strengthen environmental governance through robust monitoring, data transparency, and enforcement.

Action Steps:

- Expand the network of air quality monitoring stations across Dhaka's key traffic corridors and neighborhoods.
- Introduce low-emission zones where only clean vehicles are permitted.

- Enforce emissions testing and phasing out of old vehicles, especially diesel-fueled buses and trucks.
- Require dust control measures at all construction sites.

Expected impact:

Improved air quality, fewer respiratory illnesses, and greater accountability of pollution sources.

6. Behavioral Change and Public Participation

Policies and plans often fail due to low public awareness, lack of buy-in, or resistance from vested interest groups. Sustainable transformation requires behavioral shifts and participatory processes.

Recommendation:

Implement citizen-centered planning and education initiatives to foster sustainability culture.

Action Steps:

- Launch mass awareness campaigns on the health costs of pollution and the benefits of public transport.
- Create neighborhood-based planning committees to gather local input on infrastructure needs.
- Use digital platforms and apps to crowdsource traffic data and citizen complaints.
- Involve schools and universities in green city awareness programs and urban activism.

Expected impact:

Greater community ownership of policies, improved compliance, and sustainable civic habits.

7. Institutional Coordination and Governance Reform

A major hurdle to implementation in Dhaka is fragmented governance involving multiple agencies with overlapping mandates.

Recommendation:

Establish a Unified Urban Transport and Environment Authority with decision-making power and adequate funding.

Action Steps:

- Clarify the roles of Rajuk, Dhaka Transport Coordination Authority (DTCA), and Dhaka City Corporations to reduce institutional conflict.
- Promote data sharing and joint planning across agencies.
- Allocate specific budgets for integrated mobility and green infrastructure projects under the same authority.
- Build technical capacity through international partnerships and training.

Expected impact:

Streamlined project implementation, reduced bureaucratic delay, and better resource allocation.

8. Smart City Technologies and Innovation

Technology can play a pivotal role in managing urban systems, monitoring environmental indicators, and optimizing transport efficiency.

Recommendation:

Adopt smart mobility and pollution control technologies for data-driven urban management. Adoption of technology should prioritize low-cost, scalable, and interoperable systems that are compatible with current institutional capacity in order to avoid technology-driven but context-insensitive smart city initiatives.

Action Steps:

- Use real-time traffic management systems to reduce congestion during peak hours.
- Introduce mobile apps for multimodal trip planning to encourage public transport use.
- Implement electronic road pricing or congestion charges in critical zones to reduce peak traffic volumes.
- Use satellite data and AI to forecast air quality and inform planning decisions.

Expected impact:

Improved traffic flow, reduced emissions, better emergency response, and citizen empowerment through data access.

The solutions and recommendations presented in this chapter aim to confront Dhaka's urban crises through an integrated approach that connects spatial planning, transportation reform, green infrastructure, technological innovation, and citizen participation. These interventions are not only theoretically sound but also empirically and practically grounded in Dhaka's specific urban condition. If implemented with political will, public support, and inter-agency coordination, these proposals could pave the way for a sustainable and livable Dhaka, serving as a blueprint for other megacities in the Global South undergoing similar transformations.

9. Economic Feasibility and Financing Mechanisms

Recommendation:

- Develop diversified financing models combining public investment, land value capture, PPPs, and climate funds.

Action Steps:

- Land value capture near MRT/BRT corridors
- PPPs for electric bus systems
- Phased investment aligned with fiscal capacity
- Lifecycle cost analysis

Expected Impact

Financially sustainable and scalable urban transformation.

10. Inclusive Urban Planning and Social Equity

Recommendation

- Implement inclusive planning to guarantee fair access to transportation and services.

Action Steps

- Provide subsidized public transportation
- Gradually formalize informal transportation systems
- Develop affordable housing in sub-centers
- Incorporate gender-sensitive design principles

Expected Impact

Decreased inequality and enhanced social acceptance.

CONCLUSION

Dhaka is facing serious problems due to severe traffic congestion and environmental pollution. Because of these problems, Dhaka is becoming less attractive and it is at risk of not having enough economic resources and not being healthy enough for its people to survive in the long-term (the future). With this in mind, researchers gathered information to identify the major issues impacting the quality of life in Dhaka through the lens of sustainable urban transformation. Researchers have also used this information to develop reasonable solutions.

There is an abundance of evidence in academic and policy literature to show that unplanned development and increases in motor vehicle usage have increased the severity of traffic congestion and environmental pollution in Dhaka; Newman and Kenworthy, (1999), Banister (2008), and Beatley (2011) claim that cities that have compact mixed use and a strong public transport system are more energy-efficient, resilient and socially equitable than urban centres that do not.

Global case studies of cities including Curitiba, Singapore, and Seoul suggest that polycentric development, transit-oriented planning, and green infrastructure are all important factors in addressing urban congestion and pollution. The literature has also indicated that successful transformation requires participatory governance, innovations in technology, and changes in behavior.

Many illustrations and instances of location-based urban upgrades exist worldwide; nevertheless, applying these methods to Dhaka may result in more failures than successes. Urban revitalization practices on a sustainable basis will involve considering the socio-political dynamics of their communities, recognizing the strengths and limitations of their existing government institutions, and identifying the resources to implement change (UN-HABITAT, 2013; Parnell; Robinson, 2012). It is necessary for planners and policymakers involved in bringing about urban transformation and improvement based on what has been identified in the literature to become acclimated to their unique qualifications prior to attempting the use of global models of urban change on the communities below. Overall, importantly, the proposal made in this document consists of:

• A shift from monocentric to polycentric spatial planning, allowing for more balanced economic activity and reduced travel demand.

- The need for institutional coordination among overlapping urban agencies to address inefficiency and policy fragmentation.
- The integration of non-motorized transport (e.g., walking and cycling) into mainstream infrastructure investments, which is often sidelined in Dhaka's planning discourse.
- The application of smart city tools not only for mobility management but also for real-time pollution monitoring and citizen engagement.
- The prioritization of behavioral and cultural change alongside physical infrastructure, acknowledging that technology alone cannot deliver sustainability.

Additionally, the research advocates for a grassroots approach to engage the community, local government, and the private sector in co-creating a sustainable future for our cities. It stresses the importance of raising awareness through public awareness programs, implementing localized solutions, and creating an inclusive planning process that incorporates the opinions of typical urban residents.

This study's main conclusion is that the process of creating a sustainable urban environment in Dhaka must be holistic, spatially equitable, and involve participation from all segments of the population. Solutions must not be limited to transport infrastructure (such as road and rail development), nor should they be based solely upon short-term reactions to current events. In contrast, rethinking traffic congestion and environmental degradation will require thorough systemic changes.

The growth of cities, their ability to transport their citizenry effectively, their sustainability and the equitable distribution of power and responsibility need to be rethought comprehensively.

"Dhaka's transition from planning to implementation of sustainable urban development will require a transition toward decentralized and collaborative approaches to urbanization." This study provides an important link between the conceptual framework of sustainable urbanism and the actual context of the city of Dhaka. By demonstrating the potential for sustainable transformation within the constraints of a megacity with limited resources, political complexity, and environmental challenges, this research contributes to our overall understanding of urban sustainability.

The rapid rate of urbanization taking place in many cities across the Global South highlights the importance of addressing the unique circumstances that can lead to sustainability within megacities like Dhaka. As demonstrated through the severe challenges posed by urban development, Dhaka presents an excellent opportunity for rethinking traditional urban planning and governance practices. Through integrating mobility, greening infrastructure, reforming institutions, and increasing civic engagement, Dhaka can adapt and thrive as it continues to expand into a future that is resilient, inclusive, and sustainable.

The findings of this study call on all stakeholders - planners, policymakers, scholars, and citizens - to work together to achieve the difficult but possible task of implementing sustainable urban development in Dhaka. We have the frameworks and methodologies available to do so; we need only to have the collective courage and commitment to implement them in a manner that authentically represents both the hopes and realities of the people of Dhaka.

REFERENCES

Ahsan, A., Rahman, M., & Rahman, M. M. (2016). Traffic congestion in Dhaka city: Causes and possible solutions. *International Journal of Transportation Engineering*, *4*(1), 23–34. DOI: 10.12691/ijte-4-1-3

Alves, A., & Ferreira, J. (2021). Integrated urban planning approaches for sustainable mobility: A systematic review. *Sustainability*, *13*(7), 3748. DOI: 10.3390/su13073748

Arnstein, S. R. (1969). A ladder of citizen participation. *Journal of the American Institute of Planners*, *35*(4), 216–224. DOI: 10.1080/01944366908977225

Arnstein, S. R. (1969). A ladder of citizen participation. *Journal of the American Institute of Planners*, *35*(4), 216–224. DOI: 10.1080/01944366908977225

Banister, D. (2008). The sustainable mobility paradigm. *Transport Policy*, *15*(2), 73–80. DOI: 10.1016/j.tranpol.2007.10.005

Batty, M. (2013). *The new science of cities*. MIT Press. DOI: 10.7551/mitpress/9399.001.0001

Beatley, T. (2011). *Biophilic cities: Integrating nature into urban design and planning*. Island Press. DOI: 10.5822/978-1-59726-986-5

Cervero, R. (2013). Transport infrastructure and urban growth: The emergence of polycentric urban regions. *Journal of Transport and Land Use*, *6*(1), 7–25. DOI: 10.5198/jtlu.v6i1.425

Cervero, R., & Kockelman, K. (1997). Travel demand and the 3Ds: Density, diversity, and design. *Transportation Research Part D, Transport and Environment*, *2*(3), 199–219. DOI: 10.1016/S1361-9209(97)00009-6

Cohen, T., Saxe, S., & Soltani, A. (2014). Behavioral approaches to sustainable urban transport. *Sustainability*, *6*(5), 2739–2753. DOI: 10.3390/su6052739

Dodman, D., Hayward, B., Pelling, M., Castán Broto, V., Chow, W., Chu, E., Dawson, R., Khirfan, L., McPhearson, T., Prakash, A., & Ziervogel, G. (2022). *Cities, settlements and key infrastructure*. In IPCC Sixth Assessment Report (AR6): Climate change 2022 – Impacts, adaptation and vulnerability. Intergovernmental Panel on Climate Change. https://www.ipcc.ch/report/ar6/wg2/

Dodson, J., & Sipe, N. (2008). *Shaping urban development: The role of planning and policy in managing urban sprawl*. University of Queensland Press.

Downs, A. (2004). *Still stuck in traffic: Coping with peak-hour traffic congestion*. Brookings Institution Press.

Ewing, R., & Cervero, R. (2010). Travel and the built environment: A meta-analysis. *Journal of the American Planning Association*, *76*(3), 265–294. DOI: 10.1080/01944361003766766

Gill, S. E., Handley, J. F., Ennos, A. R., & Pauleit, S. (2007). Adapting cities for climate change: The role of the green infrastructure. *Built Environment*, *33*(1), 115–133. DOI: 10.2148/benv.33.1.115

Gössling, S., Cohen, S., & Haurie, A. (2016). *Sustainable transport and urban planning: Policies, practices and implications*. Springer.

Graells-Garrido, E., Serra-Burriel, F., Rowe, F., Cucchietti, F. M., & Reyes, P. (2021). *A city of cities: Measuring how 15-minute urban accessibility shapes human mobility in Barcelona*. arXiv. https://arxiv.org/abs/2102.12315

Gurjar, B. R., Butler, T. M., Lawrence, M. G., & Lelieveld, J. (2010). Evaluation of emissions and air quality in megacities. *Atmospheric Environment*, *44*(35), 4156–4166. DOI: 10.1016/j.atmosenv.2010.07.028

Gwilliam, K. (2011). Urban transport in developing countries. *Transport Reviews*, *31*(2), 151–174. DOI: 10.1080/01441647.2010.519887

Hanna, E., & Comín, F. A. (2021). Urban green infrastructure and sustainable development: A review. *Sustainability (Basel)*, *13*(20), 11498. DOI: 10.3390/su132011498

Harris, P., & De Leeuw, E. (2022). Integrating health into urban infrastructure planning for equity. *Cities & Health*, *6*(1), 1–4. DOI: 10.1080/23748834.2021.1938570

Healey, P. (1997). *Collaborative planning: Shaping places in fragmented societies*. Macmillan. DOI: 10.1007/978-1-349-25538-2

Hossain, M. S., Hasan, M. M., & Islam, M. R. (2020). Urban transportation and environmental sustainability: Case study of Dhaka city. *Sustainable Cities and Society*, *56*, 102084. DOI: 10.1016/j.scs.2020.102084

Islam, S., & Hasan, M. (2021). Urban green space and social equity in Dhaka: Challenges and opportunities. *Urban Forestry & Urban Greening*, *61*, 127102. DOI: 10.1016/j.ufug.2021.127102

Karim, M. R., & Hasan, M. (2013). Traffic congestion in Dhaka: Causes, impacts and policy implications. *Journal of the Bangladesh Institute of Planners*, *6*, 67–78. DOI: 10.3329/jbip.v6i0.17547

Li, Y., Svenning, J.-C., Zhou, W., & Schell, C. J. (2023). Global inequality in cooling benefits from urban green spaces and its climate adaptation potential. *Nature Communications, 14*, 6205. DOI: 10.1038/s41467-023-41846-5

Litman, T. (2021). *Evaluating transportation land use impacts*. Victoria Transport Policy Institute. https://www.vtpi.org/landuse.pdf

Louf, R., & Barthélemy, M. (2013). Modeling the polycentric transition of cities. *Physical Review Letters, 111*(19), 198702. DOI: 10.1103/PhysRevLett.111.198702 PMID: 24266493

Ma, J., Zhang, L., & Song, Y. (2021). Economic loss assessment of traffic congestion: A review and framework. *Transportation Research Part A, Policy and Practice, 143*, 1–15. DOI: 10.1016/j.tra.2020.10.007

Moreno, C., Allam, Z., Chabaud, D., Gall, C., & Pratlong, F. (2021). Introducing the "15-Minute City": Sustainability, resilience and place identity in future post-pandemic cities. *Smart Cities, 4*(1), 93–111. DOI: 10.3390/smartcities4010006

Newman, P., & Kenworthy, J. (1991). Transport and urban form in thirty-two of the world's principal cities. *Transport Reviews, 11*(3), 249–272. DOI: 10.1080/01441649108716787

Newman, P., & Kenworthy, J. (1999). *Sustainability and cities: Overcoming automobile dependence*. Island Press.

Newman, P., & Kenworthy, J. (2015). *The end of automobile dependence: How cities are moving beyond car-based planning*. Island Press. DOI: 10.5822/978-1-61091-613-4

Nowak, D. J., Hirabayashi, S., Bodine, A., & Greenfield, E. (2014). Tree and forest effects on air quality and human health in the United States. *Environmental Pollution, 193*, 119–129. DOI: 10.1016/j.envpol.2014.05.028 PMID: 25016465

OECD. (2020). *Developing an integrated approach to green infrastructure*. Organisation for Economic Co-operation and Development. https://www.oecd.org

OECD. (2020). *Nature-based solutions for cities*. Organisation for Economic Co-operation and Development.

Pangbourne, K., Mladenović, M. N., Stead, D., & Milakis, D. (2020). Questioning mobility as a service: Unanticipated implications for society and governance. *Transportation Research Part A, Policy and Practice, 131*, 35–49. DOI: 10.1016/j.tra.2019.09.033

Papageorgiou, M., Diakaki, C., Dinopoulou, V., Kotsialos, A., & Wang, Y. (2003). Review of road traffic control strategies. *Proceedings of the IEEE, 91*(12), 2043–2067. DOI: 10.1109/JPROC.2003.819610

Rahman, M., & Hossain, M. (2020). Urban planning challenges and solutions for Dhaka: Towards sustainable mobility. *Journal of Urban Affairs*, *42*(4), 509–526. DOI: 10.1080/07352166.2019.1619802

Revi, A., Satterthwaite, D., Aragón-Durand, F., Corfee-Morlot, J., Kiunsi, R., Pelling, M., Roberts, D., & Solecki, W. (2014). Urban areas. In C. B. Field et al. (Eds.), *Climate change 2014: Impacts, adaptation, and vulnerability (IPCC Fifth Assessment Report, Working Group II)*. Intergovernmental Panel on Climate Change. https://www.ipcc.ch/report/ar5/wg2/

Roy, S., Byrne, J., & Pickering, C. (2012). A systematic quantitative review of urban tree benefits, costs, and assessment methods. *Urban Forestry & Urban Greening*, *11*(4), 351–363. DOI: 10.1016/j.ufug.2012.06.006

Santos, G., Behrendt, H., & Teytelboym, A. (2010). Part II: Policy instruments for sustainable road transport. *Research in Transportation Economics*, *28*(1), 46–91. DOI: 10.1016/j.retrec.2010.03.002

Satterthwaite, D. (2016). *The urban crisis in the global South.* Routledge.

Stevens, M. (2023). Does Transit-Oriented Development (TOD) deliver on its transit promises? A meta-analysis. *Journal of the American Planning Association*, *89*(2), 214–228. DOI: 10.1080/01944363.2022.2086385

UN-Habitat. (2022). *World cities report 2022: Envisioning resilient urban futures.* United Nations Human Settlements Programme.

World Bank. (2020). *World development report 2020: Trading for development in the age of global value chains.* World Bank.

World Bank. (2023). *Bangladesh urbanization review: Managing cities for inclusive growth.* World Bank. https://www.worldbank.org/en/country/bangladesh/publication/bangladesh-urbanization-review

World Health Organization. (2017). *Urban green spaces and health: A review of evidence.* WHO Regional Office for Europe.

Zhang, L., Tan, P. Y., Diehl, J. A., & Tan, H. T. W. (2022). Integrating biodiversity into urban green infrastructure planning: The case of Singapore. *Urban Forestry & Urban Greening*, *74*, 127650. DOI: 10.1016/j.ufug.2022.127650

Zhao, P., Park, H., & Ni, J. (2018). Urban spatial structure and residents' travel behavior: A case study in Beijing. *Transportation Research Part D, Transport and Environment*, *61*, 72–84. DOI: 10.1016/j.trd.2018.01.018

KEY TERMS AND DEFINITIONS

Environmental Pollution: The presence of harmful substances (air, noise, and water pollutants) in urban environments that degrade ecosystems, threaten public health, and reduce overall urban quality of life.

Green Infrastructure: Networks of natural and semi-natural elements- such as parks, green corridors, and urban vegetation- that mitigate heat, improve air quality, enhance biodiversity, and support public health.

Integrated Land-Use and Transport Planning: A coordinated strategy aligning spatial development with mobility systems to reduce vehicle dependence, lower emissions, and improve accessibility.

Participatory Governance: Decision-making processes that actively involve citizens and stakeholders to enhance policy legitimacy, inclusiveness, and implementation effectiveness.

Polycentric Urban Development: An urban form characterized by multiple activity centers instead of a single dominant core, aimed at reducing commuting distances, redistributing travel demand, and improving accessibility.

Sustainable Urban Planning: An integrated planning approach that balances environmental protection, economic efficiency, and social equity while promoting long-term urban resilience.

Traffic Congestion: A condition where transport demand exceeds road capacity, causing delays, economic losses, higher fuel consumption, and increased emissions, particularly in rapidly urbanizing cities.

Chapter 2
Contribution of Plant Species in Noise and Air Pollution Control: A Case Study in Dhaka City

Md. Shafiqul Islam
http://orcid.org/0000-0002-5547-046X
International University of Business Agriculture and Technology, Bangladesh

Hafizur Md Rahman
State University of Bangladesh, Bangladesh

ABSTRACT

Noise and air pollution significantly affect human health, the environment, and reduce the quality of life. In Dhaka city, air and noise pollution rank as the third most severe form of environmental pollution. This study seeks to evaluate the role of plant species in reducing temperature and mitigating air and noise pollution. Data were collected on multiple air and noise parameters in both high-traffic areas during peak and off-peak hours, weekend and free days, using a Class 1 Sound Level Meter and a portable air quality meter. It was recorded an average highest noise level of 87.7 dB(A) was recorded in Bijoy Shoroni, and a minimum of 65.8 dB(A) was recorded in Ramna Park. Air quality (including PM10, PM2.5, CO, CO_2, O_3, and temperature) showed maximum values of 864 μg/m³, 321 μg/m³, 7.0 ppm, 630 ppm, 0.0 μg/m³, and 29°Cat Bijoy Shoroni, while minimum values of 62 μg/m³, 34 μg/m³, 1.0 ppm, 429 ppm, 0.0 μg/m³, and 23.3°C were recorded at Ramna Park. The presence of plant species has shown lower noise levels, lower air pollution, and lower temperatures in Ramna Park. On the other hand, Bijoy Shoroni experienced higher levels of noise and air pollution due to a scarcity of vegetation. Therefore,

DOI: 10.4018/979-8-3373-4422-5.ch002

enhancing urban greenery through plantation and implementing policies that promote eco-friendly transportation are vital steps toward mitigating noise and air pollution in Dhaka city.

1.0 INTRODUCTION

Environmental pollution is made worse and spread by cities with expanding populations, urbanization and industrialization (Ozdemir et al., 2014; Iqbal et al., 2022; Dilawar et al., 2022). Agarwal and Sharma (1980) emphasized that the phrase "air pollution" is vague, complex and truly encompasses a wide range of issues. These include particles, radio-nuclides, smog, acid rain, industrial and household smoking, construction works greenhouse gas emissions and impact, and ozone layer depletion. The major pollution parameters include carbon monoxide, carbon dioxide, ozone, and particulate matter (PM 2.5 & PM 10). Plants serve as a fundamental foundation for all ecosystems and are particularly susceptible to airborne pollution, which is recognized as having the greatest potential to influence these organisms due to ambient air pollution. The effects of such pollution are most frequently visible on the leaves, which typically act as the most prevalent and conspicuous primary receptors for a wide array of air pollutants (Wei et al., 2017). The biomonitoring of plants is a crucial method for assessing the effects of air pollution. Consequently, in recent years, urban vegetation has gained significance not only for social considerations but primarily for its impact on local and regional air quality. There is an increasing need to alter the strategy of planting trees and other plant species in light of the current state of urban environmental pollution. Including beautiful plants with the capacity to reduce pollution in the landscape design will accomplish two goals: it will make the cities environmentally friendly and pollution-free over time (Kumar et al., 2013).Noise can be defined in different ways. People, who study acoustics, define noise as complex sound waves that are a periodic, in other words, sound waves with irregular vibrations and no definite pitch. It is a pollutant and a hazard to human health and hearing. Noise also has psychological and social implications and affects our wellbeing and quality of life (Chepesiuk, 2005). Inhabitants of Dhaka City are being exposed to high level of noise pollution. Although there are many sources of noise which include industries, construction works, indiscriminate use of loudspeakers, people gathering. The main source of noise in Dhaka City is motorized traffic.

Air Pollution is widespread problem in Bangladesh. They cause a huge damage in ecosystem and also affected the human life of the people. Because of the enhanced anthropogenic activities in Bangladesh, air pollution problems have become a topic of intense debate at all platforms, Bangladesh has the worst air pollution in the entire world (Begum and Hopke, 2018). Air pollution in Bangladesh has increased rapidly because of intensive population growth, increase in the numbers of vehicles, use of fuels with poor environmental performance, badly mentioned transportation systems, industrialization, and ineffective environmental regulations. Air quality has deteriorated both due to human activities, and due to natural phenomenon, such as windblown dust particles etc. Particulate emission derived from road traffic is primarily responsible for increased death rate and respiratory problems for the urban population. This problem is acute in Dhaka being the capital of Bangladesh. Hence, estimation and understanding of the present status and the level of air pollution at different sites is deemed necessary.

A comprehensive study on air pollution and its impact needs to be undertaken to find out the degree of air pollution. In order to understand the air quality of an area, measurements of air quality parameters are required (Sokhi et al., 2021). The standard used for Bangladesh is explained in general, and in some cases, results are also compared with WHO guideline for better representation from global context. Air emission inventories can be considered as a structured collection of both information about emissions and technological, economical and territorial data. By organizing data, inventories permit the individuation of pollution sources and their localization. The quantitative emissions estimates provided by an inventory promote a better understanding of the actual emissions and help to raise the awareness of both policy makers and the general public. An emission factor is a representative value that relates the quantity of a pollutant released to the atmosphere with an activity associated with the release of that pollutant. These factors are usually expressed as the weight of pollutant divided by a unit weight, volume, distance, or duration of the activity emitting the pollutant. To calculate emission rates for different types of vehicles, various emission factor models are used throughout the world. Density, flow, vehicle composition, v/c ratio, the number of traffic lights per mile, signal coordination, and the number of stops per mile are traffic-related variables. Congested traffic conditions increase emissions and reduce speed compared to free flow conditions. The air quality of a region depends on the concentrations of air pollutants, including PM (PM2.5, PM10), Carbon Dioxide ($CO_{2)}$, Carbon Monoxide (CO), and Ozone ($O_{3)}$, over a specified averaging period.

1.1 Problem Statement

Noise pollution significantly affects human life and urban environment. Noise has become one of the pollutants that have become the concern of the international community. Sound pollution affects human health and behavior. Road traffic plays a dominant role in causing environmental noise which has a negative impact on public health. The main contributors to noise include aircraft, motor vehicles, and loud music. Unwanted noises that interfere with peace, such as barking dogs or the sounds of vehicles and airplanes, are examples of noise. One of the health implications of noise is hearing loss. This occurs due to extended, continuous exposure to loud sounds. Prolonged exposure to high decibel levels can lead to increasingly severe symptoms of deafness (Pienkowski, 2021). Noise pollution is thought to have the most detrimental effects on sleep. Sleep disorders are becoming more common in western nations as a result of lifestyle choices. It increased dramatically as a result of the growing number of automobiles and industrial equipment. Urban noise levels have risen to extremely unsettling levels, particularly along arterial roadways.

The World Health Organization ranks noise pollution as the third most dangerous type of environmental contamination. In developing countries, 70% of the world's population resides in urban areas (Jarosińska et al., 2018). Most urban dwellers are vulnerable to the detrimental effects of noise, which leads to public health concern. The increasing urbanization and population growth of today have resulted in deteriorating ambient air quality. Air pollution is one of the largest environmental hazards. It is commonly recognized urban air pollution is caused by automobiles alone. The rapid rise in motor vehicle ownership and activity in Dhaka city have a significant health, environmental, and social implications (Hossain and Huggins, 2021). The effects of air pollutants, such as gaseous and particulate matter, on plants are calculable. Noise and air pollution have been increasing daily in Dhaka city. The researcher is attempting to analyze the differences between Ramna Park and the Bijoy Shoroni traffic signal, which are among the most polluted locations. We visited and collected data from these two pre-selected sites on different dates, times, and days. The average noise and air quality levels were recorded, with a maximum of 87.7 dB(A) at the monitoring site (Bijoy Shoroni) and a minimum of 65.8 dB(A) at the monitoring site (Ramna Park). The air quality parameters (PM10, PM2.5, CO, CO2, O3, and temperature) showed a maximum of (864 µg/m3, 321 µg/m3, 7.0 ppm, 630 ppm, 0.0 µg/m3, 29°C) at the monitoring location (Bijoy Shoroni) and a minimum of (62 µg/m3, 34 µg/m3, 1.0 ppm, 429 ppm, 0.0 µg/m3, 23.3°C) at the monitoring site (Ramna Park). The Bijoy Shoroni area is the most polluted.

During mid of the day, the level of noise and air quality are elevated due to increased temperature (weather conditions) and traffic in the Bijoy Shoroni area. Clean air is vital for maintaining the delicate balance of life on this planet, not only for humans but also for wildlife, vegetation, water, and soil. Poor air quality results from various factors, including emissions from both natural and anthropogenic sources (Mallik, 2019). It occurs when pollutants reach concentrations that pose risks to human health and the environment. Therefore, it is essential to monitor air quality to ensure it remains at acceptable levels in the divisional cities of Bangladesh. To know the conditions of the air quality in any densely urbanized region, air quality measurement is essential. It is important to gather data from multiple air quality monitoring stations that may be assessed.

1.2 Objective of the Study

The main objective of this study was to measure the air and noise pollution status in the study areas. Specific objectives can be outlined as below:

- Know the present noise level and air quality parameters in the study areas.
- Identify the sources of noise and air pollution.
- Elicit the causes and sources of noise and air pollution.
- Explore the contribution of plant species in reducing temperature, noise and air pollution.
- Compare the noisiest and air-pollutedareas between the study areas in relation to the impact of plant species, peak and off-peak hours, weekdays and weekends.
- Recommended the measures to mitigate noise level, air pollution, and temperature.

2.1 LITERATURE REVIEW

Numerous studies on noise levels have been carried out by many researchers from different countries in the world (Tang & Ong, 1988; Onuu, 2000; Zannin et al., 2002; Li et al., 2002). The green belt acoustic effect of trees along the highway has been a popular research topic for the past 40 years (Gidlöf-Gunnarsson & Öhrström, 2007; Mishra et al., 2010; Aylor, 1972; Cook & Haverbeke, 1977; Pal et al., 2000; Rao et al., 2004; Fang & Ling, 2003; Tyagi et al., 2006; Pathak et al., 2008). The level of urban noise has become a disaster for the community. Plants are more economical to prevent traffic noise than concrete walls. Plants are preferred because they contribute to the beauty of the environment. Green belt of vegetation with a spaced tree layout makes it impractical to dim noise (Kragh, 1979).Tang and Ong (1988) conclude that the distance between trees of 3 meters to 25 meters only has little effect on noise. Green belts of trees with a length of more than 30 meters are better able to reduce noise compared to open grass (Reethof, 1973).

Noise spectrum was attenuated by the green belt of a pine tree with a length of 100 meters (Allen, 2005). Rows of green belts of trees and shrubs are very important to reduce noise effectively. Random arrangement of tree sequences the ability to reduce noise is smaller than a well-planned array of trees (Byun and Ching, 1999). Rows of green belts of specially regulated trees can reduce the sound (Chow et al., 1996). Chow et al., (1999) conducted research by applying a different 3D time method to predict sound propagation in a green belt of vegetation with a length of 15 meters along the highway. The findings demonstrated that the green vegetation belt might lower road noise by three decibels. The noise level drops as the tree's diameter increases. Numerous studies on air quality have been conducted by scientists, organizations, and researchers. Additionally, they computed emission rates for various vehicle types; several emission factor models are employed globally. Traffic-related factors include density, flow, vehicle composition, v/c ratio, number of traffic lights per mile, signal coordination, and number of stops per mile. The aim of this study is to present an extensive overview of the literature covering air pollution, global air pollution from power generation, emission estimation, and the effects of air pollutants on human health.

The introduction of chemicals, particles into the atmosphere that harm other living things, including food crops, or harm the ecosystem or natural environment is known as air pollution. The most current average Air Quality Index (AQI) for Dhaka was 264; in February of this year, it reached a peak of 361.When the AQI deems to be anything over 200 as being "very unhealthy" and anything over 300 as being "extremely unhealthy," a proactive approach to the problem becomes imminent. It is no wonder then that air pollution has become a primary cause for illnesses and death, with about 123,000 losing their lives to it in 2017 alone. WHO (2021) reported that almost 90 percent of people living in the cities are being exposed to dangerous levels of air pollution. Outdoor air pollution killed 3.7 million people in 2012 and it is now the world's largest single environmental health risk (WHO, 2021).

At an emergency inter-ministerial meeting, it was decided that steps would be taken to stop burning garbage at waste dump sites and shut down all illegal brick kilns around the capital. It also directed the Dhaka north and south city corporations to sprinkle water on the city roads to reduce dust in the air. Md Shahab Uddin, minister for environment, forest and climate change, who presided over the meeting held at the conference room of the ministry. He stated that total emissions from the brick manufacturing in the Greater Dhaka region, has been estimated and found 23,300 tons of PM2.5, 15,500 tons of SO_2, 302,000 tons of CO, 6,000 tons of BC, and 1.8 million tons of CO_2 emissions from the clusters of brick kilns, to produce 3.5 billion bricks per year.

In the city wise assessment, Narayanganj has been marked as the 17th city with worst air quality whereas Gazipur and Dhaka have been ranked 21st and 23rd respectively. The air is elastic, invisible and tasteless mixture of gases that surrounds the earth. Despite being a vital component of all living things, air pollution results from a number of natural processes as well as human activity. Air pollution has global impacts that transcend national borders. Air pollution is the presence in ambient atmosphere of the substances, generally resulting from the activity of man, in sufficient concentration, present for a sufficient time and under circumstances, which interfere significantly with the comfort, health or welfare of persons or with the full use or enjoyment of the property. The air that we breathe can consist of many contaminants including ozone (O_3), the particulate matter, volatile organic compounds (VOCs). Each of these contaminants has specific impacts on human health and the environment.

Health effects due to the polluted air include breathing and respiratory problems, aggravation of existing respiratory and cardiovascular disease, increased strain on pulmonary and respiratory systems, reduction of lung function, asthma, eye and nose irritation, alterations in the body's defense system against foreign materials, damage to lung tissue, carcinogens and premature death. More than one-third of the population of Bangladesh lives in cities. More than half of Bangladesh's population is expected to live in cities by the middle of the century. In and around Dhaka and other Bangladeshi divisional cities, the majority of the population growth is concentrated. The concentration of particulate matter, the primary air pollutant of concern, has been gradually rising in Dhaka and other large cities due to over population. On an annual average, this concentration is far greater than what the World Health Organization (WHO) recommends. Air pollution and greenhouse gas (GHG) emissions are growing daily in the largest cities. Based on AQI, the air in cities of Bangladesh has serious consequences for the people living in these cities (Fang & Ling, 2003). Numerous studies emphasize the intricate relationships between individual plants and the quality of ambient air, indicating that various species can exert both beneficial and detrimental impacts (Sæbø et al., 2012; Yang et al., 2015; Churkina et al., 2015).Air pollution is one of the main environmental problems facing mankind. Now in many urban areas large concentration of human activities induces considerable amounts of pollutants to be accumulated (Liang et al., 2019). As it is well known that the main pollution sources are motor vehicle traffic, which heavily affects the air quality in the densely populated cities.

Air Pollution is detrimental to the buildings, technical devices and may cause serious health damage, as well. The nature and importance of the air quality problems depend on the geographical (climate, local meteorological conditions, position, relief) and the social (existing environmental regulations, urban planning choices) factors. Vegetation can be employed to create a roadside barrier, thereby reducing pedestrians' exposure to air pollution (Abhijit et al., 2017; Abhijit & Kumar, 2019; Baldauf, 2017; Tremper, 2015).Urban trees play a significant role in lowering the temperature of their immediate environment, thereby creating a green canopy along roadside (Guo et al., 2023).

Numerous studies have demonstrated that trees can decrease temperatures by influencing transpiration, providing shade, and other mechanisms, which in turn alleviates adverse effects on residential living and energy usage (Bartesaghi-Koc et al., 2019).In urban settings, trees and impervious surfaces, such as asphalt and concrete, display varying temperature profiles. Research has consistently shown that trees contribute to a reduction in land surface temperature by offering shade (Armson et al., 2012; Sabrin et al., 2021). Additionally, urban trees help lower land surface temperature by releasing water vapor through the process of transpiration and by absorbing significant amounts of latent heat (Rahman et al., 2020).

3.1 METHODOLOGY

The methodology is the organization of collection and analysis of relevant and significant data for the purpose of the research. The study is Bijoy Shoroni and Ramna Park area, Dhaka city will be designed to use quantitative method of study. This design was chosen because of the nature of monitoring data that is needed for the study. Air and Noise testing is carried out in high traffic areas as well as being a garden area in Bangladesh. The tool used SL – 4010, IAQ-2500TH, COH-9902SD, GC-2028, Q858166, XLA-BX-O3, AS847 (Class 1 Sound Level Meter and portable air quality meter) shows a value of Average noise and air quality level has been measured maximum 87.7 (dB A) at monitoring location (Bijoy Shoroni) and measured minimum 65.8 (dB A) at monitoring locations (Ramna Park), Air quality (PM10, PM2.5, CO, CO_2, O_3 and Temperature) has measured maximum (864 µg/m3, 321 µg/m3, 7.0 ppm, 630 ppm 0.0 µg/m3 29^{0}C) at the monitoring location (Bijoy Shoroni) and measured minimum (62 µg/m3, 34 µg/m3, 1.0 ppm, 429 ppm 0.0 µg/m3 23.3^{0}C) at the monitoring location at the monitoring location (Ramna Park).All the hand devise is calibrated at site prior to sampling and placed at a height of approximately 1.0 – 1.5 m from ground level. To prevent disturbance from reflecting surfaces, the noise meter microphone facing towards the noise source with clearance of around 3 meters from any structures is ensured.

3.2 Study Area

The noise level and air quality survey conducted in and around the Dhaka city area during winter season. Monitoring of the changes of Noise and Air pollution was done in two locations Bijoy Shoroni and Ramna Park (23°45'52.43"N, 90°23'18.44"E and 23°44'30.83"N, 90°23'54.30"E) at Dhaka city.

Figure 1. Location of monitoring and activity

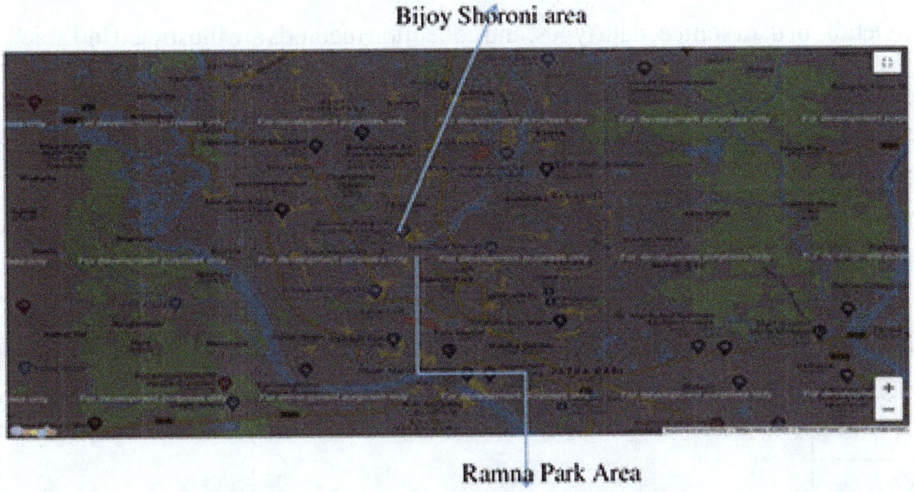

This is the schematic and highly visualized representation of the steps required to achieve the research objective. The research framework was developed through a list of steps that were used to achieve the research objectives. Data analysis was performed using Microsoft Office Excel and for the data presentation bar chart, pie chart, histogram and line graph were used. This research was conducted through comparative analysis method and questionnaire survey to estimate the casual impact of an intervention on its target population without random assignment and using quantitative data. Air quality and noise levels were measured at two locations using potable monitoring equipment (hand device), the monitor was set approximately 1.5 meters above the ground and no closer than 3 meters to any reflecting surface. Three (3) readings were taken in every hour at an interval of 20 minutes and therefore average Air quality and noise level was recorded hourly. Average hourly results from office holidays, mid of the office day, end of the office time and office hour. Out of the two locations (Bijoy Shoroni and Ramna Park area) are collect in-situ data in the office holidays, mid of the office day, end of the office time and office hour. Expert suggestion, consultations and cross-validation of 3 KIIs were taken for getting perception. In-situ data on noise and air quality (PM10, PM2.5, CO, CO_2, O_3 and Temperature) was collected from Bijoy Shoroni and Ramna Park (Figure 1). Collected data was analyzed using quantitative method in order to come up with necessary information for the study and quantitative data was transcribed and processed in themes and presented and discussed in the light of objective of the study.

3.3 Data Collection Method

Details of data source, data types, and collection methods are illustrated in Table 1.

Table 1. Data collection, measurement and analysis method

Parameters	In-situ Method	Analysis Method
Noise	Before sampling, the noise is calibrated on location and positioned between 1.0 and 1.5 meters above the ground. The noise meter microphone faces the noise source and is placed around three meters away from any structures to avoid disturbing reflected surfaces.	Class 1 Sound Level Meter complete with microphone (Model: SL – 4010
PM10	A portable gas sensor was used for the sampling. Sampling was done for less than an hour at 20-minute intervals. After then, the readings are averaged.	Electrochemical Sensor
PM2.5		
CO_2		
CO		
O_3		

3.4 Instruments of data collection

Data collection instruments were mentioned with model was presented in Table 2.

Table 2. Data collection instrument

Sl. No	Instruments name	Model Number	Monitoring parameter Name
1	Class 1 Sound Level Meter	SL – 4010	Noise Level
2	Particle counter	IAQ-2500TH	PM10 & PM2.5
3	CO Meter	COH-9902SD	CO
4	CO2 Meter	GC-2028, Q858166	CO_2
5	ES20 portable gas detector	XLA-BX-O3	O_3
6	Humidity and Temperature meter	AS847	°C

3.5 Sampling Procedure

Air Quality and Noise levels were assessed at two monitoring sites utilizing spontaneous Type potable monitoring equipment (hand device). The monitor was positioned approximately 1.5 meters above the ground and at least 3 meters away from any reflective surfaces. Three readings were recorded every hour at 20-minute intervals, resulting in an average Air Quality and Noise level documented as the respective hourly outcome. The average of hourly results was derived from Office holidays, the middle of the office day, the end of the office time, and the air and noise levels during office hours.

3.6 Sample Approach

Data was collected from the two locations (Bijoy Shoroni and Ramna Park area) on office holidays, the middle of the office day, the end of the office time, and during office hours. Expert suggestions, consultations, and cross-validation from three Key Informant Interviews (KIIs) were utilized to develop a perception questionnaire and to expedite the online survey process. The objective was to gather in-situ data on Noise and Air Quality (PM10, PM2.5, CO, CO2, O3, and Temperature) from the Bijoy Shoroni and Ramna Park areas.

3.7 Data Management

The data management derived from the study was analyzed using quantitative methods to extract essential information for the research. The quantitative data was transcribed, organized into themes, and presented and discussed in relation to the study's objectives. Following the data collection, the analytical methods employed to derive specific results are detailed in Table 2.

4.0 RESULTS

4.1 Description of the study areas

Description of the study areas with different aspects including locations, connectivity, mode of transportation, latitude and altitude and plant species were discussed in the subsections.

4.1.1 Ramna Park area

71 species of flowering trees, perennials, and annuals, 36 species fruit bearing plant, 33 species medicinal plant and 41 species of forest tree species and 11 other species were found in Ramna Park. This is the main road for communication in Dhaka city connecting Hare Road, Ramna, and Minto Road. The modes of communication are CNG, Auto Rickshaw, Motorcycle, bicycle, Heavy Truck, Medium Truck, Small Truck, Mini Bus, Large Bus, Car, and Microbus. A survey was conducted on 29-12-

2021 at a location inside the Ramna just before Minto Road (GPS: 23°44'30.83"N, 90°23'54.30"E) to count movement traffic from Vehicle movement pattern from Kazi Nazrul Islam Avenue towards Bir Uttam Ziaur Rahman road.

4.1.1.1 Plants species in Ramna Park area

Acacia auriculiformis; Albizia procera; Albizia richardiana; Alostania scholaris; Azadirachta indica; Candida voigt; Borassus flabellifer; Butea monosperma; Caesalpinia pulcherrima; Cassia fistula; Delonix regia; Ficus benghalensis; Nageshwor; champa; Sornochampa; Sapindus mukorossi; Cinnamomum camphor; Couroupita guianensis; Terminalia arjuna; Madhuca longifolia; Schleichera oleosa; Hopea odorata; and Saraca indica was found in Ramna Park.

4.1.2 Bijoy Shoroni area

In Bijoy shoroni, only flowering plants were found. The main road of communication in Dhaka city is Kazi Nazrul Islam Avenue - Bijoy Sarani - Bir Uttam Ziaur Rahman road. The mode of communications is CNG, Auto Rickshaw, Motorcycle, bicycle, Heavy Truck, Medium Truck, Small Truck, Mini Bus, Large Bus, Car, and Microbus. A survey was conducted on 29-12-2021 at a location inside the Bijoy Sarani just before Bir Uttam Ziaur Rahman road (GPS: 23°45'52.43"N, 90°23'18.44"E) to count movement traffic from Vehicle movement pattern from Kazi Nazrul Islam Avenue towards Bir Uttam Ziaur Rahman road.

4.1.2.1 Plants species at Bijoy Shoroni area

Ficus benghalensis; Caesalpinia pulcherrima; Borassus flabellifer; Delonixregia; Acaciaauriculiformis; Albiziaprocera; Azadirachtaindica; Candidavoigt; and Butea monosperma was found in Bijoy Shoroni.

4.2 Status of Noise level in the study areas

4.2.1 Noise level at Ramna Park

Noise monitoring wasdonein thestudy areas using tools (Class 1 Sound Level Meter and portable air quality meter) and average noise level was found with a minimum 65.8 (dB A) during Office holidays and maximum 77.7 (dB A) during Office hourin Ramna Park area (Figure 2).

Figure 2. Noise quality at Raman Park area

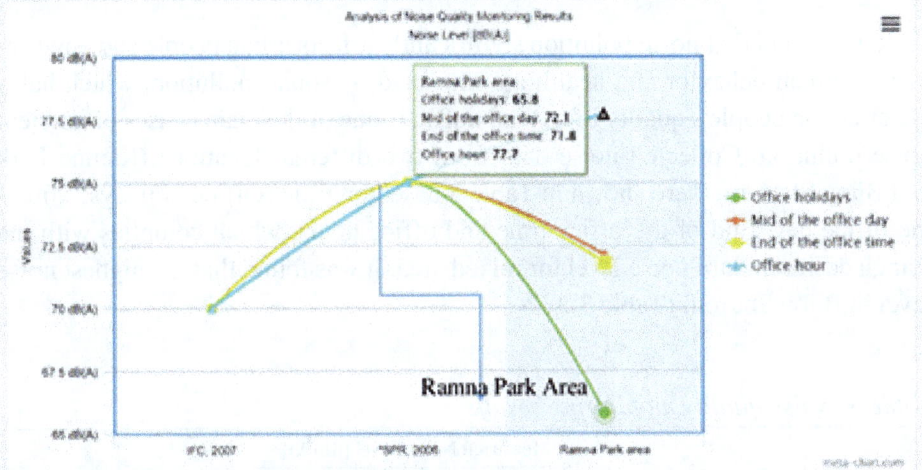

4.2.2 Noise level at Bijoy Shoroni

Results showed that the average noise level with a minimum 80.2 (dB A) during Office holidays and maximum 81.0 (dB A) during mid of the Office Dayin Bijoy Shoroni (Figure 3).

Figure 3. Noise quality at Bijoy Shoroni area

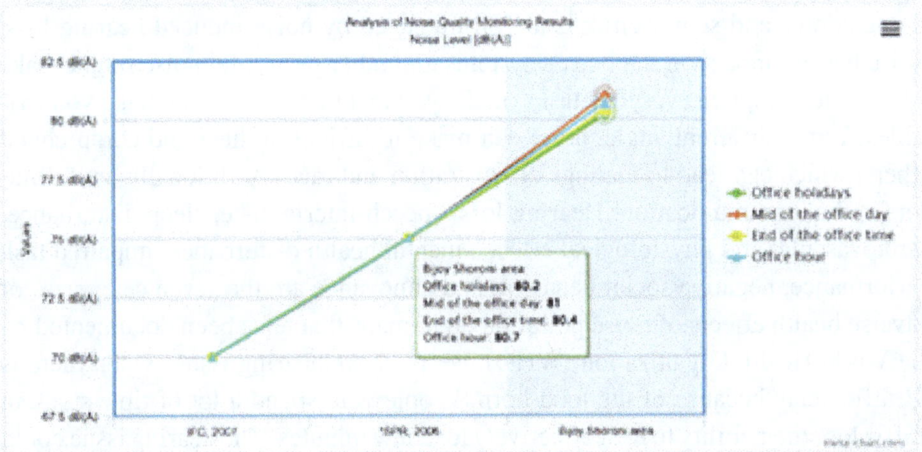

4.3.1 Noise Quality

It was found that noise pollution significantly affects urban people and environment. Human behavior and health are impacted by sound pollution, which has a big effect on people's quality life. Road traffic is one of the main causes of ambient noise pollution. Collected noise data from two different locations (Ramna Park and Bijoy Shoroni) were shown in Table 3 based on time (Office holidays, Mid of the office day, End of the office time and office hour), which complies with the Bangladesh standard noise level for mixed area. It was found that the highest noise level in Bijoy Shoroni (Table 3).

Table 3. Noise quality monitoring results

Date	Time	Measured Noise Level [dB(A)]		*SPR, 2006	IFC, 2007
		Ramna Park	Bijoy Shoroni		
04.12.2021	Office holidays	65.8	80.2	75	70
16.12.2021	Mid of the office day	72.1	81.0	75	70
20.12.2021	End of the office time	71.8	80.4	75	70
29.12.2021	Office hour	77.7	80.7	75	70

4.3.2 Impacts of noise pollution on health

Hearing loss is the physical consequence of noise exposure. Children, teenagers, young adults, and senior citizens are all impacted by noise-induced hearing loss. Noise has an impact on the body's systems and other parts, and noise might make it harder to complete everyday tasks, make you feel more tired, and make you irritable. From a different angle, noise can make it difficult to hear and comprehend others, which can lead to feelings of frustration and rage. Additionally, noise hinders verbal communication. Hearing loss, speech interference, sleep disturbance, cardiovascular and physiological effects, mental health disturbance, impaired task performance, negative social behavior, and annoyance are the seven categories of adverse health effects of noise pollution on humans that have been documented by the World Health Organization (WHO). People have hearing issues when there is a traffic delay because of the loud horn. People who spend a lot of time stuck in traffic lose their ability to hear effectively for a few minutes. The hearing issue could become irreversible if this keeps up.

4.4.1 Air Quality

Various pollutants are affected in distinct manners by diverse species and their combinations, as well as by specific conditions at each site. Across the monitoring locations of survey areas, three samples were taken at each monitoring location and concentration of particulate matter (PM10, PM2.5), Ozone (O_3), Carbon Monoxide (CO) and Carbon dioxide (CO_2) were measured and it is insignificant during November to December 2021. Across the monitoring locations, measured concentrations of air quality monitoring parameters PM10, PM2.5, O_3 and CO have been found within the standard limits expect PM10 and PM2.5 prescribed in the 'amended Schedule-2 of Bangladesh Environment Conservation Rules 1997' and WHO/World Bank Standard refer to the 'Air Quality Guideline Global Updated, 2005/ WB Guideline - EHS General 2007'. There is no standard limit set for the monitoring parameter CO_2. Considering all factors, the situation is worst in Bijoy Shoroni (Table 4). For instance, vegetation serves as an ineffective sink for various gaseous pollutants, such as nitrogen dioxide (NO2), whose deposition is restricted to stomatal uptake and consequently tends to happen during the warmer months, when NO2 poses a reduced risk. Likewise, nitric acid (NO) is inadequately addressed by vegetation, in part because of the substantial NO emissions from vegetated soils relative to bare soils.

Table 4. Air quality data

Monitoring area	Date	Time	Air Pollutants Concentration					
			PM10	PM2.5	CO	CO2	O3	Tem:
Ramna Park	04.12.2021	Office holidays	86	34	1.0	506	0.0	24.9
Bijoy Shoroni			455	225	6.0	**630**	0.0	26.1
Ramna Park	16.12.2021	Mid of the office day	94	47	1.0	555	0.0	28.9
Bijoy Shoroni			**864**	**321**	5.0	579	0.0	**29.0**
Ramna Park	20.12.2021	End of the office time	87	35	2.0	440	0.0	23.6
Bijoy Shoroni			620	315	**7.0**	560	0.0	24.5
Ramna Park	29.12.2021	Office hour	62	34	1.0	429	0.0	23.3
Bijoy Shoroni			579	251	4.0	534	0.0	24.5
Unit			µg/m3	µg/m3	ppm	ppm	µg/m3	°C
Bangladesh Standard (according to the amendment of Schedule-2, July'05 of ECR'97)			150	65	9	-	235	-
IFC, 2007			150	75	-	-	160	-

4.4.2 Coarse Particle (PM10)

It represents particles with an aerodynamic diameter of 10µm or less, or, more strictly, particles, which pass through a size selective inlet with a 50% efficiency cut-off at 10µm aerodynamic diameter. Across the survey area, three samples were taken at each monitoring location and concentration of particulate matter (PM10) was measured insignificant during December 2021. 1-hour average concentration of PM10 has measured maximum 864 µg/m3 at the monitoring location N&A-02 (Bijoy Shoroni) and measured minimum 62 µg/m3 at the monitoring in Ramna Park (Figure 4).

Figure 4. Coarse Particle (PM10) monitoring results

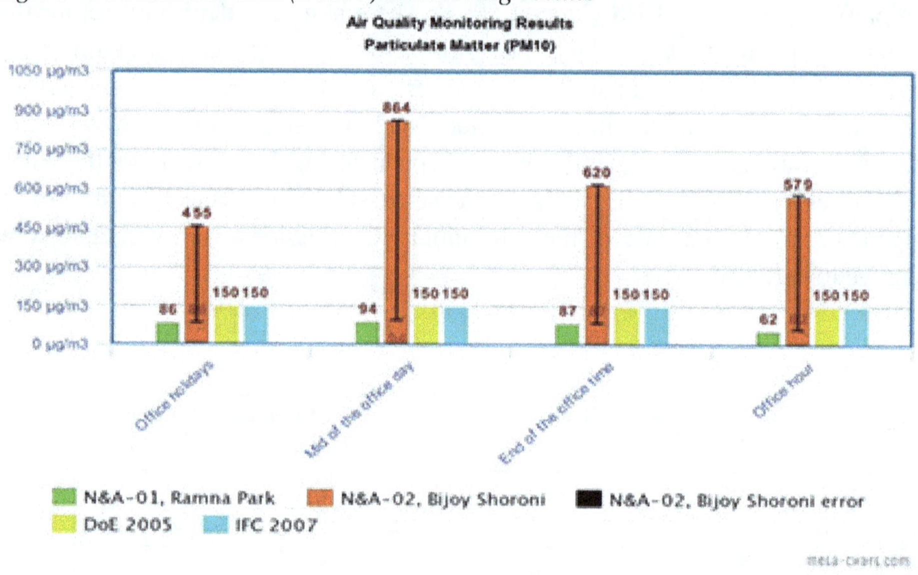

4.4.3 Fine Particle (PM2.5)

The size of PM2.5 is considered only one hair on our head. The largest fine particle is 30 times smaller than the normal human hair, which has a diameter of roughly 70 micrometers. Three samples were collected at each of the survey area's monitoring points in December 2021, the particulate matter (PM2.5) concentration was found to be negligible. The monitoring location N&A-02 (Bijoy Shoroni) recorded the highest 1-hour average PM2.5 concentration at 321 µg/m3, while the Ramna Park recorded the lowest at 34 µg/m3 (Figure 5).

Figure 5. Result of fine particulate matter (PM2.5)

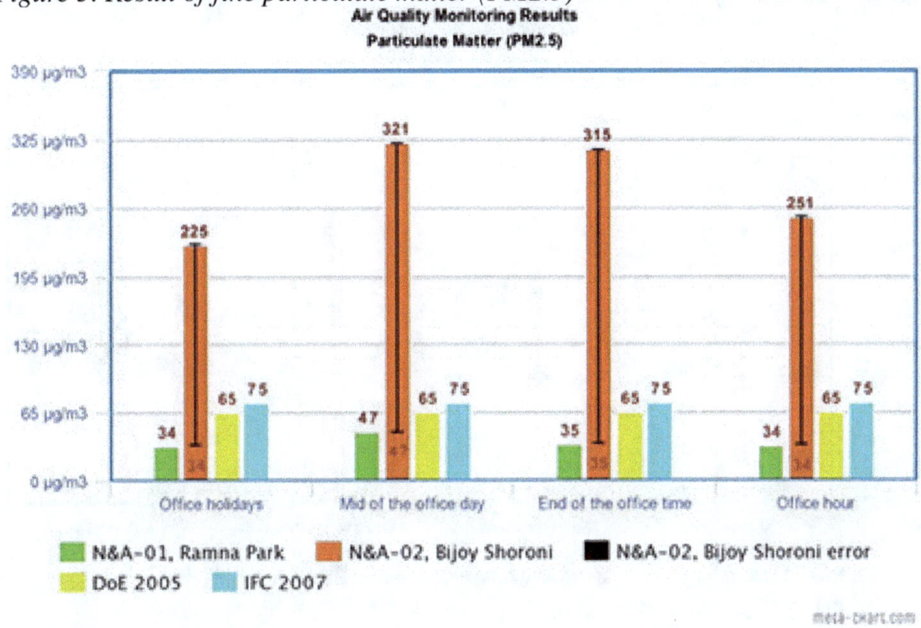

4.4.4 Carbon Monoxide (CO)

Air contains the gas carbon monoxide. Humans are poisoned by high concentrations of carbon monoxide, which is sadly invisible to the naked eye and has no taste or smell. Human health is unaffected by the natural concentration of carbon monoxide in the air, which is about 0.2 parts per million (ppm). The primary sources of extra carbon monoxide are the exhaust from motor vehicles and certain industrial processes, such the production of steel. One of the primary indoor sources of carbon monoxide is tobacco smoking. The recorded CO concentrations in December 2021 ranged from 1.0 to 7.0 ppm and it was negligible at both locations (Figure 6).

Figure 6. Results of Carbon Monoxide (CO)

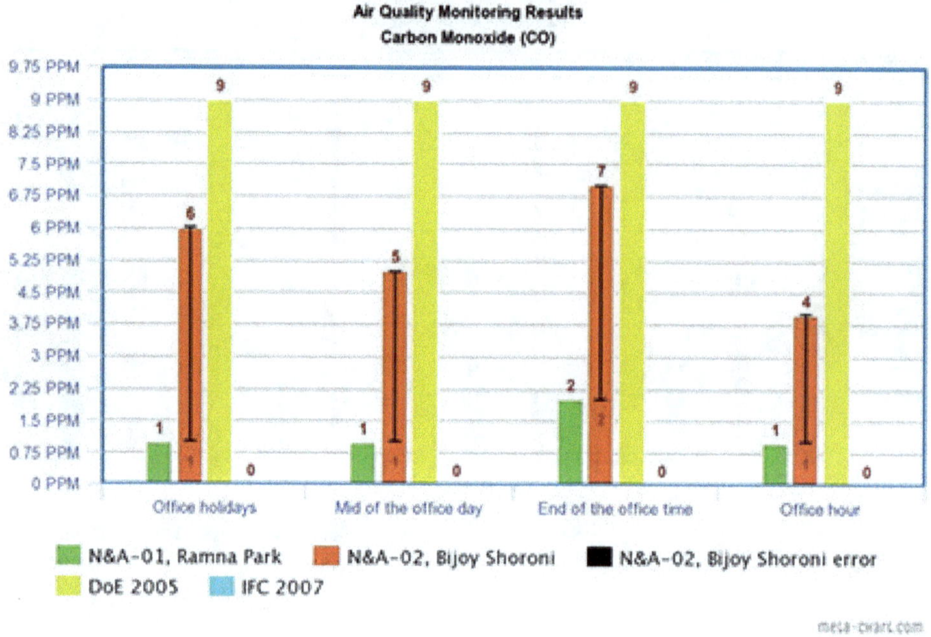

4.4.5 Carbon dioxide (CO_2)

Carbon dioxide is a chemical compound consisting of one part carbon and two parts oxygen and represented by the chemical formula CO_2.Carbon dioxide is one of the most significant gases on Earth for several reasons. Through a process called photosynthesis, plants use carbon dioxide to create carbohydrates, or sugars and starches. The globe remains warm enough for plant and animal life, including human life, because of its heat. Long-term changes in Earth's climate could be caused by rising atmospheric carbon dioxide level. It was found that CO_2 concentrations insignificant and ranges 429 - 630 ppm during December 2021 (Figure 7).

Figure 7. Monitoring result of Carbon dioxide (CO_2)

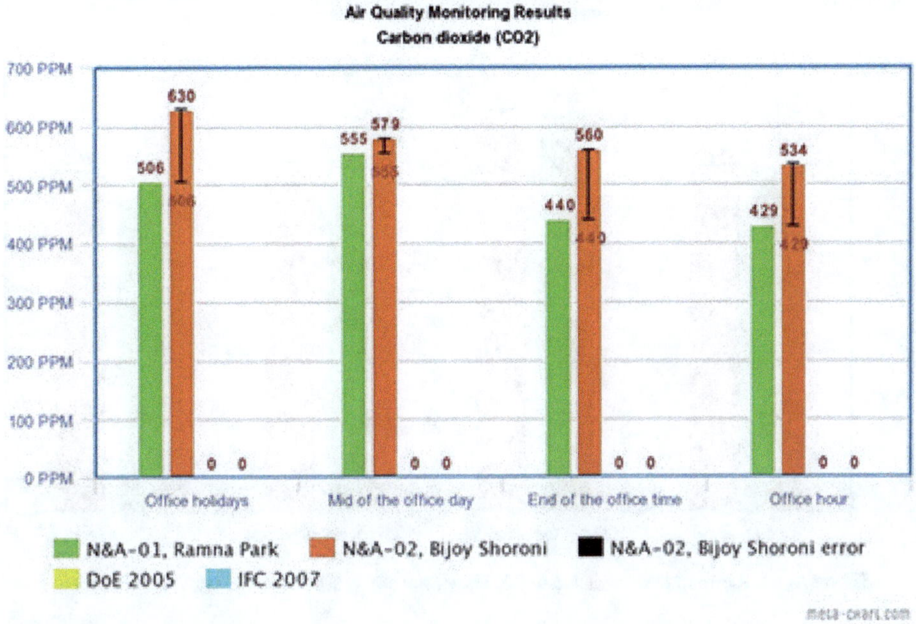

4.4.6 Ozone (O_3)

Ozone, a colorless gas with a strong smell and created by the interaction of hydrocarbons and nitrogen oxides in the presence of sunlight rather than being released straight into the atmosphere. In December 2021, it was noticed that $O_3$was negligible at both locations.

4.4.7 Temperature (^0C)

Temperature concentrations in December 2021 ranged from 23.3 to 29^{0}C, and it was found that the temperature is more in Bijoy Shoroni (Figure 8)

Figure 8. Temperature (^0C) monitoring result

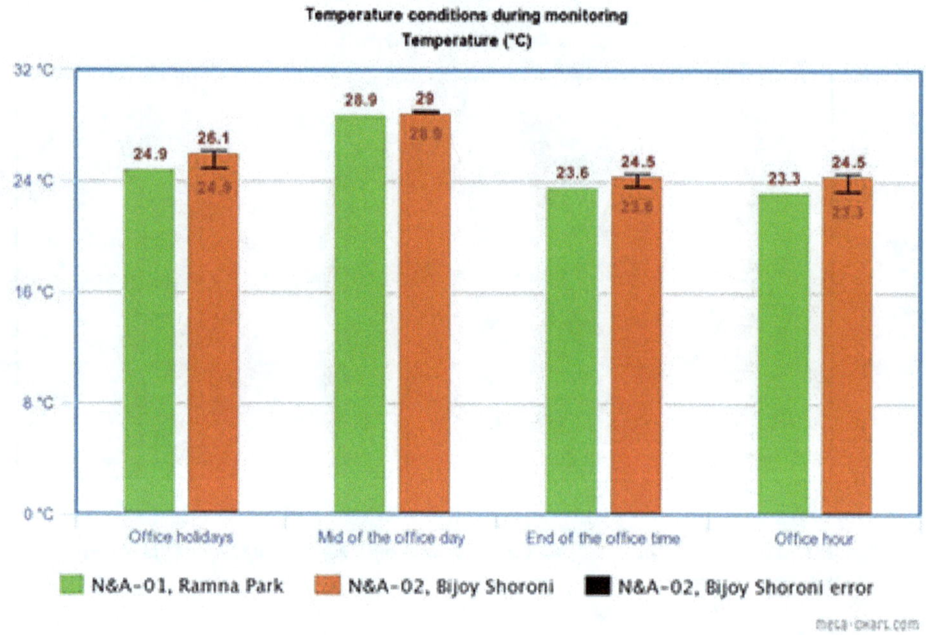

4.5.1 Impacts of Air pollution on Health

PM10 and PM2.5 include inalienable particles that are small enough to penetrate the thoracic region of the respiratory system. Inhalable PM has a well-established negative impact on health. These include the following and are caused by exposure over both short-term (hours, days) and long-term (months, years) periods: cardiovascular and respiratory morbidity, including worsening asthma symptoms, respiratory symptoms, and a rise in hospitalizations. Mortality from lung cancer as well as respiratory and cardiovascular conditions, irritation of the eyes, nose and throat, premature death in people with heart or lung disease, coughing, chest tightness and shortness of breath irregular heartbeat and heart attacks. Carbon monoxide affects healthy and unhealthy people. Increased levels of carbon monoxide reduce the amount of oxygen carried by hemoglobin around the body in red blood cells. Pollen is a term that encompasses individual pollen grains, which are responsible for carrying the male reproductive cells of plants. In angiosperms, pollen is generated within the anther, while gymnosperms produce it in the male cone. Allergic reactions in humans to airborne pollen are 20% more prevalent in urban settings compared to rural ones, partly due to the uniformity of species and the interactions between pollen emissions and pre-existing air pollutants.

4.6 Mostly noisy and air polluted area

According to findings, noise level was found at Ramna Park during Office holidays, Mid of the office day, end of the office time and office hour as 65.8 dBA, 72.1 dBA, 71.8 dBA and 77.7 dBA respectively (Table 5).In Bijoy Shoroni noise level was found duringoffice holidays, mid of the office day, end of the office time and office hour as 80.2 dBA, 81.0 dBA, 80.4 dBA and 80.7 dBA respectively. Average noise level has been measured maximum 87.7 (dB A) inBijoy Shoroni and was minimum 65.8 (dB A) in Ramna Park. Air quality (PM10, PM2.5, CO, CO_2, O_3 and Temperature) was found with the maximum (864 µg/m3, 321 µg/m3, 7.0 ppm, 630 ppm 0.0 µg/m3 29^{0}C in Bijoy Shoroni and minimum (62 µg/m3, 34 µg/m3, 1.0 ppm, 429 ppm 0.0 µg/m3 23.3^{0}C) in Ramna Park (Table 6). More noise and air pollution were recorded in Bijoy Shoroniin the mid of office day.

4.7 Source of noise and air pollution

The most point source of disturbance and pollution or contamination comes from the CNG, Auto Rickshaw, Cruiser, bike, Overwhelming Truck, Medium Truck, Little Truck, Smaller than expected Transport, Expansive Transport, Car, and Microbus.

4.8 Reasons of noise and air pollution

The most reason of upheaval pollution from transport are need of driving information, need of law enforcement, vehicle speed within the same street, need of activity run the show, unintended stoppage, poor signaling system, lack of mindfulness, huge number of vehicles within the street, and careless driving.

4.9 Role of plant species in controlling Noise and Air pollution.

Results showed that ozone concentration level is negligible in both locations. It means that there is no significant contribution of plant species on ozone concentration regulation. On the other hand, plant species contributed in noise level regulations, PM 10, PM 2.5 concentration carbon monoxide and carbon dioxide regulations (Table 5). The list of species ought to be broadened to encompass a growing number of species as additional research data is obtained. The variety present not only in species but also in types and structures of vegetation plays a crucial role in maintaining air quality, facilitating carbon sequestration, and ensuring the sustainable delivery of a range of ecosystem services.

Table 5. The contribution of plant species in controlling noise and air quality percentage analysis result

Data Collection Date	Monitoring area	Data collection Time	The contribution of plant species in controlling Noise and Air pollution					
			Noise Level [dB(A)]	**PM_{10} (µg/m3)**	**$PM_{2.5}$ (µg/m3)**	**CO (ppm)**	**CO_2 (ppm)**	**O_3 (µg/m3)**
04.12.2021	Ramna Park	Office holidays	**18%**	81.10%	84.90%	**83.30%**	19.70%	0.00%
	Bijoy Shoroni							
16.12.2021	Ramna Park	Mid of the office day	11%	89.10%	85.40%	80%	4.10%	0.00%
	Bijoy Shoroni							
20.12.2021	Ramna Park	End of the office time	10.70%	86%	**88.90%**	71.40%	**21.40%**	0.00%
	Bijoy Shoroni							
29.12.2021	Ramna Park	Office hour	3.70%	**89.30%**	86.50%	75.00%	19.70%	0.00%
	Bijoy Shoroni							
			Max 18%	Max 89.30%	Max 88.90%	Max 83.30%	Max 21.40%	Max 0.00%
			Min 3.70%	Min 81.10%	Min 84.90%	Min 71.40%	Min 4.10%	Min 0.00%

4.10 The contribution of plant species in temperature (^{0}C).

A greater number of plant species were identified in Ramna Park, and the following low temperatures were recorded: during office holidays (4.6%), in the middle of the office day (0.3%), at the end of office hours (3.7%), and throughout the office hours (4.9%). The impact of plant species on reducing temperature (°C) is presented in Table 6.

Table 6. The contribution of plant species in temperature (^{0}C) lowering

Data collection Date	Monitoring area	Data collection Time	The contribution of plant species in temperature (0C)
04.12.2021	Ramna Park	Office holidays	4.6%
	Bijoy Shoroni		
16.12.2021	Ramna Park	Mid of the office day	0.3%
	Bijoy Shoroni		

continued on following page

Table 6. Continued

Data collection Date	Monitoring area	Data collection Time	The contribution of plant species in temperature (0C)	
20.12.2021	Ramna Park	End of the office time	3.7%	
	Bijoy Shoroni			
29.12.2021	Ramna Park	Office hour	4.9%	
	Bijoy Shoroni			
			Max	4.9%
			Min	0.3%

5.1 MITIGATION MEASURES

Noise pollution is mostly related to occupational health impacts, to which due attention has to be provided. It is possible to reduce noise by shielding by barriers, monitoring of noise quality at the traffic area, replacing noisy machines with low noise/ noiseless vehicle, regular maintenance of vehicles, and establishing green belt especially surrounding Bijoy Shoroni traffic area. The diverse effects of different species on air quality are illustrated by the extensive variety of phenotypes. Furthermore, effective plant selection aimed at mitigating air pollution requires an understanding of the balance between the beneficial and detrimental traits of plants at both the species and subspecies levels, while also considering the planting environment. In the following sub-sections, we will conduct a critical evaluation of significant findings concerning the cost-benefit relationships associated with species selection for urban air quality management.

In open-road environments, woody plants can be employed to establish a continuous barrier. Research suggests that these barriers may improve air quality on the pedestrian side; however, it is crucial to take into account important factors such as the barrier's height, width, and porosity. Nevertheless, researchers frequently overlook the positive and negative impacts of built structures, impermeable surfaces, and green areas surrounding trees. In general, numerous contemporary studies examining the capacity of urban trees to mitigate land surface temperature fail to account for the three-dimensional vertical attributes of trees on a large scale, and often disregard the influence of their environment, particularly at the national level.

5.2 Discussion

The Bijoy Shoroni area is impacted by noise and air pollution, as illustrated in the reference tables (1 & 2). The primary contributors to noise and air pollution are vehicles in motion (Table 3). Factors leading to noise and air pollution have been identified, including a lack of awareness, insufficient knowledge, and violations of traffic regulations. The analysis indicates that the presence of various plant species plays a crucial role in mitigating noise pollution in the Ramna Park area, owing to the abundance of plant species. A study showed that the temperature in tree-covered regions was approximately 2.23 °C lower than that of the adjacent urban areas during the summer (Guo et al., 2023). Additionally, these plant species remarkably contribute to reducing the temperature (°C) in the Ramna Park area. A study reported that three major causes of noise pollution include construction sites, airplanes, and vehicles on the roads, according to a research (Hashim et al., 2024). With an average noise level of 119 dBA, Dhaka, the capital of Bangladesh, was declared the world's noisiest metropolis by the United Nations Environment Program (UNEP) in March 2022 (Javaid, 2022).This exceeds the recommendations set forth by the World Health Organization (WHO), which indicate that prolonged exposure to noise levels above 70 dBA can raise the risk of hearing impairment (WHO, 2021).Carbon monoxide is discharged into the environment as a result of increased traffic (Abdulgafar, 2021). Despite not being a direct driver of climate change, carbon monoxide affects the concentration of greenhouse gases such as carbon dioxide and methane. On the other hand, climate mitigation techniques aim to address the root causes of climate change while reducing any potential negative effects of noise and air pollution (Iberdrola, 2023). It has been shown that plants with thick leaves, a wider surface and width, and a longer length should be selected in order to maximize noise absorption in the mitigation of noise pollution (Hashim et al., 2024). Given the highly stressful planting conditions present in urban settings and along roadways, we propose a rationale of negative selection to establish an initial list of appropriate candidates, prior to implementing positive selection for species anticipated to improve deposition. This rationale aligns with the principle that we can more reliably identify what would be harmful rather than what would be advantageous, considering the context-dependent nature of beneficial vegetation characteristics. The species thus necessitate evaluation in field conditions. Various pollutants are affected in distinct manners by different species and their combinations, as well as by site-specific factors. Ensuring that the chosen species possesses a blend of advantageous characteristics can greatly enhance deposition, and where feasible, integrating a variety of species can assist in addressing different types of pollutants (Leonard et al., 2016).Additional investigation into each species is necessary prior to implementation to prevent possible trade-offs with ecosystem services beyond the mitigation of air pollution (Salmond et al., 2016). Air quality is an integral component of a multifaceted socio-ecological system, rendering the isolation and manipulation of functions for a particular goal (such as enhanced pollution mitigation) quite challenging. Nevertheless, by considering the spatial scale of the intervention, the context and conditions of the planting site, as well as the specific type of air pollutant targeted, it is possible to select plants with particular biophysical characteristics that can enhance the dispersive and depositional effects of a roadside vegetation barrier (Grote et al., 2016).Moreover, it is underpinned by the idea that maintaining a lack of substantial decline in air quality is a crucial prerequisite for the effective selection of species aimed at enhancing air quality (Tallis et al., 2011). A study mentioned that the impact of urban vegetation on the negative consequences of climate change cannot be easily classified as either beneficial or detrimental, as the function of urban greenery is intricately connected to the structure, composition, and distribution of the vegetation, in addition to the management criteria applied. Consequently, this understanding may enhance our comprehension of urban green spaces as a multifaceted system and lay the groundwork for subsequent research (Ferrini et al., 2020).In response to the deterioration of air quality, it is important to highlight that, as far as we are aware, no artificial devices have been developed as a means of control. Rather, the implementation of a natural heterogeneous vegetation layer, which possesses the capability to cleanse the air, is employed as the most effective strategy in urban settings to alleviate the effects of air pollution. The selection of the species should be informed by various parameters, including the percentage of pollutants eliminated, the daily release of volatile organic compounds, the generation of pollen and allergens, as well as the impact on urban heat in adjacent areas. It is essential to consider all these elements in accordance with the principle of “the right plant in the right place and with the right management”; only ensuring the survival of the plants is insufficient (Ferrini et al., 2020). They must also exhibit, for instance, elevated rates of photosynthesis and growth, thereby enhancing their environmental contributions. Consequently, the choice of plants is a critical aspect of a sustainable initiative aimed at maintaining the health and vitality of our urban environments. Although significant efforts are still required to ascertain the most effective configuration of green infrastructures within urban settings, there exists ample information to initiate proactive measures for climate change mitigation. A research indicates that the land surface temperature associated with tree cover during the summer months was lower compared to that of adjacent building sites, which were predominantly affected by the process of transpiration (Manickathan et al., 2018). Tree transpiration represents a crucial physiological mechanism and acts as the principal method for temperature regulation within the tree itself. The leaves of plants emits water vapor into the ambient atmosphere through the process of transpiration, interacting with surrounding air molecules and absorbing heat from the environment, which results in a reduction of leaf temperature (Konarska et al., 2016).

5.3 CONCLUSION

Plant species have the ability to reduce pollution including air (carbon dioxide, carbon mono oxide, ozone, and particulate matters), noise pollution and reduction of temperature. The high volume of traffic in Dhaka city has led to a rise in both air and noise pollution. Within the educational sector, noise pollution has surged considerably, disrupting both residents and students. Such disturbances can adversely affect physical health and mental well-being. It is crucial to introduce greenery that can mitigate noise along the roadways in Dhaka city, particularly in heavily trafficked areas. Plants have potential capacity to reduce noise and air pollution in the crowded cities. Due to the factors outlined in the sections titled 'Environmental pollution' and 'Species selection', the sole criterion for short listing species for inclusion was their documented tolerance to air pollution. Several characteristics present on leaf surfaces are believed to enhance deposition. As a result, plants that display one or more of these traits, in conjunction with any of the previously mentioned biophysical attributes, are considered to be relatively efficient. Vegetation with extremely low porosity can impede the removal of pollutants by restricting infiltration, which leads to air pollutants either flowing above and around the barrier or recirculation and accumulating on the upwind or source side. This research serves as a reference for managers to assess pollution-related factors of selected tree species and to develop an effective green barrier aimed at alleviating air pollution in urban environments. In the present study, we have identified the floral traits influenced by atmospheric air pollution, which can serve as ecosystem drivers for both pollution reduction and the floral dynamism of a specific area. The results of this research suggest that plants have the potential to function as effective quantitative and qualitative indicators of pollution, considering that the biomonitoring of plants is a vital approach for evaluating the impacts of air pollution on vegetation. This research demonstrates that the primary factors affecting tree cooling intensity offer insights for enhancing urban tree management to mitigate land surface temperature. In summary, these results enhance our comprehension of the role trees play in lowering land surface temperature.

5.4 RECOMMENDATIONS

The following recommendations should be considered in research, operations and policy implications:

- During license renewal process, BRTA should measure noise and emissions levels of all vehicles to ensure that respective vehicles are in compliance with the national limits prescribed in control rules 2006 & ECR 1997 for sound pollution and Bangladesh Standard according to the amendment of Schedule-2, July'05 of ECR'97 for air pollution.
- During driving license issued. BRTA should be provided day long awareness training on the impact of noise and air pollution on health and environment.
- The concepts of 'Trade-offs in plant selection' and 'Sustainability and species suitability' can serve as foundational guidelines for practitioners engaged in the selection of suitable plants when creating vegetation barriers, especially in open-road settings, aimed at enhancing urban air quality.
- Need to a lot of tree plantation surrounding at the Bijoy Shoroni area.
- Further research is required to know the types of plant species contribute more in air pollution control.
- Future research avenues have also been recognized, encompassing: the effects of air pollution on the dispersion of pollen and the allergenic potential linked to various plant species; the impact of specific plants and their characteristics (such as coniferous leaf morphology) on air pollution, especially concerning sub-micrometer particulate matter and gaseous pollutants.
- Furthermore, it is advisable that in instances where research is conducted on the impact of micromorphology on processes such as deposition, the relevant species-specific data should, whenever possible, be made easily accessible.

5.5 STUDY LIMITATIONS

Our study has several limitations. Firstly, we focused solely on the contribution of plants during daytime without examining the distinctions between trees serving different purposes, such as those in parks versus those along roads. Secondly, we were unable to acquire more detailed data on tree cover and land surface temperature, which led to diverse issues that influenced our findings.

AUTHORS CONTRIBUTION

Islam Md. Shafiqul contributed in conceptualization, methodology, writing review & editing.

Rahman Hafizur contributed in conceptualization, methodology, data collection, analysis and writing draft

DECLARATION OF COMPETING INTEREST

The authors affirm that they possess no recognized competing financial interests or personal relationships that might have seemed to affect the work presented in this paper.

REFERENCES

Abdulgafar, T. (2021). *Noise pollution & global warming [What science says]*. KRISP blog.

Agarwal, V. P., & Sharma, V. K. (1980). *Today and tomorrow's printers and publishers*.

Allen, P. (2005). Personal communication. *(Note: In APA 7, personal communications are cited only in the text and are not included in the reference list. The entry above is provided for completeness.)*

Armson, D., Stringer, P., & Ennos, A. R. (2012). The effect of tree shade and grass on surface and globe temperatures in an urban area. *Urban Forestry & Urban Greening*, *11*(3), 245–255. DOI: 10.1016/j.ufug.2012.05.002

Aylor, D. E. (1972). Noise reduction by vegetation and ground. *The Journal of the Acoustical Society of America*, *51*(1B), 197–205. DOI: 10.1121/1.1912830

Baldauf, R. (2017). Roadside vegetation design characteristics that can improve local, near-road air quality. *Transportation Research Part D, Transport and Environment*, *52*, 354–361. DOI: 10.1016/j.trd.2017.03.013

Bartesaghi-Koc, C., Osmond, P., & Peters, A. (2019). Spatio-temporal patterns in green infrastructure as driver of land surface temperature variability: The case of Sydney. *International Journal of Applied Earth Observation and Geoinformation*, *83*, 101903. DOI: 10.1016/j.jag.2019.101903

Begum, B. A., & Hopke, P. K. (2018). Ambient air quality in Dhaka Bangladesh over two decades: Impacts of policy on air quality. *Aerosol and Air Quality Research, 18*(7), 1910–1920. DOI: 10.4209/aaqr.2017.11.0465

Byun, D. W., & Ching, J. K. S. (1999). Science algorithms of the EPA Models-3 Community Multiscale Air Quality (CMAQ) modeling system. United States Environmental Protection Agency, Office of Research and Development., EPA/600/R-99/030.

Chepesiuk, R. (2005). Decibel hell: The effects of living in a noisy world. *Environmental Health Perspectives, 113*(1), A35–A41. DOI: 10.1289/ehp.113-a34

Chow, J. C., Watson, J. G., Lowenthal, D. H., Hackney, R., Magliano, K., Lehrman, D., & Smith, T. (1999). Temporal variations of PM2.5, PM10, and gaseous precursors during the 1995 integrated monitoring study in central California. *Journal of the Air & Waste Management Association, 49*(9), 16–24. DOI: 10.1080/10473289.1999.10463909

Chow, J. C., Watson, J. G., Lu, Z., Lowenthal, D. H., Frazier, C. A., Solomon, P. A., Thuillier, R. H., & Magliano, K. (1996). Descriptive analysis of PM2.5 and PM10 at regionally representative locations during SJVAQS/AUSPEX. *Atmospheric Environment, 30*(12), 2079–2112. DOI: 10.1016/1352-2310(95)00402-5

Churkina, G., Grote, R., Butler, T. M., & Lawrence, M. (2015). Natural selection? Picking the right trees for urban greening. *Environmental Science & Policy, 47*, 12–17. DOI: 10.1016/j.envsci.2014.10.014

Cook, D. I., & Haverbeke, D. F. (1977). Suburban noise control with planting and solid barrier combinations []. University of Nebraska–Lincoln.]. *Research Bulletin (International Commission for the Northwest Atlantic Fisheries)*, •••, 100.

Dilawar, A., Chen, B., Ashraf, A., Alphonse, K., Hussain, Y., Ali, S., Jinghong, J., Shafeeque, M., Boyang, S., Sun, X., & Hussain, S. (2022). Development of a GIS based hazard, exposure, and vulnerability analyzing method for monitoring drought risk at Karachi, Pakistan. *Geomatics, Natural Hazards & Risk, 13*(1), 1700–1720. DOI: 10.1080/19475705.2022.2090863

Fang, C. F., & Ling, D. L. (2003). Investigation of the noise reduction provided by tree belts. *Landscape and Urban Planning, 63*(4), 187–195. DOI: 10.1016/S0169-2046(02)00190-1

Ferrini, F., Fini, A., Mori, J., & Gori, A. (2020). Role of vegetation as a mitigating factor in the urban context. *Sustainability (Basel), 12*(10), 4247. DOI: 10.3390/su12104247

Gidlöf-Gunnarsson, A., & Öhrström, E. (2007). Noise and well-being in urban residential environments: The potential role of perceived availability to nearby green areas. *Landscape and Urban Planning, 83*(2–3), 115–126. DOI: 10.1016/j.landurbplan.2007.03.003

Grote, R., Samson, R., Alonso, R., Amorim, J. H., Cariñanos, P., Churkina, G., Fares, S., Le Thiec, D., Niinemets, Ü., Mikkelsen, T. N., & Paoletti, E. (2016). Functional traits of urban trees: Air pollution mitigation potential. *Frontiers in Ecology and the Environment, 14*(10), 543–550. DOI: 10.1002/fee.1426

Guo, A., He, T., Yue, W., Xiao, W., Yang, J., Zhang, M., & Li, M. (2023). Contribution of urban trees in reducing land surface temperature: Evidence from China's major cities. *International Journal of Applied Earth Observation and Geoinformation, 125*, 103570. DOI: 10.1016/j.jag.2023.103570

Hashim, W., Jamaludin, S. M., Mohamed, R. M. A. R., Gunasekaran, S. S., Thabit, R., & Gohel, H. (2024, July). Analysing type of plants for reducing noise pollution in climate change adaptation. []. IOP Publishing.]. *IOP Conference Series. Earth and Environmental Science, 1372*(1), 012056. DOI: 10.1088/1755-1315/1372/1/012056

Hossain, M. A., & Huggins, R. (2021). The environmental and social impacts of unplanned and rapid industrialization in suburban areas: The case of the greater Dhaka region, Bangladesh. *Environment & Urbanization Asia, 12*(1), 73–89. DOI: 10.1177/0975425321990319

Iberdrola. (2023). *Adapting to climate change: What will the earth look like in 2030?* Nature.

Iqbal, M. M., Li, L., Hussain, S., Lee, J. L., Mumtaz, F., Elbeltagi, A., Waqas, M. S., & Dilawar, A. (2022). Analysis of seasonal variations in surface water quality over wet and dry regions. *Water (Basel), 14*(7), 1058. DOI: 10.3390/w14071058

Jarosińska, D., Héroux, M. È., Wilkhu, P., Creswick, J., Verbeek, J., Wothge, J., & Paunović, E. (2018). Development of the WHO environmental noise guidelines for the European region: An introduction. *International Journal of Environmental Research and Public Health, 15*(4), 813. DOI: 10.3390/ijerph15040813

Javaid, A. (2022). *Top 15 noisiest cities in the world: Check rankings of Moradabad, Kolkata, and Asansol.* Jagran Josh. https://www.jagranjosh.com/general-knowledge/top-15-noisiestcities-in-the-world-1648473437-1

Konarska, J., Uddling, J., Holmer, B., Lutz, M., Lindberg, F., Pleijel, H., & Thorsson, S. (2016). Transpiration of urban trees and its cooling effect in a high latitude city. *International Journal of Biometeorology*, *60*(1), 159–172. DOI: 10.1007/s00484-015-1014-x

Kragh, J. (1979). Pilot study on railway noise attenuation by belts of trees. *Journal of Sound and Vibration*, *66*(3), 407–415. DOI: 10.1016/0022-460X(79)90859-9

Kumar, S. R., Arumugam, T., Anandakumar, C., Balakrishnan, S., & Rajavel, D. (2013). Use of plant species in controlling environmental pollution. *Bulletin of Environment, Pharmacology and Life Sciences*, *2*(2), 52.

Leonard, R. J., McArthur, C., & Hochuli, D. F. (2016). Particulate matter deposition on roadside plants and the importance of leaf trait combinations. *Urban Forestry & Urban Greening*, *20*, 249–253. DOI: 10.1016/j.ufug.2016.09.008

Li, B., Tao, S., & Dawson, R. W. (2002). Evaluation and analysis of traffic noise from the main urban roads in Beijing. *Applied Acoustics*, *63*(10), 1137–1142. DOI: 10.1016/S0003-682X(02)00024-5

Liang, L., Wang, Z., & Li, J. (2019). The effect of urbanization on environmental pollution in rapidly developing urban agglomerations. *Journal of Cleaner Production*, *237*, 117649. DOI: 10.1016/j.jclepro.2019.117649

Mallik, C. (2019). Anthropogenic sources of air pollution. In *Air pollution: Sources, impacts and controls* (pp. 6–25). CAB International. DOI: 10.1079/9781786393890.0006

Manickathan, L., Defraeye, T., Allegrini, J., Derome, D., & Carmeliet, J. (2018). Parametric study of the influence of environmental factors and tree properties on the transpirative cooling effect of trees. *Agricultural and Forest Meteorology*, *248*, 259–274. DOI: 10.1016/j.agrformet.2017.10.014

Mishra, R. K., Parida, M., & Rangnekar, S. (2010). Evaluation and analysis of traffic noise along bus rapid transit system corridor. *International Journal of Environmental Science and Technology*, *7*(4), 737–750. DOI: 10.1007/BF03326183

Onuu, M. U. (2000). Road traffic noise in Nigeria: Measurements, analysis and evaluation of nuisance. *Journal of Sound and Vibration*, *233*(3), 391–405. DOI: 10.1006/jsvi.1999.2832

Ozdemir, B., Bayramoglu, E., & Demirel, O. (2014). Noise pollution and human health in Trabzon parks. *Studies on Ethno-Medicine*, *8*(2), 127–134. DOI: 10.1080/09735070.2014.11917627

Pal, A. K., Kumar, V., & Saxena, N. C. (2000). Noise attenuation by green belts. *Journal of Sound and Vibration, 234*(1), 149–165. DOI: 10.1006/jsvi.2000.2863

Pathak, V., Tripathi, B. D., & Mishra, V. K. (2008). Dynamics of traffic noise in a tropical city Varanasi and its abatement through vegetation. *Environmental Monitoring and Assessment, 146*(1-3), 67–75. DOI: 10.1007/s10661-007-0060-1

Pienkowski, M. (2021). Loud music and leisure noise is a common cause of chronic hearing loss, tinnitus and hyperacusis. *International Journal of Environmental Research and Public Health, 18*(8), 4236. DOI: 10.3390/ijerph18084236

Rahman, M. A., Stratopoulos, L. M., Moser-Reischl, A., Zölch, T., Häberle, K. H., Rötzer, T., Pretzsch, H., & Pauleit, S. (2020). Traits of trees for cooling urban heat islands: A meta-analysis. *Building and Environment, 170*, 106606. DOI: 10.1016/j.buildenv.2019.106606

Rao, P. S., Gavane, A. G., Ankam, S. S., Ansari, M. F., Pandit, V. I., & Nema, P. (2004). Performance evaluation of a green belt in a petroleum refinery: A case study. *Ecological Engineering, 23*(2), 77–84. DOI: 10.1016/j.ecoleng.2004.06.013

Reethof, G. (1973). Effect of plantings on radiation of highway noise. *Journal of the Air Pollution Control Association, 23*(3), 185–189. DOI: 10.1080/00022470.1973.10469763

Sabrin, S., Karimi, M., Nazari, R., Pratt, J., & Bryk, J. (2021). Effects of different urban-vegetation morphology on the canopy-level thermal comfort and the cooling benefits of shade trees: Case-study in Philadelphia. *Sustainable Cities and Society, 66*, 102684. DOI: 10.1016/j.scs.2020.102684

Sæbø, A., Popek, R., Nawrot, B., Hanslin, H. M., Gawronska, H., & Gawronski, S. W. (2012). Plant species differences in particulate matter accumulation on leaf surfaces. *The Science of the Total Environment, 427*, 347–354. DOI: 10.1016/j.scitotenv.2012.03.084

Salmond, J. A., Tadaki, M., Vardoulakis, S., Arbuthnott, K., Coutts, A., Demuzere, M., Dirks, K. N., Heaviside, C., Lim, S., Macintyre, H., McInnes, R. N., & Wheeler, B. W. (2016). Health and climate related ecosystem services provided by street trees in the urban environment. *Environmental Health : A Global Access Science Source, 15*(S1, Suppl. 1), 95. DOI: 10.1186/s12940-016-0103-6

Sokhi, R. S., Moussiopoulos, N., Baklanov, A., Bartzis, J., Coll, I., Finardi, S., Friedrich, R., Geels, C., Grönholm, T., Halenka, T., Ketzel, M., Maragkidou, A., Matthias, V., Moldanova, J., Ntziachristos, L., Schäfer, K., Suppan, P., Tsegas, G., Carmichael, G., & Kukkonen, J. (2021). Advances in air quality research—Current and emerging challenges. *Atmospheric Chemistry and Physics Discussion, 2021*, 1–133. DOI: 10.5194/acp-2021-581

Tallis, M., Taylor, G., Sinnett, D., & Freer-Smith, P. (2011). Estimating the removal of atmospheric particulate pollution by the urban tree canopy of London, under current and future environments. *Landscape and Urban Planning, 103*(2), 129–138. DOI: 10.1016/j.landurbplan.2011.07.003

Tang, S. H., & Ong, P. P. (1988). A Monte Carlo technique to determine the effectiveness of roadside trees for containing traffic noise. *Applied Acoustics, 23*(4), 263–271. DOI: 10.1016/0003-682X(88)90040-0

Tremper, A. H. (2015). *Impact of green screens on concentrations of particulate matter and oxides of nitrogen in near road environments* [Doctoral dissertation, King's College London].

Tyagi, V., Kumar, K., & Jain, V. K. (2006). A study of the spectral characteristics of traffic noise attenuation by vegetation belts in Delhi. *Applied Acoustics, 67*(9), 926–935. DOI: 10.1016/j.apacoust.2005.09.002

Wei, X., Lyu, S., Yu, Y., Wang, Z., Liu, H., Pan, D., & Chen, J. (2017). Phylloremediation of air pollutants: Exploiting the potential of plant leaves and leaf-associated microbes. *Frontiers in Plant Science, 8*, 1318. DOI: 10.3389/fpls.2017.01318

World Health Organization. (2021). *Compendium of WHO and other UN guidance on health and environment.*

Yang, J., Chang, Y., & Yan, P. (2015). Ranking the suitability of common urban tree species for controlling PM2.5 pollution. *Atmospheric Pollution Research, 6*(2), 267–277. DOI: 10.5094/APR.2015.031

Zannin, P. H. T., Diniz, F. B., & Barbosa, W. A. (2002). Environmental noise pollution in the city of Curitiba, Brazil. *Applied Acoustics, 63*(4), 351–358. DOI: 10.1016/S0003-682X(01)00052-4

KEY TERMS AND DEFINITIONS

Air: is the refreshing and breathing element that should be available for every citizen of this earth.

Green space: is the protective zone of urban life by lowering the temperature, air pollution and controlling the noise.

Huge noise: is not acceptable and bearable when it exceeds the limit especially in the huge trafficked zone in the city and industrial areas.

Pollution: is the disintegration of the element from the origin and causes serious problem to the human being, animals and to the environment.

Chapter 3
Assessing the Impact of Government Policies and Practices for Pollution Control and Transforming the Greater Cities of India

R. Kathaian
http://orcid.org/0000-0001-6137-0847
Vel Tech Rangarajan Dr. Sagunthala R&D Institute of Science and Technology, India

Rajani H. Pillai
Mount Carmel College, India

Roopa Adarsh
Mount Carmel College, India

M. Thenkovan
Guru Nanak College, Chennai, India

P. Kumarasamy
Dhanalakshmi Srinivasan University, India

U. S. Syama
Kristu Jayanti University, India

ABSTRACT

Cities in India are making a pivotal contribution to economic growth and substantially driving the gross domestic product (GDP). The top cities, namely Delhi, Mumbai, Kolkata, Bangalore, Chennai, Hyderabad, Pune, Ahmedabad, Surat, and Jaipur, contribute to industrial growth, employment generation, and economic development. Subsequently, urbanisation also creates numerous issues and challenges, including environmental pollution and sustainability concerns. This chapter aims to assess the

DOI: 10.4018/979-8-3373-4422-5.ch003

various positive impacts of pollution control practices and policies of the Government of India. The government has taken many efforts in the form of policy to arrest the pollution, such as industrial pollution, vehicle pollution, and traffic congestion. The government of India's policies, such as the National Clean Air Programme, measures for control of vehicular emissions, measures for control of emissions from vehicles. Based on the research results and conclusions, an appropriate suggestion will be given for the sustainable transformation of cities in India.

INTRODUCTION

India is facing rapid urbanisation of cities, with an important segment of its population in urban areas. The major Indian urban areas, such as Delhi, Mumbai, Bangalore, Kolkata, Hyderabad, and Chennai, are now facing enormous population growth and are the main contributors to this growing trend. Though urbanisation brings various economic opportunities, it also creates certain challenges, such as infrastructure strain and inequity. India's economic climate has been greatly impacted by the country's recent rapid urbanization (Basu & Mishra, 2024). Urban areas like Bangalore, Chennai, Mumbai, and Delhi have developed into major economic hubs as a result of improved infrastructure, more job opportunities, and greater living standards. Urbanization serves as a gauge of the shift from traditional rural providence to contemporary factory providence. This procedure is long-term (Savithiri et al. 2025). The result of poverty and population growth is urbanization, which leads to rural-urban migration. Urbanization is happening because of rural pushes rather than urban pulls. The ongoing population growth and urbanization of large cities in India are indicators of the country's urbanization trend. Economic power and social conversion are significantly influenced by urbanization. It might worsen inequality, generate new opportunities, or all three. Cities could be a powerful tool for reducing poverty within their borders as well as through the easy, human, and commercial flow between India's urban and rural areas (Goyal et al. 2018). By concentrating on the poor, they may potentially worsen poverty and replicate the current societal inequality. In a nutshell, urbanization is a conversion that can be measured. Economic growth and the development of urban infrastructure are related. This is particularly true for India, a country that is gradually becoming more urbanized. The establishment of necessary manufacturing facilities as a result of the infrastructure investments instantly boosts the economy (Praharaj & Han, 2019). Additionally, it reduces the costs of companies and transactions, increasing their competitiveness and facilitating market integration. In addition to benefiting already-existing enterprises, this infrastructure improvement attracts new investments, which boost the economy (Vani et al. 2025). It is additionally noted that a nation's weak and insufficient infrastructure poses a significant barrier to its commercial development and economic expansion. The nation's inadequate physical infrastructure hinders attempts to boost productivity and efficiency and keeps it from reaching its full economic potential (Vedant, 2024).

Background of the study

Urbanization, or the expansion of cities, has a significant effect on quality-of-life metrics by altering numerous facets of people's happiness and health (Ahvenniemi et al. 2017). It highlights how a greater rate of urbanization signifies the economy's sectoral shift from agriculture to industry and services, which boosts growth. In Indian society, urbanization has become widespread. Cities have grown as a result of the expansion of industries. People have been migrating to industrial areas in pursuit of work as a result of industrialization (Javidroozi et al. 2023). Towns and cities have grown as a result of this. The process of population concentration in a specific area is another definition of urbanization. The process of developing into urban area, relocating to a city, and switching from agricultural to other city-specific activities is known as urbanization. As per the World Bank, India's urban population as a percentage of the country's overall population was last recorded as 32.75 per cent in 2015. The primary cause of urban issues is urban sprawl, or the actual growth of cities in terms of both population and geographic area in fast-expanding cities (Savithiri et al. 2026). The economic foundation of the majority of cities is unable to handle the issues brought on by their excessive growth. metropolis have grown in size as a result of the nearly constant large-scale immigration from rural regions and small towns into large metropolises. The predicament of overpopulation is a result of urbanization. One effect of urban overpopulation is overcrowding. Naturally, overcrowding must be a problem in cities with a huge population crammed into a tiny area. Nearly all of India's major cities are excellent examples of this. For instance, Mumbai has six per cent of an acre of vegetation for every thousand people, even though the Greater Mumbai Master Plan recommends four acres. India's metropolitan areas are overcrowded in both “absolute” and “relative” senses. These cities' high population densities are absolute, but the issue of supplying the city with services and other amenities is relative, even if the population numbers are not extremely high. It contributes to the ongoing issue of housing scarcity in cities (Hassan & Lee, 2015). This issue is particularly severe in urban locations where a significant number of underemployed and unemployed immigrants from nearby towns join the town without a place to live (Alamelu Sathyabama et al. 2025). According to a 1959 Indian Sample Survey, 44% of urban homes had one room or fewer. Sixty-seven per cent of households in major cities lived in one room or less. The issue is further aggravated by the extremely slow rate of development of housing that is now taking place. Approximately 2.5 million new homes are needed in Indian cities each year, but less than 15 per cent of that demand is being met. The housing issue discussed above is just as significant as the unemployment issue. It is estimated that between 15 and 25 percent of India's workforce is unemployed in urban areas. Among those with more education, this number is much higher. Four major cities are thought to house roughly half of all qualified urban unemployed people. Furthermore, considering the substantial cost of living in metropolitan regions, even while incomes in urban areas are larger than those in rural areas, they are still incredibly low. Over the past few decades, air pollution and climate change have become major global concerns (Jha et al. 2014). The current evaluation was conducted in the setting of Indian cities, where air pollution and climate change have a substantial negative influence on human health. Human health is at risk due to the growing metropolitan areas' extreme climate events, which include droughts, floods, excessive rains, and extremely high temperatures. Increased temperatures brought on by climate change-induced heat waves have caused urban dwellers to experience thermal discomfort and some health problems. The paper also discusses how air pollution levels in the majority of Indian megacities are rising above the recommended limits (Garau & Pavan, 2018).

Objectives

To assess the various environmental impacts, such as pollution control and the transformation of cities through government policies.

To analyse the various case studies and success stories of cities in India.

Research Problem

Rapid expansion of cities in India led to significant traffic and environmental issues, including air and water pollution, inefficient waste management, reduced green spaces, and extensive carbon emissions due to excessive automobile vehicles use. In response to this issue, various government policy measures and environmental initiatives to urban areas for the control of traffic, pollution, and promotion of sustainable urban transformation. However, even after the implementation of different urban policies, cities in India have continuously exhibiting wide differences in environmental outcomes, increasing concerns about the impact, implementation processes, and scalability of those policy interventions. While many cities in India have emerged as notable success cases in adopting sustainable practices such as air quality monitoring system, integrated solid waste administering system, renewable energy system for smart city scheme. But, these urban cities remains limited and not extended across all the cities of the nation. Therefore, it requires a comprehensive assessment that would help to navigate the government policies for measurable positive outcomes. Moreover, the core research problem lies in the process of understanding how and what extent public policies contributed in an immense way to control pollution and transform the cities in India.

Methodology

The research chapter adopts an explanatory and analytical design based on certain available data, case studies, and existing literature. The related literature has been collected and reviewed from published articles from various indexed journals, government reports, policy notes, and reports published by national and international agencies and institutions to identify and understand the environmental impacts, issues relating to urbanisation, and policy initiatives of the government to make smart cities for sustainable cities. Secondary sources from urban transport policy and smart city mission reports have been used to demonstrate the urbanisation, traffic issues, environmental concerns, and policy measures.

LITERATURE REVIEW

Urbanisation

The societal transition from traditional rural cultures to contemporary urban communities is known as urbanization. The term "urbanization" refers to a process that manifests itself through transformations in the demographics, which are social, economic, technological, and environmental facets of a society throughout time, space, and sector (Sivasubramanian, 2024). The progressive concentration of people in an urban area is known as urbanization. The population is becoming more concentrated in human settlements that are larger than villages, people are becoming more involved in additional and additional production functions, and some social characteristics that are characteristic of traditional agricultural societies are gradually being adopted (Uma Maheswari et al. 2025). In addition to industrialization, urbanization is also linked to modernization, and all three of these factors might occasionally cooperate. Whereas urbanization in developing nations is rapid and does not coincide with industrialization but rather by the rapid expansion of services as an industry in the economy, urbanization in developed nations has been slow but steady and was fueled by the agricultural and mechanical revolutions, higher per capita income, and a higher standard of living. Modernization and industry have not kept pace with urbanization in third-world countries. The contemporary urbanization process is a relatively new phenomenon that is still evolving in the majority of emerging nations. This process is exposing unique characteristics as it continues to evolve in underdeveloped nations (Shruti et al. 2021).

Smart City Mission

The Indian government's Smart Cities Mission is one of several programs aimed at enhancing urban living conditions. Launched on June 25, 2015, the mission's original five-year duration was later extended to accommodate ongoing activities. It is anticipated that the mission will end on June 30, 2024. One hundred cities with different populations were selected from India's states and union territories as part of the project (Angelidou, 2017). The Ministry of Housing and Urban Affairs is in charge of the mission, and a special purpose vehicle (SPV) made up of a few chosen municipal employees and other interested parties acts as the city-level executing authority. The SPV is in charge of all facets of creating and carrying out different projects and activities, which are determined by conducting stakeholder interviews and demand assessment surveys. This initiative's unique characteristics include the use of new technology, a participatory approach, and the creation of clever solutions for urban issues. Area-Based Developments (ABD) and Pan-City Actions are the two main intervention areas upon which the Smart Cities Mission (SCM) is built. ABD focuses on upgrading existing infrastructure and turning specific portions of the city into pilot areas with clever solutions. Three methods, including retrofitting, reconstruction, and green-field development, were used to initially conceptualize ABD (Sultaana & Saha, 2025). Among the 100 major cities, 51 areas, including Visakapatnam, Delhi, Surat, Trivandrum, has selected to upgrade their present infrastructure through retrofitting. A concept that included retrofitting and rebuilding was adopted by the next forty cities, including Raipur, Amristar, Jalandhar, and Indore. Ahmedabad has focused on smart transportation, whereas Surat has adopted smart water meters and flood alert devices (Trindade et al. 2017). From the time of selection to the end of the project period, each city chosen for the project has received an average of 500 crore. States and urban local bodies (ULBs) raised the matching money. Resources also come from both external and internal financial sources. This figure, which was made public by the government of India in July 2024, is used to a degree of 93% in various implementation projects. All told, 8,016 projects totalling 1,64,223 crore that were started nationwide were finished (Saha & Singh, 2017).

Smart Cities for Sustainability

New workplaces have been established in cities, and working conditions have improved. For example, fishermen can now use Kakinada's restored boat-building yard. With the addition of amenities like vendor seating stalls, drinking water, garbage and sewage management, and parking, Indore's classic weekly market has been transformed. An additional instance is Belagavi, which has developed a bustling street market that is pedestrian-friendly (Banerjee et al. 2024). To help different population groups, create revenue, efforts are being made to help them develop their skills. To help artists and craftspeople prepare and share new creations, the online handicraft incubation center has been established in Tirupati. The Jabalpur incubation centre hosts events including networking, intellectual property rights seminars, and entrepreneurship camps to assist individuals in locating employment and launching a business. Likewise, the Rozgar Training Center in Lucknow provides training for a range of occupations. Both the federal and state governments have implemented housing-specific initiatives, like the Pradhan Mantri Awas Yojana (2006) (Urban). Furthermore, as seen by the building of more than 44,000 subsidised housing complexes and 6,300 rooms under communal housing projects in Bhubaneswar, the Smart Cities Mission tackles social equality issues in the housing sector. (Bibri, 2018). Additionally, registered patients' medical records have been digitized, and the information stored within the framework will help track, observe, and manage disease outbreaks (Farid et al. 2021).

Government Policies on Pollution Control

Action Plan for Improvement of Air Quality

The goal of Clean Air 2025 was to minimise

The main goal of Clean Air 2025 is to reduce the emission of hazardous air pollutants while reducing indoor and household air pollution. A suitable regulation, along with the identification, measurement, and monitoring of air pollutants, as well as an efficient stakeholder participation process, is highly required to reach this goal (Das & Chadchan, 2023).

Digital Technologies and AI for Smart Cities

Urban cities around the world are using modern, specialized technologies to solve social, economic, and other problems as a result of enhanced urbanization. The potential and significant AI to revolutionise urban development has gained recognition as a transformative instrument. AI could present various opportunities for creating and running smart cities due to its capacity for learning, prediction, and independent operation. Its uses are numerous, including better and improved public services (Rautela et al. 2024). AI is a technological tool that can greatly aid in the integration of various key aspects of smart cities, including people, economy, mobility, environment, and government. AI and other innovative methods have demonstrated significant promise in generating the good intricate issues that are inherent to the development of smart cities, starting from cybersecurity and intelligent transportation monitoring systems to energy grids and smart healthcare provisions (Das, 2025). By 2050, it is predicted that 60 percent of the global population will live in urban areas, raising issues about the effects on the environment, management, and security (Musa et al. 2023) & (Han et al. 2022). It is relatively impossible to overestimate the importance of AI in processing the volumes of various data produced by IoT, a prominent element of smart city applications. To make the best decisions, urban cities could use AI technology (Yang et al. 2022) & Das, 2024).

Impacts of the policy

Cities across the world are struggling with the need to transform their conventional traffic management systems in line with growing urbanisation and the result of an increase in vehicle traffic. AI is an important technological powerhouse that has the potential to completely transform the dimensions of urban mobility and is at the forefront of this revolutionary journey (Iyer, 2021). The symbiotic relation between AI and traffic management is thoroughly assessed in this chapter with a particular emphasis on smarter solution reviews and their multiple effects on the fabric of the urban world (Wang et al. 2024). Unprecedented population pressure is occurring across various urban areas, leading to worsening traffic congestion, longer commutes, and reduced road safety. AI becomes a ray of hope as an answer to this urban quandary, providing intellectual and dynamic results that transform with traffic patterns (Ramadan et al. 2024). Through various adoptable systems, the incorporation of AI in traffic management holds the promise of maximising traffic flow. Machine learning smart traffic signal networks could instantly respond to provide real-time traffic solutions (Dureja et al. 2024). With gridlock jeopardizing not only the impact of public transit but also the quality of life for urban people. This context creates an ideal environment for analysing novel solutions, with a particular emphasis on the potentially transformative capabilities of AI (Yigitcanlar et al. 2020).

Table 1. Long-term action plan

Long-term Action Plan	Implementation (Days)	Responsible Department
Launch of electric buses for public transport	360	State Transport
Expressway roads	360	National Highways
Peripheral roads	360	National Highways

continued on following page

Table 1. Continued

Long-term Action Plan	**Implementation (Days)**	**Responsible Department**
Multilevel parking	360	City Development Authority
Development of Bike and Bicycle Zone	360	City Development Authority
Use of Bio-Ethanol for vehicles	360	State Transport
Strict compliance with polluting vehicles	Regular Activity	RTO & Traffic Police
Public awareness campaign for the reduction of Air Pollution	Regular Activity	RTO & Traffic Police
Widening the roads	90	State Transport
Motion bridges	180	State Transport
Maintaining 33 per cent of the forest area in the city	360	Forest Department

Source: National Urban Transport Policy.

Figure 1. Count of long-term action plan by responsible department

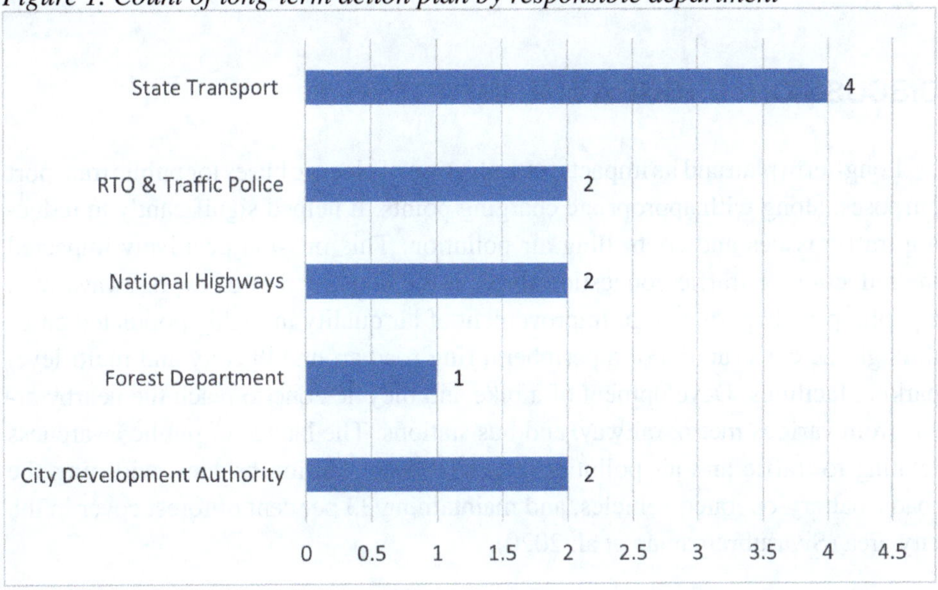

Figure 2. Implementation days (long term)

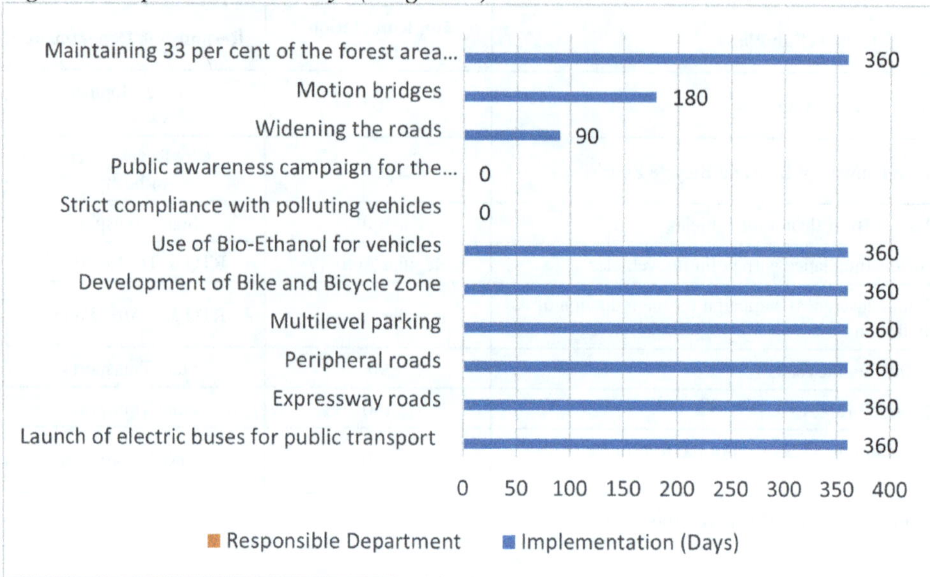

DISCUSSION: PROGRAMME IMPACTS

Long-term plan and its impacts: introduction of electric buses for public transport purposes, along with appropriate charging points. It helped significantly in reducing traffic issues and controlling air pollution. This mission positively impacted the reduction of traffic congestion through the construction of expressways with a public-private partnership. Improvement of air quality in highly populated cities through the construction of a peripheral ring road around the city and multi-level parking facilities. Development of a bike and bicycle zone to reach the nearby areas from various metro, railway, and bus stations. The launch of public awareness relating to traffic and air pollution. Installation of motion bridges, widening the roads, battery-operated vehicles, and maintaining 33 per cent of forest cover in the city area (Sivasubramanian et al. 2020).

Table 2. Short-term action plan

Short-term Action Plan	Implementation (Days)	Responsible Department
Plantation of specific plants for pollution control	90	Forest Department
Potholes free roads to control traffic & pollution	Regular Activity	City Development Authority
Greening of Open Areas	90	Forest Department
Implementation of water fountains at major traffic signals	90	City Corporation
Wall-to-Wall dust control	180	City Corporation

Source: Same as the previous table.

Figure 3. Implementation days (short-term)

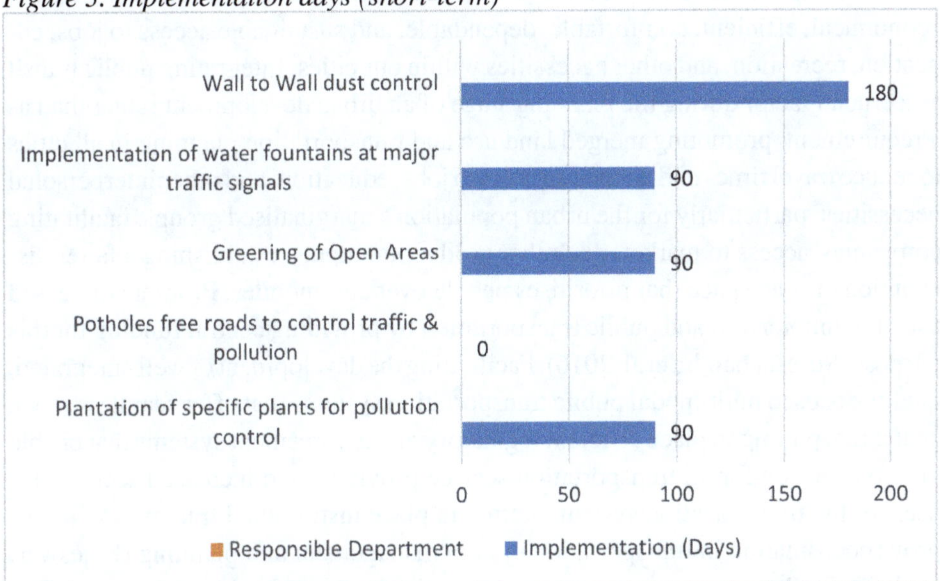

National Clean Air Programme

The metro and non-metro cities have performed well due to this action plan. This programme helped and contributed to strengthening air quality monitoring through the installation of new air quality monitoring stations. Based on this programme implementation, capacity building, city planning, and enhanced inter-departmental actions have increased. With a revised goal of a 40 per cent decrease in PM10 levels from 2017 to 2026, it was subsequently prolonged (NCAP, 2019). The previous

CREA "Tracing the Hazy Air" assessments from the years 2022 to 2024 noted the program's difficulties and flaws. This most recent analysis examines external air quality trends until 2024 and evaluates the improvements made under NCAP. To reach the forty percent decrease target by 2026, it is imperative to assess how the present progress compares to the initial goals. There are still 28 cities without continuous ambient quality of air monitoring stations (CAAQMS) out of 130. Just 97 of the 102 cities with monitoring stations reported having 80 or greater coverage of PM10 data. Twenty-nine of the 97 NCAP cities have elevated PM10 levels. Out of 97 cities, only 41 met the original NCAP goal of reducing PM10 levels by 20–30% (Central Pollution Control Board. (2011) & (Ganguly et al. 2020).

National Urban Transport Policy, 2006.

This policy aims to provide the increasing number of city dwellers with safe, economical, efficient, comfortable, dependable, and sustainable access to jobs, education, recreation, and other necessities within our cities. Integrating public transit as a crucial factor during the planning stage of an urban development rather than as a requirement. promoting merged land use and transportation planning in all cities to reduce travel times and enhance access to jobs, education, and other interpersonal necessities, particularly for the urban population's marginalised groups. facilitating companies' access to markets and other production inputs. establishing a fairer distribution of road space that prioritises people over automobiles. Promote increased use of nonmotorized and public transportation by providing central funding for this purpose (Sureshchandra et al. 2016). Facilitating the development of well-integrated, quality-focused multimodal public transportation systems that offer seamless mode switching. putting in place efficient regulatory and enforcement systems that enable fair competition for all transportation service providers and increased security for users of the transportation system. putting in place institutional frameworks to improve coordination in transportation system management and planning (Rajeswari et al. 2025). Presenting automated transportation systems for the control of traffic. addressing issues related to trauma response and road safety, lowering pollution levels through improved enforcement, more stringent regulations, altered travel habits, advancements in technology, etc. Establishing a knowledge management system that would meet the needs of all urban transport professionals, including planners, researchers, teachers, students, etc., and increasing institutional and human capacity to plan for sustainable urban transport (Shehab et al. 2021). India's economy is expected to experience rapid growth. Both the secondary and tertiary sectors of the economy, or the service and industrial sectors, will account for the majority of this future growth. The condition of our cities and towns is essential to India's future development because economic activity in these sectors mostly occurs in urban areas (Balasubramanian et al. 2025). Furthermore, about 30 per cent of India's population currently lives in urban areas. Rapid urbanisation raises this percentage to over 60 per cent before it starts to stabilise, according to global experience. As a result, it is predicted that India's urban population will increase from 285 million in 2001 to roughly 473 million in 2021 and 820 million by 2051. Cities must thus address the needs of those who have not yet joined the urban population as well as the mobility requirements of the present population. Urban areas must facilitate the simple and sustainable movement of people and goods to sustain the necessary level of economic activity. Inappropriately, though, there have been some issues with this flow of people and goods. Travel expenses have significantly increased, particularly for the impoverished. This is mostly because using less expensive non-motorised modes, such as walking and cycling, has become very dangerous because they must share the same space as motorised modes (Chaturvedi et al. 2024).

Additionally, cities have tended to expand as a result of population growth, and longer commutes have rendered non-motorised modes of transportation impractical.

This has significantly increased the difficulty of obtaining livelihoods, especially for the impoverished. With accident rates rising from 1.6 lakh in 1981 to over 3.9 lakh in 2001, travelling in the city has become riskier. Over the same period, the number of people killed in traffic accidents has also increased, rising from 28,400 to exceed 80,000. Again, because many of the people killed or injured are cyclists, pedestrians, or pavement dwellers, this has tended to affect the poor more severely (Das, 2020). Poor mobility can significantly impede economic growth and lower the quality of life if the aforementioned issues are not resolved. Therefore, a policy that provides precise guidance and an arrangement for future action is required in order to address this quickly expanding issue (Kumar & Dahiya, 2017). India's cities differ greatly in terms of their topography, population, urban form, income levels, economic activity, growth constraints, and other factors. As a result, these city-specific characteristics will need to inform the transportation system's design. Furthermore, land use planning and transportation planning are inextricably linked, and they must be developed jointly in a way that minimizes travel needs while serving the entire population. In summary, the characteristics of sustainable transportation systems must be incorporated into an integrated master plan (Tavares et al. 2021). Instead of creating a transportation infrastructure after unchecked sprawl has occurred, consideration should be given to directing a city's future growth around a planned-out transport network when creating such plans. Therefore, transportation plans should allow a city to adopt an urban form that best supports the main economic and social operations of its citizens while also fitting the geographic constraints of its location. Unfortunately, when creating plans for land use and strategic development, transportation planning has not gotten the attention it deserves (Dankan et al. 2024).

Sustainable Urban Traffic Management & Air Pollution Control

The Central Government would thus support policies that prioritise people and more fairly distribute road space, such as allocating lanes and corridors only to public transportation and alternative forms of transportation or allocating lanes to cars carrying more than three people (Sethi & Kosmopoulos, 2025). A physically populated area that incorporates both tangible and intangible resources for its residents to use is known as a "smart city." To become a system that can provide residents with the services they expect without lowering or altering their quality of life, ICT-based infrastructures must be compatible with built-in infrastructures (Sharma et al. 2025). By meeting the six pillars of sustainability economy, people, governance, mobility, circumstances, and living, it is also recognised as an innovation that aims to provide sustainable growth and services. Mobility should receive particular and significant attention since it is a crucial component and one of the six dimensions for creating smart cities (Kumar et al. 2017). Reducing the direct and indirect effects of traffic congestion is the foundation of the concept of decongested cities through traffic regulations (De Palma & Lindsey, 2011). Road accidents, environmental pollution from exhaust, fuel waste, etc., are examples of these direct impacts; travel time loss is an example of an indirect impact. The foundation of sustainable transportation is the creation of cleaner transportation systems and decongested roads (Rocha Filho et al. 2020). By enforcing traffic laws and utilising cutting-edge technology, decongestion lowers the number of traffic accidents and fosters a more accessible environment with less risk of pollution and degradation (Bigazzi & Rouleau, 2017). The primary cause of air pollution from traffic emissions is a rapidly and significantly growing urban population; records show that over 50% of fuels produced worldwide are used by transportation industries, which greatly contribute to environmental pollution (Singh et al. 2023). This percentage is expected to rise in the years that follow due to the unprecedentedly high rate of continuous vehicle ownership worldwide (Erhardt et al. 2021). When vehicles release harmful gaseous substances like carbon dioxide, methane, organic compounds with volatile compounds, nitrogen oxides and particulate matter, etc., both intermittent and recurring traffic congestions contributes to environmental pollution. If left unchecked, this could defeat the goals of sustainable transportation (Venkatesh, 2014; Tang et al., 2020).

Promotion of Electric Vehicles and Pollution Control

One of the main causes of air pollution in cities is the toxic emissions from internal combustion engines. Besides being energy-efficient, electric vehicles (EVs) can lower greenhouse gas emissions, air pollution, and noise pollution from transportation-related activities (Tu & Zan, 2025). EVs can help lessen reliance on non-renewable energy sources, carbon dioxide emissions, and oil demand. They might make it easier to move toward the production of clean energy, which would lower the nation's pollution levels (Jia et al. 2023). India is probably going to fulfil its commitments made in Paris in 2015 by 2030, even though its absolute greenhouse gas emissions have doubled. These emissions would be below the current global averages on a per-capita basis. These projections, however, make the assumption that the current estimates of electricity demand will decline and that the switch to renewable energy sources will happen more quickly (Moriarty, 2023). Infrastructure problems, financial obstacles, behavioural problems, technical problems, and outside influences are the elements that hinder the adoption of EVs (Selvakumar et al. 2018). These obstacles can be addressed with the aid of an environmental policy designed to promote the use of EVs. Additionally, a variety of appealing vehicles that cater to consumers' social needs help to boost EV sales in addition to incentives (Son et al. 2025). It is necessary to examine the economic features and societal patterns of each nation before deciding on its EV policy components. Incentives for the manufacturing and use of electric vehicles (EVs) and prohibitions or disincentives for the production as well as use of polluting autos would make up the ideal policy mix. Due to their comparatively low prices, two-wheelers would be simple to target in the EV market. This would also benefit two-wheeler users, who are more vulnerable to and impacted by air pollution (Guttikunda et al. 2019).

Case Storie 1

Numerous companies in India have started a variety of projects.

For instance, HCL Infosystems feels that electronic manufacturers must enable environmentally friendly disposal of products when their useful lives are coming to an end. HCL Infosystems backs India's continuous push for distinct laws pertaining to e-waste. In India, HCL has been creating a simple, practical, and safe e-waste recycling program. Mahendra is significantly contributing to the reduction of air pollution through electric vehicles for sustainability. It is offering zero CO_2 emissions by substituting traditional fuel-powered vehicles. The electric vehicles of Mahindra are not only helpful for cost-effectiveness, but also help to promote air quality and reduce noise pollution in the cities (Usharani, S. et al. 2025).

As one of pioneer in India to introduce a full-fledged range of restriction of dangerous substances compliance and also take back policy of e-waste for the recycling process. In 2025, the HCL group of companies introduced a 25k crore value of the ClimaForce fund to help decrease carbon emissions.

Table 3. HCL 2025 data highlights

Resource Consumption	Efficiency
Less Energy Consumption	23%
Energy from RE	34%
Emission Reduction	46%
Reduction of Water Use	32.20%

Source: HCL Info Report

Table 4. Prospective target and progress in align with SDGs

Target	Progress
To achieve net zero by 2040	Achieved 46 percent in Scope 1 & 2and 22 percent in Scope 3
Achieve zero waste	
Reduce the absolute scope 50% by 2030	Achieved 46 percent

Source: HCL Info Report

In the year 2025, HCL Info Technologies completed a corporate carbon footprint evaluation and publicly disclosed its greenhouse gas discharges in accordance with the GHG protocol, emphasizing Scope one (direct emissions owned and controlled), Scope two (indirect emissions from outside purchased energy sources), and Scope three (indirect emissions from the value chain). However, the company has not revealed a category level for its Scope three break-ups.

Case Story 2

Asian Paints Limited

Asian Paints Ltd demonstrates a strong and significant commitment to environmental protection by integrating the principles and policies of sustainable development into all its functions. The policy of the company emphasises highly responsible sourcing of raw materials, following of regulations, and efficient resource utilization. Operational commitments such as safe handling of various materials, optimising through 3R processes, and plastic waste management. Adapting to Water Shortage through wastewater recycle: Imagine treating and recycling od wastewater not as one option but as an integral practice. Asian paints climate action idea and plan focuses on decarbonising functions and building sustainable and responsible supply partnerships. Through consistent use of renewable energy sources and aimed efficiency measures, the company has significantly achieved Scope one and Scope two emissions from baseline. The company's sustainable supply chain initiatives foster active collaboration across the value chain for consistent progress.

Table 5. Pollution control measures and progress

Particulars	Percentage Achieved	SDGs
Renewable Energy	57.60%	SDG7, SDG11, & SDG12
Recycled Plastics	20150 MT	SDG7, SDG11, SDG12, & SDG17
Recycled Water Use	478%	SDG7, SDG11, & SDG12

Source: Asian Paints Website Report

Achieving 57.6 percent of renewable sources of energy usage, indicating that more than half of the company's energy consumption comes from solar energy, wind energy, or related renewable sources. It contributes to achieving and promoting clean energy, reaching sustainable cities and communities in terms of reducing urban pollution and better energy systems, and it encourages efficient use of available resources. Similarly, recycling of 20150 Metric Tons of plastics demonstrates and practices good approaches for the circular economy. It helps to reduce environmental degradation and plastic use.

Case Story 3

Reliance Industries

Reliance Industries aims to enable village communities to be more income-earning, nutrition-rich, and climate-resilient. Reliance covered 1346 villages for water resource development and effective utilisation of water through CSR activities. We are adamant that the world must quickly achieve complete emissions reductions; being carbon-neutral is insufficient. Fossil fuels cannot be used for much longer. As a result, we are quickly switching to renewable fuels to meet our captive energy needs. To reduce emissions and make the transition to clean energy sources and materials easier, we are making the most of biodiesel and bio-pathways. Increasing material recycling and maximizing circularity throughout value chains will be key components of this shift. At competitive prices, we hope to capture, store, and convert CO2 into valuable chemicals and materials.

Figure 4. Reduction of air emission

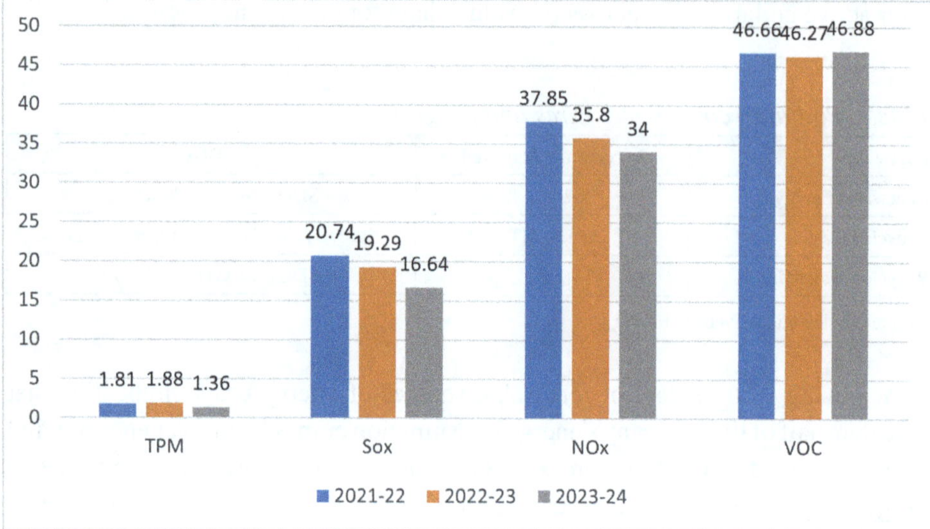

Reliance is aggressively moving towards renewable sources, consuming almost 6.85 million units of green energy in the financial year 2024. The company is focusses to reach net-zero emissions by 2035 through an investment of Rs. 5.9 lakh crore to make 100 Giga Watts of solar energy capacity.

Waste Management and Pollution Control in Cities

Rapid urbanisation, industrialisation, and the improvement of living standards are all linked to environmental pollution. Many nations needed industrialisation to grow, and today it is still required to increase a nation's economy and achieve self-sufficiency. On the other hand, industrialisation has led to significant problems with environmental pollution. Waste thus seems to be a consequence of development. A nation such as India has no option but to discard them (Adarsh et al. 2025). On the other hand, non-renewable resources are running out every day due to the rising demand for raw materials for industrial production. Therefore, by turning these undesirable wastes into usable beginnings for a range of advantageous uses, attempts will be made to control pollution arising from waste disposal (Peris-Ortiz et al. 2016). Industrial solid waste management is outside the purview of local governments. Under applicable regulations, solid waste-generating industries are required to manage their waste and secure authorisation from the corresponding State Pollution Control Boards (Mikucioniene et al. 2024). Every day, individuals dispose of their garbage according to standard procedures. Waste disposal, however, is turning into a significant and confusing problem for all human settlements worldwide. Putting solid waste out of sight doesn't fix the issue; instead, it makes it worse and eventually gets out of control. This practice has well-known negative effects on the environment, human health, food, water, and air pollution, and the loss of precious resources that might otherwise be recovered from solid waste. For this reason, waste management needs to be a global priority. Waste management is now a national and international issue (Roy & Srinivasan, 2025). Waste management gets more complicated as human settlements get more developed. The search for good solutions to this issue is ongoing, but it grows more and more obvious that technologically based solutions that do not involve human intervention will not last long and will only cause more issues (Venkatesan et al. 2025). The best method for handling solid waste is solid waste management, which typically comprises appropriate component segregation and scientific recycling. The term "solid waste management" (SWM) is widely used to describe the use of strategies to guarantee the efficient performance of the different tasks involved in the collection, transport, processing, administration, and disposal of solid waste (Khlie & Benmamoun, 2024). Every single one of these four categories of waste is subject to a distinct set of rules and regulations due to the differences in their nature. Waste generation, storage for primary collection, secondary gathering, and transport, recycling, treatment, and disposal are the cornerstones of waste management procedures in India. In India, waste management in every city is largely the responsibility of municipal corporations working with the municipality's health department (Khadke & Waghmare, 2018).

Challenges

India's cities differ greatly in terms of their topography, population, urban form, income levels, economic activity, growth constraints, and other factors. As a result, these city-specific characteristics will need to inform the transportation system's design. Furthermore, land use planning and transportation planning are inextricably linked, and they must be developed jointly in a way that minimises travel needs while serving the entire population (Han et al. 2022). In summary, the characteristics of sustainable transportation systems must be incorporated into an integrated master plan. Instead of creating a transport system once unchecked sprawl has occurred, consideration should be given to directing a city's future growth around a pre-planned transport network when creating such plans (Batty et al. 2012). Therefore, transportation plans should allow a city to adopt an urban form that best supports the main economic and social activities of its citizens while also fitting the geographic constraints of its location. Unfortunately, when creating plans for land use and strategic development, transportation planning has not gotten the attention it deserves (Guttikunda et al. 2023). Even before settlements are established, these corridors would need to be physically guarded against encroachment by installing barriers along designated routes and building roads on brief sections. This would suggest that in order to direct the settlements' locations and prevent excessive sprawl, sections of the passageway would rise first (Ghosh & Arora, 2022).

CONCLUSION AND RECOMMENDATIONS

The widespread movement of individuals from rural to urban regions is one of the main reasons for urban unemployment. Migration from rural to urban areas has been happening for a while, but it hasn't always been as problematic as it is now. Rural residents' widespread poverty forces them to relocate to cities in search of employment and the promise of a better quality of life. This chapter assessed the various positive impacts of pollution control practices and policies of the Government of India (Adhisekar et al. 2026). The government has taken many efforts in the form of policy to arrest the pollution, such as industrial pollution, vehicle pollution, and traffic congestion. The government of India's policies, such as the National Clean Air Programme, measures for control of vehicular emissions, measures for control of industrial emissions, and control of emissions from stubble burning, actions from the Central Pollution Control Board, and technical interventions include monitoring, implementation, stakeholder consultation, media outreach, and regulatory actions. Based on the research results and conclusions, an appropriate suggestion will be given for the sustainable transformation of cities in India. While PM2.5 management should be given priority in policy and activities, PM10 measurement may still be gathered for more comprehensive air quality evaluations (Segnestam, 2002). The list of failure cities shall be updated every two years by NCAP 2.0, as more and more cities are surpassing air quality criteria. The static list that has been in place from the beginning of NCAP does not allow prosperous towns to be eliminated from the program, and does not include newly developing non-attainment cities. NCAP 2.0 must guarantee that these plans are carried out successfully, with transparent accountability procedures and sufficient financial resources, to fully address transboundary pollution issues throughout the area (Joss, 2011). We require a more rigorous, integrated, and strategic framework for waste prevention to effectively address waste issues. Instead of merely replacing present ones with models from developed nations, we must improve upon them. India desperately requires a clear strategic disposal plan and robust implementation to prevent epidemics and make every city both economically and ecologically healthy. An efficient waste management system within the nation can be created with the help of numerous stakeholders by methodically analysing the strengths and weaknesses of the community and municipal corporation in order to meet financial sustainability, socioeconomic, and environmental targets within the arena of waste management (Mishra et al. 2022).

Implications

The outcome of this chapter could help to design policy by the central government and strengthen smart plans for urban areas. It will help to identify successful policy implications to control pollution and build sustainable urban transformation models applied across the cities of India. Evidence from data, literature, and case examples can support the refinement of various India's flagship programmes such as National Clean Air Initiatives, Smart City Scheme and AMRUT by the union government. This study provides various policy implications to state governments for actionable insights for effective city plans, waste and traffic management.

REFERENCES

Adarsh, R., Jothi, E., Sivasubramanian, K., & Madhusudhanan, R. (2025). Analysing the Status of Agricultural Finance in Rural India in Promoting Agriculture. In Al Mubarak, M. (Ed.), *Sustainable Digital Technology and Ethics in an Ever-Changing Environment. Studies in Systems, Decision and Control* (Vol. 237). Springer., DOI: 10.1007/978-3-031-86708-8_52

Adhisekar, K., Lenin, S., Ilankadhir, M., Vani, H., & Kumarasamy, P. (2026). A review-based analysis on the application of AI in retail management for sustainable business in India. In *Studies in Systems* (Vol. 620, pp. 323–330). Decision and Control. DOI: 10.1007/978-3-032-02056-7_26

Ahvenniemi, H., Huovila, A., Pinto-Seppä, I., & Airaksinen, M. (2017). What are the differences between sustainable and smart cities? *Cities (London, England), 60*, 234–245. DOI: 10.1016/j.cities.2016.09.009

Alamelu Sathyabama, N., Ramesh, K., & Sivasubramanian, K. (2025). Waste to Wealth: Impact of Solid Waste Management Strategies for Achieving SDGs in India. In Alareeni, B. (Ed.), *Big Data in Finance: Transforming the Financial Landscape. Studies in Big Data* (Vol. 169). Springer., DOI: 10.1007/978-3-031-80656-8_39

Alamelu Sathyabama, N., Ramesh, K., & Sivasubramanian, K. (2025). Economic and Social Issues of Women Domestic Workers in India: An Analysis. In Al Mubarak, M. (Ed.), *Sustainable Digital Technology and Ethics in an Ever-Changing Environment. Studies in Systems, Decision and Control* (Vol. 236). Springer., DOI: 10.1007/978-3-031-86712-5_49

Angelidou, M. (2017). The role of smart city characteristics in the plans of fifteen cities. *Journal of Urban Technology, 24*(4), 3–28. DOI: 10.1080/10630732.2017.1348880

Balasubramanian, G.. (2025). Digital Payment Systems and Business Development of Small and Petty Traders in India: A Monetary Analysis. In Hamdan, A. (Ed.), *Achieving Sustainable Business through AI, Technology Education and Computer Science. Studies in Big Data* (Vol. 158). Springer., DOI: 10.1007/978-3-031-70855-8_2

Balasubramanian, G., Jeyaratha, V. C., Sudha, T. S., Thamilmani, R., Kumarasamy, P., & Sivasubramanian, K. (2025). Scaling the Progress of Sustainable Business Practices Among Small Scale Industries Relating to India: A Review. In Hamdan, A., & Braendle, U. (Eds.), *Harnessing AI, Machine Learning, and IoT for Intelligent Business. Studies in Systems, Decision and Control* (Vol. 555). Springer., DOI: 10.1007/978-3-031-67890-5_27

Balasubramanian, G., Saravanan, M., Kasthuri, R., Thamilmani, R., Vinothina, V., & Sivasubramanian, K. (2025). Application of Digital Marketing in Retail Sector Through Innovative Technology: An Evaluative Review. In Alareeni, B. (Ed.), *Big Data in Finance: Transforming the Financial Landscape. Studies in Big Data* (Vol. 169). Springer., DOI: 10.1007/978-3-031-80656-8_33

Banerjee, A., Costa, B., Forkan, A. R. M., Kang, Y.-B., Marti, F., McCarthy, C., Ghaderi, H., Georgakopoulos, D., & Jayaraman, P. P. (2024). 5G-enabled smart cities: A real-world evaluation and analysis of 5G using a pilot smart city application. *Internet of Things : Engineering Cyber Physical Human Systems*, *28*, 101326. DOI: 10.1016/j.iot.2024.101326

Basu, D., & Mishra, S. (2024). Mine reclamation practices and effects of stakeholder perception — A case study of Saoner mines, Maharashtra, India. *Journal of Engineering and Applied Sciences (Asian Research Publishing Network)*, *71*(1), 62.

Batty, M., Axhausen, K. W., Giannotti, F., Pozdnoukhov, A., Bazzani, A., Wachowicz, M., Ouzounis, G., & Portugali, Y. (2012). Smart cities of the future. *The European Physical Journal. Special Topics*, *214*(1), 481–518. DOI: 10.1140/epjst/e2012-01703-3

Bibri, S. E. (2018). The IoT for smart sustainable cities of the future: An analytical framework for sensor-based big data applications for environmental sustainability. *Sustainable Cities and Society*, *38*, 230–253. DOI: 10.1016/j.scs.2017.12.034

Bigazzi, A. Y., & Rouleau, M. (2017). Can traffic management strategies improve urban air quality? A review of the evidence. *Journal of Transport & Health*, *7*, 111–124. DOI: 10.1016/j.jth.2017.08.001

Central Pollution Control Board. (2011). *Air quality monitoring, emission inventory, and source apportionment study for Indian cities*. Government of India.

Chaturvedi, V., Ghosh, A., Garg, A., Avashia, V., Vishwanathan, S. S., Gupta, D., Sinha, N. K., Bhushan, C., Banerjee, S., Datt, D., Bansal, J., Pathak, M., Dhar, S., Singh, A. K., Khan, N., Rashmi, R. R., Agrawal, S., Agarwal, D., Singh, A., & Prasad, S. (2024). India's pathway to net zero by 2070: Status, challenges, and way forward. *Environmental Research Letters*, *19*(11), 112501. DOI: 10.1088/1748-9326/ad7749

Dankan Gowda, V., Patil, S., Srinivas, V., Prasad, K., Jagtap, M. M., & Kaur, M. (2024). Smart urban ecosystems with IoT-based strategies for traffic optimization and pollution control. In Proceedings of the 2024 2nd International Conference on Advancement in Computation & Computer Technologies (InCACCT) (pp. 777–782). IEEE. DOI: 10.1109/InCACCT61598.2024.10551250

Das, D. (2020). Perspectives of smart cities in South Africa through applied systems analysis approach: A case of Bloemfontein. *Construction Economics and Building*, *20*(2), 65–88. DOI: 10.5130/AJCEB.v20i2.6657

Das, D. K. (2024). Exploring the symbiotic relationship between digital transformation, infrastructure, service delivery, and governance for smart sustainable cities. *Smart Cities*, *7*(2), 806–835. DOI: 10.3390/smartcities7020034

Das, D. K. (2025). Digital technology and AI for smart sustainable cities in the Global South: A critical review of literature and case studies. *Urban Science (Basel, Switzerland)*, *9*(3), 72. DOI: 10.3390/urbansci9030072

Das, D. K., & Chadchan, J. (2023). A proposed framework for an appropriate governance system to develop smart cities in India. *Territory, Politics, Governance*, •••, 1–22. DOI: 10.1080/21622671.2023.2229872

De Palma, A., & Lindsey, R. (2011). Traffic congestion pricing methodologies and technologies. *Transportation Research Part C, Emerging Technologies*, *19*(6), 1377–1399. DOI: 10.1016/j.trc.2011.02.010

Dureja, A., Dureja, A., Kumar, V., & Sabharwal, S. (2024). Combining digital twin technology and intelligent transportation systems for smart mobility. In Springer (Ed.), [Title of the book] (pp. 281–296). Cham, Switzerland: *Springer*. ISBN 978-3-031-58522-7. DOI: 10.1007/978-3-031-58523-4_14

Erhardt, G., Mucci, R., Cooper, D., Sana, B., Chen, M., & Castiglione, J. (2021). Do transportation network companies increase or decrease transit ridership? Empirical evidence from San Francisco. *Transportation*, *49*(2), 313–342. DOI: 10.1007/s11116-021-10178-4

Farid, A. M., Liner, B., Alshareef, M., Singh Badhesha, P., Boccaletti, C., Alessandro, N., Cacho, A., Carlier, C.-I., Corriveau, A., & Khayal, I.. (2021). Smart city drivers & challenges in urban mobility, healthcare, and interdependent infrastructure systems. *IEEE Potentials*, *40*(1), 11–16. DOI: 10.1109/MPOT.2020.3011399

Ganguly, T., Selvaraj, K. L., & Guttikunda, S. K. (2020). National Clean Air Programme (NCAP) for Indian cities: Review and outlook of clean air action plans. *Atmospheric Environment: X*, *8*, 100096. DOI: 10.1016/j.aeaoa.2020.100096

Garau, C., & Pavan, V. M. (2018). Evaluating urban quality: Indicators and assessment tools for smart sustainable cities. *Sustainability (Basel)*, *10*(2), 575. DOI: 10.3390/su10030575

Ghosh, B., & Arora, S. (2022). Smart as (un)democratic? The making of a smart city imaginary in Kolkata, India. *Environment and Planning C. Politics and Space*, *40*(1), 318–339. DOI: 10.1177/23996544211027583

Goyal, S., Sahni, A., & Garg, T. (2018). Identification and prioritisation of barriers in building smart cities. *International Journal of Mechanical Engineering and Technology*, *9*, 819–827.

Guttikunda, S. K., Dammalapati, S. K., Pradhan, G., Krishna, B., Jethva, H. T., & Jawahar, P. (2023). What is polluting Delhi's air? A review from 1990 to 2022. *Sustainability (Basel)*, *15*(5), 4209. DOI: 10.3390/su15054209

Guttikunda, S. K., Nishadh, K. A., Gota, S., Singh, P., Chanda, A., Jawahar, P., & Asundi, J. (2019). Air quality, emissions, and source contributions analysis for the Greater Bengaluru region of India. *Atmospheric Pollution Research*, *10*(5), 941–953. DOI: 10.1016/j.apr.2019.01.002

Han, L., Zhou, W., Li, W., & Qian, Y. (2022). Challenges in continuous air quality improvement: An insight from the contribution of the recent clean air actions in China. *Urban Climate*, *46*, 101328. DOI: 10.1016/j.uclim.2022.101328

Hassan, A. M., & Lee, H. (2015). The paradox of the sustainable city: Definitions and examples. *Environment, Development and Sustainability*, *17*(6), 1267–1285. DOI: 10.1007/s10668-014-9604-z

Iyer, L. S. (2021). AI-enabled applications towards intelligent transportation. *Transportation Engineering*, *5*, 100083. DOI: 10.1016/j.treng.2021.100083

Javidroozi, V., Carter, C., Grace, M., & Shah, H. (2023). Smart, sustainable, green cities: A state-of-the-art review. *Sustainability (Basel)*, *15*(6), 5353. DOI: 10.3390/su15065353

Jha, I. S., Sen, S., & Kumar, R. (2014). Smart grid development in India—A case study. In *Proceedings of the Eighteenth National Power Systems Conference (NPSC)* (pp. 1–6). IEEE. DOI: 10.1109/NPSC.2014.7103866

Jia, S., Bi, L., Zhu, W., & Fang, T. (2023). System dynamics modeling for improving the policy effect of traffic energy consumption and CO_2 emissions. *Sustainable Cities and Society*, *90*, 104398. DOI: 10.1016/j.scs.2023.104398

Joss, S. (2011). Eco-cities: The mainstreaming of urban sustainability; key characteristics and driving factors. *International Journal of Sustainable Development and Planning*, *6*(3), 268–285. DOI: 10.2495/SDP-V6-N3-268-285

Khadke, P. A., & Waghmare, P. B. (2018). Class and size wise distribution of urban centre and their determinants in Maharashtra. *Research Review International Journal of Multidisciplinary*, *3*, 722–730.

Khlie, K., & Benmamoun, Z. (2024). Towards smarter and greener cities: Harnessing AI and green technology for urban sustainability. *Journal of Infrastructure Policy Development*, *8*(8), 6300. DOI: 10.24294/jipd.v8i8.6300

Kumar, P., Gulia, S., Harrison, R. M., & Khare, M. (2017). The influence of odd–even car trial on fine and coarse particles in Delhi. *Environmental Pollution*, *225*, 20–30. DOI: 10.1016/j.envpol.2017.03.017

Kumar, V. T. M., & Dahiya, B. (2017). Smart economy in smart cities. In Kumar, V. T. M. (Ed.), *Smart economy in smart cities* (pp. 3–76). Springer., DOI: 10.1007/978-981-10-1610-3_1

Mikucioniene, D., Mínguez-García, D., Repon, M. R., Milašius, R., Priniotakis, G., Chronis, I., Kiskira, K., Hogeboom, R., Belda-Anaya, R., & Díaz-García, P. (2024). Understanding and addressing the water footprint in the textile sector: A review. *AUTEX Research Journal*, *24*(1), 1. DOI: 10.1515/aut-2024-0004

Mishra, P., Thakur, P., & Singh, G. (2022). Sustainable smart city to society 5.0: State-of-the-art and research challenges. *SAIEE African Research Journal*, *113*(4), 152–164. DOI: 10.23919/SAIEE.2022.9945865

Moriarty, P. (2023). Advanced communication and computational technologies in a sustainable urban context: Smart grids, smart cities, and smart health. In Power Systems: Vol. Part F1423 (pp. 389–415). *Springer*. DOI: 10.1007/978-3-031-38506-3_15

Musa, A. A., Malami, S. I., Alanazi, F., Ounaies, W., Alshammari, M., & Haruna, S. I. (2023). Sustainable traffic management for smart cities using Internet-of-Things-oriented intelligent transportation systems (ITS): Challenges and recommendations. *Sustainability (Basel)*, *15*(9859), 9859. Advance online publication. DOI: 10.3390/su15139859

Peris-Ortiz, M., Bennett, D., & Yábar, D. P. B. (Eds.). (2016). *Sustainable smart cities: Creating spaces for technological, social, and business development*. Springer.

Praharaj, S., & Han, H. (2019). Cutting through the clutter of smart city definitions: A reading into the smart city perceptions in India. *City, Culture and Society*, *18*, 100289. DOI: 10.1016/j.ccs.2019.05.005

Rajeswari, S., Savithiri, D., Usharani, S., Sujatha, P., Kasthuri, R., & Sivasubramanian, K. (2025). Economic Analysis of Trend in Agricultural Price: A Statistical Picture with Reference to India. In Hamdan, A., & Braendle, U. (Eds.), *Harnessing AI, Machine Learning, and IoT for Intelligent Business. Studies in Systems, Decision and Control* (Vol. 555). Springer., DOI: 10.1007/978-3-031-67890-5_50

Ramadan, M. N. A., Ali, M. A. H., Shin, Y. K., Alkhedher, M., & Alherbawi, M. (2024). Real-time IoT-powered AI system for monitoring and forecasting of air pollution in industrial environment. *Ecotoxicology and Environmental Safety*, *283*, 116856. DOI: 10.1016/j.ecoenv.2024.116856

Rautela, K. S., & Goyal, M. K. (2024). Transforming air pollution management in India with AI and machine learning technologies. *Scientific Reports*, *14*(1), 20412. DOI: 10.1038/s41598-024-71269-7

Rocha Filho, G. P., Meneguette, R. I., Torres Neto, J. R., Valejo, A., Weigang, L., Ueyama, J., Pessin, G., & Villas, L. A. (2020). Enhancing intelligence in traffic management systems to aid in vehicle traffic congestion problems in smart cities. *Ad Hoc Networks*, *107*, 102265. DOI: 10.1016/j.adhoc.2020.102265

Roy, P., & Srinivasan, K. K. (2025). Geographically weighted nonlinear regression for cost-effective policies to enhance bus ridership. *Sustainability (Basel)*, *17*(6), 2485. DOI: 10.3390/su17062485

Saha, A. R., & Singh, N. (2017). Smart cities for a sustainable future: Can Singapore be a model for Delhi? *International Journal of Economic Research*, *14*, 367–379.

Savithiri, D., Kasthuri, R., Rajeswari, P., Vinothina, V., & Sivasubramanian, K. (2026). Self-help group model for financial sustainability of rural women in India: An explanatory study. In B. S. Awwad (Ed.), Artificial intelligence and technology: Systems management, decisions and control for sustainability in the digital age (Studies in Systems, Decision and Control, Vol. 645, pp. 969–976). Springer, Cham. DOI: 10.1007/978-3-032-10016-0_77

Savithiri, D., Usharani, S., Sujatha, P., Rajeswari, S., Kasthuri, R., & Sivasubramanian, K. (2025). Women Labour Force Participation and Economic Development in India: An Economic Analysis. In Hamdan, A., & Braendle, U. (Eds.), *Harnessing AI, Machine Learning, and IoT for Intelligent Business. Studies in Systems, Decision and Control* (Vol. 555). Springer., DOI: 10.1007/978-3-031-67890-5_11

Segnestam, L. (2002). Indicators of environment and sustainable development: Theories and practical experience (Environmental Economics Series No. 89, pp. 1–4). *The World Bank.*

Selvakumar, M., Reddy, M. A., Sathish, V., & Venkatesh, R. (2018). Potential influence of metro on bus: A case study. *Journal of the Institution of Engineers (India)*: Series A, 99, 379–384.

Sethi, D., & Kosmopoulos, P. G. (2025). Rooftop solar photovoltaic potential in polluted Indian cities: Atmospheric and urban impacts, climate trends, societal gains, and economic opportunities. *Remote Sensing (Basel)*, *17*(7), 1221. DOI: 10.3390/rs17071221

Sharma, N., Kumar, R., Vihari, N. S., Arora, M., & Saini, J. R. (2025). Prioritising smart city themes for multi-national enterprises and United Nations Sustainable Development Goals. *Sustainability (Basel)*, *17*(10), 4251. DOI: 10.3390/su17104251

Shehab, M. J., Kassem, I., Kutty, A. A., Kucukvar, M., Onat, N., & Khattab, T. (2021). 5G networks towards smart and sustainable cities: A review of recent developments, applications and future perspectives. *IEEE Access : Practical Innovations, Open Solutions*, *10*, 2987–3006. DOI: 10.1109/ACCESS.2021.3139436

Shruti, S., Singh, P. K., & Ohri, A. (2021). Evaluating the environmental sustainability of smart cities in India: The design and application of the Indian Smart City Environmental Sustainability Index. *Sustainability (Basel)*, *13*(1), 327. DOI: 10.3390/su13010327

Singh, A. K., Kumar, V. R. P., Dehdasht, G., Mohandes, S. R., Manu, P., & Pour Rahimian, F. (2023). Investigating the barriers to the adoption of blockchain technology in sustainable construction projects. *Journal of Cleaner Production*, *403*, 136840. DOI: 10.1016/j.jclepro.2023.136840

Singh, S., Singh, J., Goyal, S. B., Sehra, S. S., Ali, F., Alkhafaji, M. A., & Singh, R. (2023). A novel framework to avoid traffic congestion and air pollution for the sustainable development of smart cities. *Sustainable Energy Technologies and Assessments*, *56*, 103125. DOI: 10.1016/j.seta.2023.103125

Sivasubramanian, K. (2024). An analysis of green retailing for sustainable business through digital technology application in Indian retail market. In *Building climate neutral economies through digital business and green skills* (pp. 1–22). IGI Global., DOI: 10.4018/979-8-3693-5673-9.ch005

Sivasubramanian, K., Pushpa, A., Raju, V., & Kumar, M. D. (2020). Women and informal employment: An analysis of socio-economic and health conditions of women home-based workers in Chennai, India. *Journal of International Women's Studies*, *21*(5), 97–106.

Son, H., Jang, J., Park, J., Balog, A., Ballantyne, P., Kwon, H. R., Singleton, A., & Hwang, J. (2025). Leveraging advanced technologies for (smart) transportation planning: A systematic review. *Sustainability (Basel), 17*(5), 2245. DOI: 10.3390/su17052245

Sultaana, M. A., & Saha, A. K. (2025). Unleashing the potential of 5G for smart cities: A focus on real-time digital twin integration. *Smart Cities, 8*(2), 70. DOI: 10.3390/smartcities8020070

Sureshchandra, M. S., Bhavsar, J. J., & Pitroda, R. J. (2016). Assessment of critical success factors for smart cities using the significance index method. *International Journal of Advanced Research in Innovative Ideas in Education, 2*(2), 802–810.

Tang, J., McNabola, A., & Misstear, B. (2020). The potential impacts of different traffic management strategies on air pollution and public health for a more sustainable city: A modelling case study from Dublin, Ireland. *Sustainable Cities and Society, 60*, 102229. DOI: 10.1016/j.scs.2020.102229

Tavares, P., Ingi, D., Araújo, L., Pinho, P., & Bhusal, P. (2021). Reviewing the role of outdoor lighting in achieving Sustainable Development Goals. *Sustainability (Basel), 13*(22), 12657. DOI: 10.3390/su132212657

Trindade, E. P., Hinnig, M. P. F., Da Costa, E. M., Marques, J. S., Bastos, R. C., & Yigitcanlar, T. (2017). Sustainable development of smart cities: A systematic review of the literature. *Journal of Open Innovation, 3*(3), 11. DOI: 10.1186/s40852-017-0063-2

Tu, G., & Zan, Z. (2025). The green paradox of new energy vehicles: A system dynamics analysis. *Sustainability (Basel), 17*(9), 3938. DOI: 10.3390/su17093938

Uma Maheswari, V., Balasubramanian, G., Sugatha Kumari, B., Ramya, C., Thamilmani, R., & Sivasubramanian, K. (2025). Business Strategies and Employment Growth of Retail Industry: An Economic Perspective in India. In Hamdan, A. (Ed.), *Achieving Sustainable Business through AI, Technology Education and Computer Science. Studies in Big Data* (Vol. 158). Springer., DOI: 10.1007/978-3-031-70855-8_41

Usharani, S.. (2025). An Evaluative Review of Solid Waste Management System in India for the Sustainable Future. In Hamdan, A., & Braendle, U. (Eds.), *Harnessing AI, Machine Learning, and IoT for Intelligent Business. Studies in Systems, Decision and Control* (Vol. 555). Springer., DOI: 10.1007/978-3-031-67890-5_47

Vani, H., Kumarasamy, P., Chand, M. V., Kasthuri, R., & Sivasubramanian, K. (2025). Women and Work: An Economic Investigation on Women Fish Vendors in Coastal Areas of Chennai District, India. In Hamdan, A., & Braendle, U. (Eds.), *Harnessing AI, Machine Learning, and IoT for Intelligent Business. Studies in Systems, Decision and Control* (Vol. 555). Springer., DOI: 10.1007/978-3-031-67890-5_34

Vedant, A. (2024). Urbanisation and Economic Development in India: A Case Study of Major Cities. *International Journal of Novel Research and Development, 9*(8).

Venkatesan, S., Rajkamal, S., Ramesh, K., & Sivasubramanian, K. (2025). An Analytical Review of the Performance of Selected SDGs in South Indian States Using NITI AAYOG Data. In Hamdan, A., & Braendle, U. (Eds.), *Harnessing AI, Machine Learning, and IoT for Intelligent Business. Studies in Systems, Decision and Control* (Vol. 555). Springer., DOI: 10.1007/978-3-031-67890-5_22

Venkatesh, G. (2014). A critique of the European Green City Index. *Journal of Environmental Planning and Management, 57*(3), 317–328. DOI: 10.1080/09640568.2012.741520

Wang, Z., Fan, M., & Fan, Y. (2024). The impact of digitally inclusive finance on environmental pollution: A case study of air pollution. *PLoS One, 19*(7), e0305963. DOI: 10.1371/journal.pone.0305963

Yang, C., Liang, P., Fu, L., Cui, G., Huang, F., Teng, F., & Bangash, Y. A. (2022). Using 5G in smart cities: A systematic mapping study. *Intelligent Systems with Applications, 14*, 200065. DOI: 10.1016/j.iswa.2022.200065

Yigitcanlar, T., Desouza, K., Butler, L., & Roozkhosh, F. (2020). Contributions and risks of artificial intelligence (AI) in building smarter cities: Insights from a systematic review of the literature. *Energies, 13*(6), 1473. DOI: 10.3390/en13061473

KEY TERMS AND DEFINITIONS

Air Programme: The National Clean Air Programme was introduced in 2019 to enhance the quality of air by reducing CO2 emissions by nearly 20 to 30 per cent in India. Further, it was extended to a revised target of 40% reduction in particulate matter level at PM10 by 2026. It was found that modest improvements have been identified in PM10 and PM2.5 levels in cities such as Delhi, Kanpur, and Lucknow by around 15 to 25 per cent compared to 2017 PM levels.

Government Policies: Government policy consists of regulatory measures, directives, and budget requirements formulated through government officials to serve and enhance the welfare of the public. It is an ever-changing process involving various stakeholders, generating policies influenced by different interests and considerations. Meanwhile, government policies and programmes evolve, responding to a changing social requirement. A helpful place to start when analyzing and placing the rights for advancement in the Indian context is the Preamble of the Indian Constitution, as well as the chapters on the basic freedoms and the Directive Principles of Government Policy. As is widely understood, the Indian Constitution, as it was interpreted in 1950, made an unambiguous differentiation between a set of basic rights that protect people from arbitrary or coercive state action and a set of regulations that should serve as the goals and aspirations for the State's actions in the interest of the welfare of its citizens.

NOx: Nitrogen Oxides

SOx: Sulphur Oxides

TPM: Total Particulate Matter

Urbanisation & Smart cities: Sustainable development and governance techniques are needed in urbanizing cities to control urban growth and provide residents with a high-quality, dignified existence. Numerous city administrations throughout the globe have worked hard in this area and have been successful in overcoming the different obstacles associated with the urbanization phenomenon. Similar conditions have not been maintained in many Indian cities, which has contributed to a lower standard of living for their residents. As the initial city in the SCM, Bhubaneswar has emerged as a model for the technological resurgence of urban governance. Bhubaneswar has been a model for other cities in the region thanks to its citizen-friendly programs, which include savvy parking systems, bicycle-sharing programs, and effective garbage management. Likewise, as part of the monsoon-related problems in that area, flood mitigation measures and drainage system improvements have been implemented under the Guwahati city planning programs. Just as crucial as the Smart Cities Mission's conceptual framework is its financial foundation. For this purpose, the centre has approved a Rs. 48,000 crore cash contribution.

VOC: Volatile Organic Compounds

Chapter 4

Air Pollution, Sustainable Air Quality Management, and Public Health

Laxmi Kant Bhardwaj
http://orcid.org/0000-0001-7518-4199
Amity Institute of Environmental Sciences, Amity University, Noida, India

Prangya Rath
http://orcid.org/0000-0002-6764-1625
Amity Institute of Environmental Sciences, Amity University, Noida, India

K. Dhanaraj
http://orcid.org/0000-0002-9121-7314
Amity Institute of Geoinformatics and Remote Sensing, Amity University, Noida, India

Made P. Raju
Amity Institute of Environmental Sciences, Amity University, Noida, India

Yogesh Kumar Sharma
http://orcid.org/0000-0001-5646-4291
Department of Chemistry, Kalindi College, University of Delhi, New Delhi, India

ABSTRACT

Air pollution is a global concern as it influences the health of humans and the environment because of increasing industrialization, urbanization, as well as other anthropogenic activities, several air pollutants like carbon monoxide (CO), nitrogen oxides (NOx), particulate matter (PM), sulfur dioxide (SO2), volatile organic compounds (VOCs), etc. are released into the atmosphere. These pollutants are associ-

DOI: 10.4018/979-8-3373-4422-5.ch004

ated with serious health problems such as cancers, premature deaths, neurological disorders, respiratory problems, as well as cardiovascular illnesses. These effects of air pollution are observed in susceptible inhabitants which including children, the elderly, and low-income groups. Sustainable air quality administration is needed to mitigate these influences and preserve public health. This chapter highlights the outline, sources, as well as impact of air pollution on the health of humans. Different air pollutants, sustainable management of air quality, recommendations, and future directions, as well as innovations, are also focused on in this chapter.

1. INTRODUCTION

Air pollution refers to the existence of toxic particles in the atmosphere that negatively affect the health of humans and the environment (Sharma et al. 2013; Manisalidis et al., 2020; Raju et al. 2026). It occurs in liquid, solid, and gaseous states. It can be natural or anthropogenic and impacts millions of people. It can be categorized on the basis of the sources and the properties of contaminants. It is the major cause of climate change (Bhardwaj 2023). Although it may begin in one country, its impact can reach other countries via atmospheric transport (Colvile et al. 2001; Fenger 2009). That is why this issue impacts the health of people and the environment on a global level. In developing countries, rapid urbanization has made air pollution worse due to industrial development, population rise, and elevated emissions from transportation, industries, and energy expenditure (Zhang et al. 2022).

Air pollution is a major concern not only because of declining air quality but also due to additional health issues, such as cardiovascular disorders, respiratory illnesses, and even premature death. Due to the comprehensive-ranging outcomes on the environment, climate change, and economic change, air pollution has become an extremely complicated problem that requires prompt consideration. This has emphasized the importance of addressing the universal catastrophe through air quality control procedures. Through a systematic study, technology, and regulatory engagements, the sustainable attempt aims to establish a balance between defending the environment and promoting economic growth by lowering contaminant releases in the environment. Public health consequences are also advanced by decreasing exposure to dangerous contaminants due to valuable air quality control approaches.

The addition of remote sensing and Geographic Information Systems (GIS) in air pollution investigations has been extensively researched, explaining the utility and role of the integration in air quality observation and assessment. Remote sensing, especially satellite-based platforms (e.g., Moderate Resolution Imaging Spectroradiometer (MODIS), Landsat, and Sentinel-5P), has been widely used to supervise atmospheric contaminants such as $PM_{2.5}$ and PM_{10}, NOx, and SO_2 at the local and global level. Studies have indicated these technologies' applicability in the detection of spatial and temporal pollution models, particularly in areas with limited ground-based monitoring setups. For example, aerosol optical depth (AOD) products, obtained from satellite observations, have been applied to estimate particulate strengths, showing interaction between satellite-based products and in situ depths. This has permitted the improvement of predictive models that link satellite imagery to air quality symbols, providing valuable data to policymakers and urban developers.

GIS shows a supporting task by giving a toolset for spatial investigation, data integration, and visualization of air pollution statistics. Data highlights the effectiveness of GIS in mapping pollution resources, modelling toxin distribution, and evaluating contamination display threats for the residents. Numerous studies have engaged GIS to evaluate the link between air pollution and health consequences, integrating demographic, land usage, and meteorological data for a comprehensive examination. Furthermore, GIS-based applications have been applied to estimate the impact of urban morphology on pollution levels, such as the effects of green spaces and the traffic flow on air quality. The integration of remote sensing and GIS has proven clearly effective in urban settings, where intricate relationships between emission sources and local relief warrant detailed spatial examination. Since the acquisition of increasingly finer-resolution satellite data and advanced GIS techniques, these technologies have become critical for air pollution studies and management.

This chapter investigates the complicated interplay between air pollution, sustainable air quality administration, and public health, highlighting the value of multidisciplinary and inclusive methods in the engagement against this concern. This chapter also aimed to describe the occurrence, types of air pollutants, sources, mitigation, and challenges of air pollution. This chapter also focuses on future directions and innovations and gives some recommendations for decreasing the dangerous consequences of air pollution. The different components of this chapter are shown in Figure 2.

Figure 1. The long-range atmospheric transport mechanism (grasshopper effect) for air pollutants

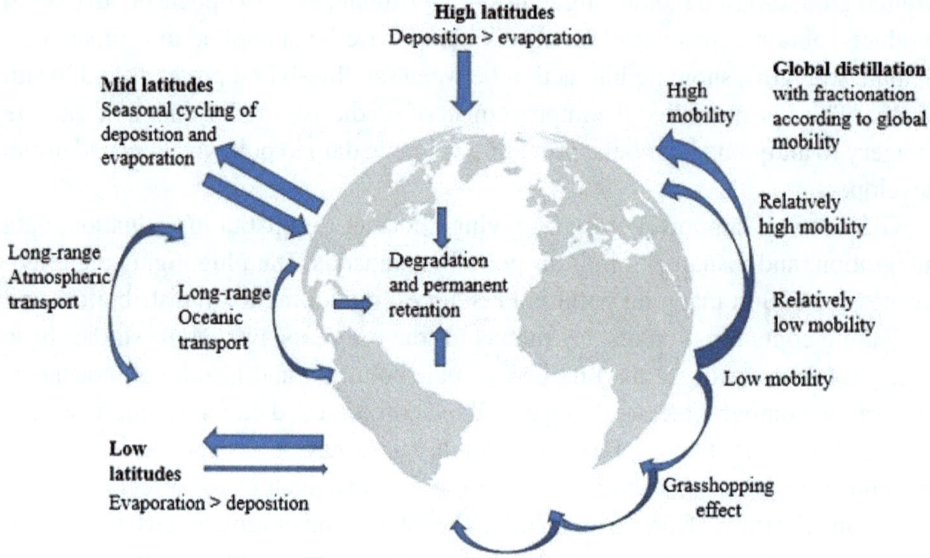

Figure 2. The study of air pollution and its components (types of air pollutants, sources, mitigation, challenges, future direction, innovations & recommendations for air pollution)

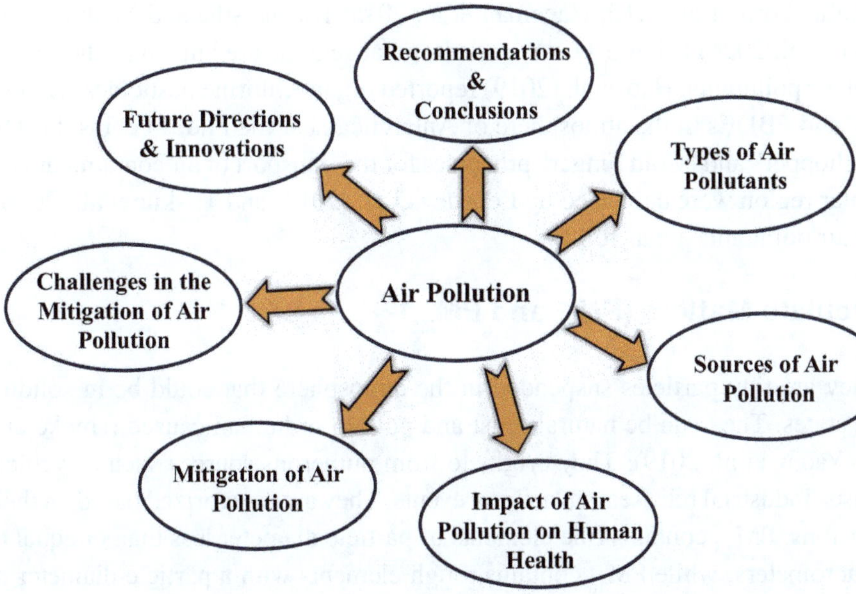

2. AIR POLLUTANTS

Air pollution is triggered by air pollutants like particulate matter ($PM_{2.5}$ and PM_{10}), NOx, SO_2, CO, and VOCs (Li et al. 2019; WHO 2021). There are different kinds of air contaminants, each originating from diverse sources and having different effects on human health and the environment. Contaminants can be classified into criteria and non-criteria contaminants (Griffin 2016; Al-Hemoud et al. 2024). Criteria contaminants are those whose application limits have been decided to recognize between poor and standard air quality, such as NO_x, PM_{10}, benzene (C_6H_6), SO_2, lead (Pb), ozone (O_3) and CO (Saxena and Sonwani 2019). Whereas, non-criteria pollutants are those designated as toxic substances by the authority (Lioy 2018). The concentration of air pollutants is increasing with the increase in population and urbanization (Lin and Zhu 2018; Qing 2018). The different types and sizes of air pollutants are shown in Table 1 (Manisalidis et al. 2020).

These air contaminants are transferred to the polar region due to human actions and long-range atmospheric transport mechanisms and can be stored up in snow/ice (Kallenborn et al. 2015; Hageman et al. 2015; Kulshrestha and Mishra 2020; Thakur et al. 2026) (Figure 1). As a result, polar regions are known as the recipients of air pollutants. Hao et al. (2019) reported organochlorine pesticides (OCPs), PCBs, and PBDEs in the atmosphere of Antarctica near the Fildes Peninsula. The "Grasshopper" and "Cold Finger" processes for the transport of air contaminants to the polar region were described by Lebedev et al. (2014) and Thakur et al. (2026). Some air pollutants are as follows:

Particulate Matters (PM_{10} and $PM_{2.5}$)

They are tiny particles suspended in the atmosphere that could be in solid or liquid states. They can be natural (dust and pollen) or human-caused (smoke and soot) (Yadav et al. 2019). They originate from numerous sources such as vehicle exhausts, industrial releases, and natural events. They are categorized based on their dimensions. $PM_{2.5}$ contains fine elements of particle diameter less than or equal to 2.5 micrometers, while PM_{10} contains rough elements with a particle diameter of up to 10 micrometers (Pikridas et al. 2018). They can enter extremely deeply into the lungs and bloodstream (Brook and Rajagopalan 2009; Thompson 2018).

Nitrogenous Oxides (NO_x)

Nitric oxide (NO) and nitrogen dioxide (NO_2) are collectively referred to as NO_X. In the atmosphere, NO is rapidly oxidized to NO_2, which plays a critical role in the formation of secondary pollutants, including O_3, photochemical smog, and secondary PM. They are emitted primarily from vehicles, power plants, industrial activities, and biomass burning (McDonald et al. 2012). They contribute to the creation of smog and acid rain, eutrophication of aquatic ecosystems, and damage to vegetation by disrupting photosynthesis. Prolonged exposure to NO_X adversely affects human health by irritating the respiratory tract, reducing lung function, and exacerbating asthma and other pulmonary diseases, particularly in children and the elderly.

Sulfur Dioxide (SO_2)

It is primarily released from the combustion of sulfur-containing fossil fuels in power plants and industrial facilities, as well as from petroleum refining, metal smelting, and volcanic eruptions (Hung and Ashner 2020). It can respond with other mixtures in the environment to form secondary contaminants, for example, sulfuric acid (H_2SO_4), nitric acid (HNO_3), etc. After exposure, sulfur dioxide (SO_2) is associated with adverse health effects, particularly respiratory irritation, bronchoconstriction, and aggravated asthma. It is seen to participate in acid rain formation.

Volatile Organic Compounds (VOCs)

Volatile organic compounds (VOCs) are characterized by their high vapor pressure and ability to easily evaporate into the atmosphere at ambient temperatures. They originate from natural as well as anthropogenic sources. Vegetation is the natural source, while industrial processes, automobile emissions, fuel combustion, solvent and pesticide use, etc., are examples of anthropogenic activities (Li et al. 2018; Wang et al. 2023). These hazardous chemicals influence the formation of ground-level O_3 and smog. They also play a critical role in photochemical reactions. They are toxic and carcinogenic and pose risks to human health by affecting the respiratory system, central nervous system, and immune function.

Carbon Monoxide (CO)

It is a colorless, tasteless, and odorless gas emitted by partially combusting carbon-containing fuels such as petrol, coal, diesel, biomass, and natural gas (Yadava and Bhatt 2021). Vehicle releases, industrial procedures, residential cooking, and heating are the main causes of the presence of CO in the atmosphere (Wang et al. 2019). It can persist for a long time once released into the atmosphere. In high amounts, it can act as a barrier to transport oxygen (O_2) throughout the body after binding with hemoglobin and forming carboxyhemoglobin.

Ozone (O_3)

It is a secondary contaminant formed in the troposphere through photochemical reactions between NO_x and VOCs in the presence of sunlight (Wang et al. 2022). Unlike stratospheric ozone, which protects life by absorbing ultraviolet radiation, ground-level ozone is harmful to human health and the environment. Short-term exposure can irritate the respiratory system, reduce lung function, and exacerbate asthma, while long-term exposure is associated with chronic respiratory diseases and increased mortality. It also damages vegetation by impairing photosynthesis, reducing crop yields, and weakening forest ecosystems, thereby posing serious ecological and economic concerns, especially in urban and industrial regions.

Heavy Metals

They are the most hazardous air pollutants due to their persistence, toxicity, and ability to penetrate deep into the human respiratory system. They are released into the natural environment due to numerous industrial processes, coal combustion, vehicular emissions, mining, waste incineration, and metallurgical processes (Bhardwaj et al. 2024a; Bhardwaj et al. 2025 a & b). Exposure to heavy metals is linked to severe health effects, including neurotoxicity, respiratory and cardiovascular diseases, kidney damage, and increased cancer risk. The presence of some metals, for example, Pb, mercury (Hg), and cadmium (Cd), is toxic even at minute concentrations if inhaled or ingested (Alam et al. 2023; Gupta et al. 2024). Pb emitted from battery manufacturing plants is responsible for children's cognitive development. Hg could be sustained in the environment for a longer duration and could lead to neurological diseases.

Persistent Organic Pollutants (POPs)

These are volatile chemical compounds that are toxic and persist in the environment for a longer duration (Boethling et al. 2009; Weber et al. 2011; Akhtar et al. 2021; Bhardwaj and Jindal 2021; Bhardwaj et al. 2026). They tend to accumulate in the organisms and lead to long-term consequences on the health of humans. They can be transported over long distances, leading to widespread contamination far from their original sources. They may induce oxidative stress, endocrine disruption, immunotoxicity, and carcinogenic effects. Dioxins, polychlorinated biphenyls (PCBs), polycyclic aromatic hydrocarbons (PAHs), polybrominated biphenyl ethers (PBDEs), and certain organochlorine pesticides are examples of POPs (Wikoff et al. 2012; Alawi et al. 2018).

Ammonia (NH_3)

It is a gaseous compound and reacts in the atmosphere with acidic gases such as sulfur dioxide (SO_2) and nitrogen oxides (NO_X) to form secondary inorganic aerosols like ammonium sulfate and ammonium nitrate. It is emitted into the environment from agricultural actions, especially the use of nitrogen-based fertilizers and livestock waste, along with emissions from biomass burning and certain industrial processes. It can affect the quality of soil and water (Bittman and Mikkelsen 2009; Behera et al. 2013). It is responsible for the creation of fine PM and contributes to nutrient inequality. Exposure to high concentration pose serious health risks and can penetrate deep into the lungs and bloodstream, leading to respiratory and cardiovascular diseases.

Table 1. Types and size of air pollutants

S. No.	Types	Air Pollutants	Size (μm)
1	Particulate toxins	Smog	0.01-1.0
		Soot	0.01-0.8
		Tobacco smoke	0.01-1.0
		Fly ash	1.0-100.0
		Cement dust	8.0-100.0
2	Biological toxins	Bacteria and bacterial spores	0.7-10.0
		Viruses	0.01-1.0
		Fungi and molds	2.0-12.0
		Allergens	0.1-100.0
3	Types of dust	Atmospheric dust	0.01-1.0
		Heavy dust	100.0-1000.0
		Settling dust	1.0-100.0
4	Gases	Different gaseous toxins	0.0001-0.01

3. REVIEW OF LITERATURE

Air pollution was first reported in the 13th century in London (Te Brake 1975). As per the Clean Air Act (CAA), two types of air quality specifications are available. In the primary standard, limits to contaminants to guard the health of children and communities are described, while in the secondary standard, limits of contaminants to guard crops, buildings, and vegetation are described. An air quality standard for suspended particulates and SO_2 is available in the study, which was done by the European Commission (EC) in 1980 (Smeets 1982). The air quality standards for O_3, CO, NOx, Pb, and PM have been set by the United States Environmental Protection Agency (USEPA) (Grambsch 2001).

Several researchers studied the air quality index in various countries. Tang et al. (2020) studied air quality in Dublin, Ireland. This study was based on the traffic situations of 2013. They included two pollutants, NO_2 and $PM_{2.5,}$ in their study and reported that traffic is the chief source of air pollution in urban spaces. Zhang et al. (2022) reviewed the history of air pollution in high, mid, and low-income countries in relation to urbanization and stated that low- and mid-income nations are more affected by air pollution in urban areas. Abdul Jabbar et al. (2022) studied air quality in six South Asian countries (India, Nepal, Bangladesh, Pakistan, Sri Lanka, and Nepal) between 1990 to 2020. They reported the greatest rise in CO_2 discharges in India (578.5 to 2441.8 million tons), methane (CH_4) discharges in Pakistan (70.4 to

151 million tons), and nitrous oxide (N_2O) emissions in Bangladesh (16.5 to 29.3 million tons). They also reported the highest level of $PM_{2.5}$ in Bangladesh (76.9 μg/m^3) during the COVID-19 pandemic. Kumar and his team reviewed the sources of major indoor air pollutants and their effects on human health (Kumar et al. 2023). They reported that indoor air quality (IAQ) is one of the fundamental elements that can affect more women and children due to long-term exposure to poor indoor air quality.

4. SOURCES OF AIR POLLUTION

The sources of air pollution can differ based on the geographical setting (Huang and Batterman 2000; Harrison 2006; Lelieveld et al. 2015). These sources include transport (cars, trucks, and airplanes), industrial resources, agricultural events, construction, waste burning, and natural causes like wildfires and volcanic explosions. Releases from thermal power plants, refineries, and petrochemicals are the main sources of air pollution (Hannun and Razzaq 2022; Tavella et al. 2025). Dust from building locations also adds to air pollution (Azarov et al. 2018; Unsal 2025). Air contaminants are mainly emitted from the burning activity (Friedrich & Reis 2004). The burning of wood can be a source of polycyclic aromatic hydrocarbons (PAHs) (Bhardwaj et al. 2026). Volatile organic chemicals (VOCs) are immediately released from human-caused and natural sources. NO_x is released from both static and mobile sources and can fall with rain in the form of HNO_3 (Abiodun et al. 2014).

Marelle et al. (2016) stated that the local traces of air pollution induce the atmospheric components. Numerous sources distribute harmful constituents into the atmosphere, leading to decreased air quality and threatening human and environmental well-being. These sources can be considered as human-caused and natural bases, with each contributing differently to air quality. Knowledge of these sources is an important requirement for developing effective control or mitigation methods. Natural sources of air pollution are those that do not involve human activity (Popescu and Ionel 2010; Guttikunda et al. 2014). Most emissions come from volcanic eruptions, which release ash, PMs, and SO_2 gases into the surroundings. Wildfires, caused by natural events such as lightning strikes, release CO, NOx, and VOCs. PM is also contributed by sea spray close to coastal locations and dust storms in desert places.

However, human activity is now the foremost source of air pollution in contemporary times. The use of fossil fuels, such as oil, coal, and natural gas, in industrial activities leads to large emissions of SO_2, NOx, and $PM_{2.5}$. This is heavily influenced by manufacturers, power plants, and refineries. Transportation is another significant source, with gasoline and diesel-run automobiles emitting CO, NOx, and unburned hydrocarbons (HCs). Within the domestic sector, the burning of biomass for heating and cooking, as well as inefficient energy use, leads to pollutants like black carbon (BC) and VOCs (Raju et al. 2020). Agricultural activities also contribute, with NH_3 being produced from fertilizers and CH_4 from animals. The worst feature of air pollution is being created by urbanization and rapid industrialization because of the emissions of informal industries, the burning of rubbish, and construction works. Therefore, the combination of various sources creates a complex air pollution scenario, requiring coordination across sectors to manage and control the system effectively. Source management will be essential to minimize health issues and achieve the long-term benefits of improvements in air quality.

5. IMPACT OF AIR POLLUTION ON HUMAN HEALTH

Air pollution significantly affects human health and other living beings, having multifaceted as well as far-reaching effects (Gul and Das 2023; Ayejoto et al. 2023; Izah et al. 2024). Its impacts include respiratory diseases, cardiovascular problems, aggravation of existing health conditions, decreased air quality, harm to crops and ecosystems, and assistance to climate change. Air pollution adversely affects individuals' health and impacts people of all ages and socioeconomic groups. A variety of acute and chronic health conditions can also arise in the act of inhaling pollutants, including CO, PM, $SO_{2,}$ NOx, and ground-level O_3. It can reduce lung function and cognitive impairments during children's development.

It increases stress and anxiety levels and reduces the quality of life in highly contaminated regions. The toxic consequences of air pollution on human health highlight the value of implementing comprehensive air quality management policies. These policies should target decreasing chemical releases, supervising air quality, and executing targeted interventions to protect vulnerable inhabitants. Such actions are necessary to preserve public health, improve the quality of life, and mitigate the long-term effects of air contamination. The alterations in the air toxins influence the equilibrium of radiation and influence polar climate heating (Shindell and Faluvegi 2009). The dangerous impacts of air contamination are far-reaching and can be demonstrated in numerous ways:

Respiratory Problems and Cardiovascular Issues

Higher levels of PM in the air have been observed to lead to different types of respiratory disorders (Wu et al. 2018; Kyung and Jeong 2020). These include bronchitis, asthma, chronic obstructive pulmonary disease (COPD), and degraded lung function. Excessive exposure to PM for a longer duration has been reported to negatively affect lung development in children. Additionally, the existence of

NOx in the air worsens respiratory infections. NOx can induce inflammation in the airways and decreased lung function (Haccuria et al. 2014; Lu et al. 2023). Yue et al. (2017) studied the harmful impact of air pollution and reported different types of cancers. PM in high concentrations leads to oxidative stress and inflammation in the body of humans and is linked with cardiovascular disorders and heart attacks (Hoek et al. 2013). The risk of air pollution is higher in elderly people, children, and those who are under medical treatment (Peled 2011; Di et al. 2017; Kingdon 2023).

Cancer Risk and Neurological Effects

Certain air toxins, such as formaldehyde (CH_2O) and C_6H_6, are known as carcinogens (Dan et al. 2020; Tesfaye et al. 2020). These chemicals can cause different types of cancer after exposure in the long term. Roy and D'Angiulli (2024) described the link between air contamination and neurological disorders. Higher levels of contaminants have been connected with increased danger of Alzheimer's and Parkinson's disease (Chin-Chan et al. 2015; Rajendran et al. 2022; Bhardwaj et al. 2024b; Calderón-Garcidueñas et al. 2024). The presence of a high concentration of Pb can negatively affect the development of the brain (Nava-Ruiz et al. 2012; Chibowska et al. 2016; Karri et al. 2016)

Adverse Pregnancy Outcomes and Affecting Children's Health

Air pollution has been reported to cause issues with respiratory function, difficulties in developmental processes, and prolonged health consequences. Lower birth weights and premature births in pregnancy are also reported to be associated with air pollution (Stieb et al. 2012; Liu et al. 2019). Sram et al. (2013) studied the development of children and reported that children are highly vulnerable to the consequences of air pollution. Air pollution can destructively affect children's nervous system development (Genc et al. 2012; Calderón-Garcidueñas et al. 2014; Sram et al. 2017).

Harm to Ecosystems and Climate Change

Air pollution can damage aquatic environments and forests (Křeček and Hořická 2006; Smith 2012). SO_2 and NO_x are the major causes of acid rain that can harm vegetation, damage oil quality, and affect water bodies (Bhardwaj 2023). Singh and Yadav (2021) studied the correlation between air pollution and climate change and reported that greenhouse gases (GHGs) such as CO_2 influence climate change and global warming.

Reduced Quality of Life and Disproportionate Impact on Vulnerable Communities

Air pollution has a significant impact on social as well as economic terms (Martinez et al. 2018; Druzhinin et al. 2020; Sivarethinamohan et al. 2021). It decreases the quality of life and productivity while increasing the overall costs of healthcare caused by an increase in illness and hospitalizations. CO interferes with the transport of O_2 in the human body. This leads to prolonged headaches, different types of severe health problems, and, in extreme cases, it could lead to death. Air pollution has been reported to exacerbate social inequalities. There have been studies that have shown that marginalized and less-income communities are affected disproportionately and thus are exposed to more health problems (Cook et al. 2021).

6. MITIGATION OF AIR POLLUTION

Regulatory actions, technological developments to decrease releases, transitioning to better energy resources, and public understanding and involvement in lowering individual contributions to air contamination are the efforts that are involved in controlling air pollution.

These efforts hold a range of approaches and act at numerous stages, such as individual, community, industrial, and governmental. Air pollution control requires interdisciplinary actions at global and local levels to address the multifaceted nature of pollutants and their impacts. Internationally, agreements (e.g., the Paris Agreement) aim to decrease emissions of GHGs and air pollutants by promoting cleaner and more sustainable energy production, sustainable industrial practices, and improvements in transportation technology. Implementing renewable energy resources (i.e., solar, wind, and hydro) has become a cornerstone of these strategies, along with coal and fossil fuel retirement efforts.

Collaborative work across the globe is evident in research and innovation for cleaner technologies, supporting knowledge transfer, and building capacity in various countries. By targeting the primary sources of transboundary pollution (industrial emissions and deforestation), these actions focus on reducing the extended environmental and climatic consequences of air pollution. At the local level, cities and communities implement targeted interventions based on the unique sources of pollution they face. In urban areas, strategies such as low-emission zones (to reduce vehicle emissions), promoting public transport (to replace vehicle trips), and integrating cycle and pedestrian facilities (to provide alternatives to vehicle trips) are used to mitigate annual vehicle output.

Local authorities enforce more stringent industrial controls and deploy air quality monitoring infrastructure to detect and quickly address hotspots. Natural ways, such as urban forests and green roofs, help eliminate toxins and improve the urban population's quality of life. Public understanding programs and community-based actions, such as controlling biomass flaming and enhancing waste management, are also important in focusing on localized contamination traces. Through the common incorporation of universal activities and local actions, the community collectively contributes to a greener and restored environment for future generations.

There are some effective ways to reduce and control air pollution:

Promotion of Clean Energy Sources and Better Urban Planning

Sources of renewable energies include power generated from solar, wind, and hydroelectric sources. These sources can substantially decrease releases from traditional fossil fuel-based energy creation. Public transport, pedestrian-friendly spaces, and effective traffic management can decrease city overcrowding and vehicular pollution. The promotion of electric and hybrid automobiles can substantially decrease vehicle emissions.

Improving Industrial Processes, Regulations, and Policy Implementation

The implementation of cleaner manufacturing equipment, utilizing low-emission fuels, and the installation of pollution control strategies in manufacturing can decrease toxic releases. The implementation and enforcement of precise release standards and protocols for manufacturing units and automobile scans efficiently monitor and decrease pollutant releases.

Waste Management, Recycling, Afforestation, and Green Spaces

Waste management procedures, such as recycling and proper removal, can decrease the risk of air pollutants from burning and landfills. Trees act as natural filters, absorb contaminants from the atmosphere, and improve the quality of the air.

Public Awareness, Education, and International Cooperation

The awareness and education of people about the consequences of air contamination can help reduce air pollution. The promotion of personal actions like applying energy-efficient applications and decreasing personal vehicle usage can significantly decrease the impact of air pollution. International agreements and collaborations can help mitigate the impact of contaminants that cross national boundaries.

Monitoring and Research

Monitoring and research are important elements in the assessment of air pollution. Continuous observation of air quality is essential to recognizing the traces and effects of air pollution. It can provide the necessary and updated data to the researchers and scientists who are working on air pollution. A satellite-based air pollution observing method permits the constant tracking of contaminants, such as $PM_{2.5}$, PM_{10}, NOx, SO_2, CO, and O_3. Investment in research and innovation can help develop successful mitigation approaches. Research activities enhance monitoring by studying air pollution sources, dispersion, and effects, contributing to the development of specific policies and practical action plans. The development of analytic instruments, including big data analytics and machine learning (ML), has optimized the use of complex environmental information to facilitate integrative air quality analysis.

Monitoring and research of air pollution have been significantly enhanced by remote sensing and GIS, which enable the visualization and analysis of spatial data. Satellite-based remote sensing instruments, such as MODIS and Sentinel-5P, provide overviews of air-pollutant atmospheric scenarios, enabling region-wide, large-scale systematic monitoring of air quality, especially in areas where ground-truth data are limited or unavailable. GIS integrates spatial and temporal data from multiple sources to generate a comprehensive map that represents pollution concentrations, trends, and affected areas. These technologies are particularly relevant for transboundary pollution and investigating the spatial correlation between emission sources and affected populations. Remote sensing and GIS are also powerful in pollutant transport modelling and risk zone identification, which can inform targeted interventions.

Better accuracy, as well as the usage of technologies required for precise monitoring, are needed to enhance our knowledge of the role of air pollution and other variables of the environment. Research studies comprising novel pollutants in categories of microplastics (MPs) and nanomaterials (NMs) are one of the key focuses of investigation (Bhardwaj et al. 2023; Bhardwaj 2024). Cooperation between different sectors of academia, government organizations, and intergovernmental institutions is necessary for carrying forward innovation along with knowledge transfer. Combining remote sensing with GIS for the purpose of monitoring and research would help in expanding the scope of air quality management. It would also simultaneously support the implementation of new mitigation strategies against air pollution. These techniques would be beneficial for protecting global public health.

7. CHALLENGES IN THE MITIGATION OF AIR POLLUTION

Decreasing air pollution is a complex problem at the universal and regional scales, driven by scientific, economic, political, and public restrictions. A significant problem is the complexity of pollution sources, which range from manufacturing processes and transportation to agriculture and domestic energy consumption. This heterogeneity is met with specific strategies adapted to address various regional and sectoral contributions. Furthermore, economic limitations often prevent the adoption of cleaner technologies, particularly in developing countries where fossil fuels and traditional technologies are heavily relied upon. The economic burden of transitioning to renewable energy, modernizing industrial processes, and introducing advanced technologies is unsustainable without adequate financing and support from the international community.

Political and institutional roadblocks also present significant challenges. These consist of inconsistently formulated strategies, ineffective enforcement of policies, and, in specific cases, an absence of worldwide consensus on reduction goals for climate change. The conflict between economic progress and environmental defense often leads to inaction. Additionally, public awareness and resistance to change may slow down progress, as individuals and communities may be reluctant to adopt cleaner practices or accept policy measures that affect their way of life. Weather conditions and transboundary air pollution add difficulties, since pollutants of one region usually reach adjacent regions, creating the need for international collaboration to address air pollution effectively. Overcoming these barriers requires concretive action, technological advancements, robust governance, and active community participation to achieve measurable, lasting improvement in air quality. Despite these challenges, efforts to address and mitigate air pollution globally must continue to safeguard public health and the atmosphere. Some of the key challenges include:

Complexity of Sources, Urbanization, and Population Growth

The sources of air pollution are diverse. The coordination efforts to control toxic gases generated from the different sources are complex. Fast urbanization and population expansion indicate intensified vehicular transport, industrial activities, and energy utilization, resulting in significant releases and exacerbating air contamination challenges.

Climate Change, Air Quality, and Lack of Awareness

Climate change is linked to air quality. A balanced methodology is necessary to address the issues of climate change and air pollution at the same time because some ways to decrease GHG releases can increase the issue of air pollution. Most people are not aware of the health impacts of air contamination or the actions that can be taken to decrease it. Learning and understanding movements are necessary to notify the public.

Political, Regulatory Hurdles, and Environmental Justice

Implementing effective policies and regulations to control emissions often faces political and economic challenges. The balancing of environmental affairs with economic consequences can be a substantial hurdle because air contamination doesn't have political or geographical limits. The transportation of air contaminants from one nation to another adds complications because addressing air quality involves international cooperation and agreements. Air pollution extremely influences low-income communities and less affluent regions. Achieving justice in addressing air quality problems is a task due to social, economic, and political components.

Technological and Infrastructural Challenges

Upgrading infrastructure and implementing cleaner technologies require substantial investment and time. This transition often faces resistance due to cost implications and technological limitations.

Monitoring, Data Collection, and Innovative Solutions

Efficient monitoring systems are necessary to track the quality of air and collect data on the level of contaminants. However, the setup of the perfect monitoring systems is challenging in developing nations. Finding innovative, cost-effective solutions to mitigate air pollution is a challenging task for developing regions.

8. CASE STUDIES

(a) Creation of Ultra Low Emission Zone (ULEZ) in London, United Kingdom

The authority of London created an Ultra Low Emission Zone (ULEZ) in 2019 to decrease the emissions of NO_2 and PM from transport. It is the largest clean air zone in the world. It is a region where a release standard-based charge is directed to non-compliant road vehicles. It manages 24 hours a day, 7 days a week, every day of the year to help clear London's air. The target of this initiative was high-polluting vehicles with daily charges to incentivize cleaner transport alternatives. A study conducted by GLA (2020) reported the reduction of the concentration of NO_2 by 44% within ULEZ. This initiative is thought to be a sustainable example for urban air quality management. Tsocheva et al. (2023) studied the effects of air pollution on children aged 6–9 years in primary schools in Central London during the first phase of the ULEZ. They reported improved lung function in the children.

(b) Implementation of "Odd-Even Vehicle Rule" in Delhi, India

The Delhi government implemented this rule between 2016 and 2020. It is also known as the vehicle rationing rule. It is a traffic management strategy. According to this rule, the private vehicles with record-keeping plates ending in odd numbers are permitted on odd-numbered dates, and those with even-numbered plates are allowed on even-numbered dates. The aim of this rule was to decrease the number of cars on the road during peak hours and to reduce the city's critical levels of PM pollution, which often exceeded WHO limits. A study conducted by Sharma and his team in 2017 on the air quality of Delhi city during Phases 1 and Phase II of the Odd-Even rule of 2016. They reported a decrease in traffic-related air pollution and a minor improvement in the air quality of Delhi city as compared to before the implementation of the scheme. Goyal and Gandhi (2016) studied the Air Quality Index (AQI) during the Odd-Even rule in 2016 and reported that PM_{10} and $PM_{2.5}$ were the main contributors of air contaminants in Delhi.

(c) Air Pollution Control Action Plan (2013–2017) in Beijing, China

The Beijing Municipal Government created and published the "Beijing Clean Air Action Plan (2013–2017). The set target of this plan was the reduction of $PM_{2.5}$ to a level lower than the threshold of 60 $\mu g/m^3$ by 2017. It was focused on the reduction of air pollution from various sources like vehicles, industries, and coal-fired power plants, as well as promoting clean energy. The government invested $17 billion in this plan to enhance the air quality. As per the study by the Chinese Ministry of Ecology and Environment in 2018, the reduction of $PM_{2.5}$ by 35.6% was reported during these five years.

(d) Long-Term Air Quality Regulations in Los Angeles, USA

The city of Los Angeles was facing severe smog and ozone pollution. The management of the city implemented the Clean Air Act in 1963, which targeted industrial emissions and vehicular pollution. This law authorizes the EPA to determine

National Ambient Air Quality Standards (NAAQS) to safeguard public health and to standardize releases of dangerous air pollutants like ozone, particulate matter, and other air toxics. A study done by Gauderman and his team in 2015 reported the respiratory health of children from 1994 to 2011 and air quality improvement. They found improved lung development and reduced air pollutants such as NO_2 and $PM_{2.5}$.

9. FUTURE DIRECTIONS AND INNOVATIONS

The future of air pollution reduction depends on the adoption of state-of-the-art technologies, international partnerships, and sustainable approaches that balance economic growth with environmental protection. Technological innovations, such as improved renewable energy, electric vehicles, and carbon capture and storage, are ripe to reduce major emission sources. Novel technologies like hydrogen fuel cells and bioenergy are emerging as attractive green technologies for clean energy generation. Furthermore, intelligent city solutions, including Internet of Things (IoT)-based air quality monitoring and data-driven urban planning, are well equipped to dynamically manage pollution reduction efforts.

On a larger scale, policy frameworks and transnational agreements will play a key role in the future of air quality management. Strengthening commitments to international agreements, such as the Paris Agreement, and fostering collaboration among countries can leverage the implementation of clean technologies and the dissemination of information. Innovations in geoengineering, such as stratospheric aerosol injection or urban greening, offer potential unconventional methods to reduce atmospheric pollution. Public engagement and education will also remain vital in promoting behavioural changes and supporting community-driven initiatives. The combination of these technological advances, strong policies, and social participation could effectively address air pollution, enhancing healthy environments and sustainable development for future generations.

One promising direction for the future is the development of artificial intelligence (AI) and ML applications for air pollution prediction, monitoring, and mitigation. These technologies can handle large volumes of environmental data to analyze pollution trends, forecast high-risk events, and develop optimal response strategies. AI-based models can be utilized to develop early warning systems, allowing policymakers and communities to proactively mitigate exposure to pollutants. Advancements in remote sensing technologies utilizing satellites could also improve the accuracy and determination of air quality monitoring and pollution distribution data across urban and rural areas. These tools have the potential to address data gaps, particularly where ground-based observation infrastructure is limited, and support fairer air quality management.

Future scope needs to focus on integrating air pollution reduction with the reduction in climate change to achieve Sustainable Development Goals (SDGs). For example, combining transit-oriented development with mixed-income communities was observed to decrease dependence on automobiles, improve the quality of air, and reduce the levels of GHG emissions. Furthermore, the circular economy model, which aims towards minimizing waste and maximizing resource usage, has significant potential to reduce pollution related to both industries and domestic. Advancements in financial instruments, like green bonds and carbon credits, have been suggested to be beneficial for building cleaner technologies and infrastructure. These programs could be made more successful by the formation of strong collaborations among governments, industries, academia, and the population to leverage innovations and policies for enhancing their effectiveness and practical approaches.

10. RECOMMENDATIONS

Researchers/scientists are continuously working towards minimizing air pollution and simultaneously providing recommendations for future problems. They suggest that regularly monitoring the parameters of air pollution would be beneficial for data generation. More advanced and sophisticated instruments are required for air pollutant analysis. Emission of hazardous CO, NOx, and HCs needs to be measured using catalytic converters. The authorities need to design more strategies for air pollution control. Addressing air pollution challenges needs more concerted efforts collectively from governments, industries, communities, and the public so as to implement better and more comprehensive strategies, regulations, and technological advancements.

Addressing the toxic and harmful effects generated due to air pollution requires strategies that would reduce emissions. These would simultaneously implement strict regulations that would promote clean energy techniques, improve transportation facilities for the public, and increase awareness relating to the significance of clean air and its importance in regulating human health and the environment. Awareness, education, and implementing mitigation measures to reduce indoor air pollution are important for ensuring better and healthier indoor environments, as they would minimize health risks that are associated with air pollutants. In India, efforts to reduce air pollution are progressing. Such policies target vehicular emissions, initiatives that promote the usage of cleaner energy, and reduce the burning of agricultural wastes. However, there is a need for more improved and comprehensive measures, strictly enforcing the regulations, as well as increased awareness and public participation, to effectively mitigate multifaceted challenges.

11. CONCLUSION

Air pollution has become a major global issue in recent decades. Both developing and developed nations are severely affected. Activities such as construction, mining, and vehicular emissions degrade air quality. Air pollution harms human health and wildlife. It is linked to severe diseases and premature deaths. Pollutants can travel long distances and reach polar regions. They accumulate in ice and persist for extended periods. Growing health concerns have led to stricter environmental policies. Clean energy technologies and electric vehicles offer hope for improvement. Research links air pollution to respiratory, cardiovascular, and neurological disorders. International cooperation addresses transboundary air pollution challenges. Temporary reductions during COVID-19 showed the impact of reduced emissions. However, lasting improvement requires structural and policy changes. Governments must enforce strong regional air quality management strategies. Combining regulation, renewable energy, sustainable transport, and public awareness is essential for cleaner air.

ACKNOWLEDGMENT

The authors thank their respective institutes for providing the platform to do this study.

REFERENCES

Abdul Jabbar, S., Tul Qadar, L., Ghafoor, S., Rasheed, L., Sarfraz, Z., Sarfraz, A., Sarfraz, M., Felix, M., & Cherrez-Ojeda, I. (2022). Air quality, pollution and sustainability trends in South Asia: A population-based study. *International Journal of Environmental Research and Public Health, 19*(12), 7534. DOI: 10.3390/ijerph19127534

Abiodun, B. J., Ojumu, A. M., Jenner, S., & Ojumu, T. V. (2014). The transport of atmospheric NO x and HNO 3 over Cape Town. *Atmospheric Chemistry and Physics, 14*(2), 559–575. DOI: 10.5194/acp-14-559-2014

Akhtar, A. B. T., Naseem, S., Yasar, A., & Naseem, Z. (2021). Persistent organic pollutants (POPs): sources, types, impacts, and their remediation. In *Environmental pollution and remediation* (pp. 213–246). Springer Singapore. DOI: 10.1007/978-981-15-5499-5_8

Al-Hemoud, A., Al-Senafi, A., Al-Khayat, A., Behbehani, W., Al-Dashti, H., Al-Dabbous, A., Al-Qadeeri, G., Gasana, J., Mihalkov, V., Petrov, P., Alahmad, B., & Koutrakis, P. (2024). The complex chemical effects of air pollutants, meteorology and dust on the concentrations of criteria and non-criteria air pollution in Kuwait. *Discover Atmosphere, 2*(1), 15. DOI: 10.1007/s44292-024-00017-3

Alam, S., Bhardwaj, L. K., Mallick, R., & Rai, S. (2023). Estimation of heavy metals and fluoride ion in vegetables grown nearby the stretch of River Yamuna, Delhi (NCR), India. *Indian Journal of Environmental Protection, 43*(1), 64–73.

Alawi, M., Masaad, M., & Al-Hussaini, M. (2018). Comparative study of persistent organic pollutant (POP)(chlorinated pesticides, PCBs, and dioxins/furans) concentrations in cancer-affected human organs with those of healthy organs. *Environmental Monitoring and Assessment, 190*(8), 470. DOI: 10.1007/s10661-018-6856-3

Ayejoto, D. A., Agbasi, J. C., Nwazelibe, V. E., Egbueri, J. C., & Alao, J. O. (2023). Understanding the connections between climate change, air pollution, and human health in Africa: Insights from a literature review. *Journal of Environmental Science and Health. Part C, Environmental Carcinogenesis & Ecotoxicology Reviews, 41*(3-4), 77–120.

Azarov, V., Petrenko, L., & Manzhilevskaya, S. (2018). The study of local dust pollution of atmospheric air on construction sites in urban areas. In *Energy Management of Municipal Transportation Facilities and Transport* (pp. 430–439). Springer International Publishing.

Behera, S. N., Sharma, M., Aneja, V. P., & Balasubramanian, R. (2013). Ammonia in the atmosphere: A review on emission sources, atmospheric chemistry and deposition on terrestrial bodies. *Environmental Science and Pollution Research International, 20*(11), 8092–8131. DOI: 10.1007/s11356-013-2051-9

Bhardwaj, L., & Jindal, T. (2021). Polar ecotoxicology: Sources and toxic effects of pollutants. In *New frontiers in environmental Toxicology* (pp. 9–14). Springer International Publishing.

Bhardwaj, L. K. (2023). A Comprehensive review on the climate change and its impact on health. https://doi.org/DOI: 10.20944/preprints202305.0159.v1

Bhardwaj, L. K., Mittal, R., Rath, P., Gupta, U., Bhardwaj, D., Dahal, Y., & Kumar, A. (2026). Persistent Organic Pollutants (POPs), A Threat to Arctic Environment and Their Impact on Human Health. In Xenobiotics in Environment (pp. 15-24). CRC Press.

Bhardwaj, L. K., Rath, P., Bajpai, S., Upadhyay, D., Jain, H., Kumar, N., & Sinha, S. (2024b). COVID-19 and the Interplay with Antibacterial Drug Resistance. In *Frontiers in Combating Antibacterial Resistance: Current Perspectives and Future Horizons* (pp. 246-273). IGI Global Scientific Publishing.

Bhardwaj, L. K., Rath, P., & Choudhury, M. (2023). A comprehensive review on the classification, uses, sources of nanoparticles (NPs) and their toxicity on health. *Aerosol Science and Engineering, 7*(1), 69–86. DOI: 10.1007/s41810-022-00163-4

Bhardwaj, L. K., Rath, P., Choudhury, M., Metzdorff, A., & Jodder, P. K. (2025a). Electronic wastes and contamination of water. In *Electronic Waste* (pp. 48–57). CRC Press. DOI: 10.1201/9781003582311-6

Bhardwaj, L. K., Rath, P., Jain, H., & Choudhury, M. (2025b). Exploring the effects of e-waste on soil, water quality and human health. *Discover Civil Engineering, 2*(1), 12. DOI: 10.1007/s44290-025-00167-2

Bhardwaj, L. K., Rath, P., & Tokas, R. (2024a). E-waste management in developing countries: current practices, challenges, disposal, and impact on human health and environment. *Sustainable solutions for e-waste and development*, 90-102.

Bittman, S., & Mikkelsen, R. (2009). Ammonia emissions from agricultural operations: Livestock. *Better Crops with Plant Food, 93*(1), 28–31.

Boethling, R., Fenner, K., Howard, P., Klečka, G., Madsen, T., Snape, J. R., & Whelan, M. J. (2009). Environmental persistence of organic pollutants: Guidance for development and review of POP risk profiles. *Integrated Environmental Assessment and Management*, *5*(4), 539–556. DOI: 10.1897/IEAM_2008-090.1

Brook, R. D., & Rajagopalan, S. (2009). Particulate matter, air pollution, and blood pressure. *Journal of the American Society of Hypertension*, *3*(5), 332–350. DOI: 10.1016/j.jash.2009.08.005

Calderón-Garcidueñas, L., Stommel, E. W., Torres-Jardón, R., Hernández-Luna, J., Aiello-Mora, M., González-Maciel, A., Reynoso-Robles, R., Pérez-Guillé, B., Silva-Pereyra, H. G., Tehuacanero-Cuapa, S., Rodríguez-Gómez, A., Lachmann, I., Galaz-Montoya, C., Doty, R. L., Roy, A., & Mukherjee, P. S. (2024). Alzheimer and Parkinson diseases, frontotemporal lobar degeneration and amyotrophic lateral sclerosis overlapping neuropathology start in the first two decades of life in pollution exposed urbanites and brain ultrafine particulate matter and industrial nanoparticles, including Fe, Ti, Al, V, Ni, Hg, Co, Cu, Zn, Ag, Pt, Ce, La, Pr and W are key players. Metropolitan Mexico City health crisis is in progress. *Frontiers in Human Neuroscience*, *17*, 1297467. DOI: 10.3389/fnhum.2023.1297467

Calderón-Garcidueñas, L., Torres-Jardón, R., Kulesza, R. J., Park, S. B., & D'Angiulli, A. (2014). Air pollution and detrimental effects on children's brain. The need for a multidisciplinary approach to the issue complexity and challenges. *Frontiers in Human Neuroscience*, *8*, 613.

Chibowska, K., Baranowska-Bosiacka, I., Falkowska, A., Gutowska, I., Goschorska, M., & Chlubek, D. (2016). Effect of lead (Pb) on inflammatory processes in the brain. *International Journal of Molecular Sciences*, *17*(12), 2140. DOI: 10.3390/ijms17122140

Chin-Chan, M., Navarro-Yepes, J., & Quintanilla-Vega, B. (2015). Environmental pollutants as risk factors for neurodegenerative disorders: Alzheimer and Parkinson diseases. *Frontiers in Cellular Neuroscience*, *9*, 124. DOI: 10.3389/fncel.2015.00124

Colvile, R. N., Hutchinson, E. J., Mindell, J. S., & Warren, R. F. (2001). The transport sector as a source of air pollution. *Atmospheric Environment*, *35*(9), 1537–1565. DOI: 10.1016/S1352-2310(00)00551-3

Cook, Q., Argenio, K., & Lovinsky-Desir, S. (2021). The impact of environmental injustice and social determinants of health on the role of air pollution in asthma and allergic disease in the United States. *The Journal of Allergy and Clinical Immunology*, *148*(5), 1089–1101. DOI: 10.1016/j.jaci.2021.09.018

Dan, S., Pant, M., Kaur, T., & Pant, S. (2020). Toxic effect of formaldehyde. *Systematic Reviews.*

Di, Q., Dai, L., Wang, Y., Zanobetti, A., Choirat, C., Schwartz, J. D., & Dominici, F. (2017). Association of short-term exposure to air pollution with mortality in older adults. *Journal of the American Medical Association, 318*(24), 2446–2456. DOI: 10.1001/jama.2017.17923

Druzhinin, P. V., Shkiperova, G. T., Potasheva, O. V., & Zimin, D. A. (2020). The Assessment of the Impact of the Economy's Development on Air Pollution. *Ekonomicheskie i Sotsialnye Peremeny, 13*(2), 125–142. DOI: 10.15838/esc.2020.2.68.8

Fenger, J. (2009). Air pollution in the last 50 years–From local to global. *Atmospheric Environment, 43*(1), 13–22. DOI: 10.1016/j.atmosenv.2008.09.061

Friedrich, R., & Reis, S. (Eds.). (2004). *Emissions of air pollutants: measurements, calculations and uncertainties.* Springer Science & Business Media. DOI: 10.1007/978-3-662-07015-4

Gauderman, W. J., Urman, R., Avol, E., Berhane, K., McConnell, R., Rappaport, E., Chang, R., Lurmann, F., & Gilliland, F. (2015). Association of improved air quality with lung development in children. *The New England Journal of Medicine, 372*(10), 905–913. DOI: 10.1056/NEJMoa1414123

Genc, S., Zadeoglulari, Z., Fuss, S. H., & Genc, K. (2012). The adverse effects of air pollution on the nervous system. *Journal of Toxicology, 2012*(1), 782462. DOI: 10.1155/2012/782462

Goyal, P., & Gandhi, G. (2016). Assessment of air quality during the 'odd-even scheme'of vehicles in Delhi. *Indian Journal of Science and Technology, 9*(48), 1–7. DOI: 10.17485/ijst/2016/v9i30/97528

Grambsch, A. (2001). *Climate change and air quality.* Department of Tansportation Centre for Climate Change and Environmental Forecasting.

Greater London Authority (GLA). (2020). ULEZ One Year Report. https://www.london.gov.uk

Griffin, R. D. (2016). *Principles of air quality management.* CRC Press. DOI: 10.1201/9781420006544

Gul, H., & Das, B. K. (2023). The impacts of air pollution on human health and well-being: A comprehensive review. *Journal of Environmental Impact and Management Policy, 36*(36), 1–11. DOI: 10.55529/jeimp.36.1.11

Gupta, U., Shukla, S., Bhardwaj, L. K., Rath, P., Upadhyay, D., Sharma, B., & Singh, H. (2024). Unraveling urban plant strategies for heavy metal uptake and detoxification. In *Urban Forests, Climate Change and Environmental Pollution: Physio-Biochemical and Molecular Perspectives to Enhance Urban Resilience* (pp. 93–119). Springer Nature Switzerland. DOI: 10.1007/978-3-031-67837-0_5

Guttikunda, S. K., Goel, R., & Pant, P. (2014). Nature of air pollution, emission sources, and management in the Indian cities. *Atmospheric Environment, 95*, 501–510. DOI: 10.1016/j.atmosenv.2014.07.006

Haccuria, A., Michils, A., Michiels, S., & Van Muylem, A. (2014). Exhaled nitric oxide: A biomarker integrating both lung function and airway inflammation changes. *The Journal of Allergy and Clinical Immunology, 134*(3), 554–559. DOI: 10.1016/j.jaci.2013.12.1070

Hageman, K. J., Bogdal, C., & Scheringer, M. (2015). Long-range and regional atmospheric transport of POPs and implications for global cycling. [). Elsevier.]. *Comprehensive Analytical Chemistry, 67*, 363–387. DOI: 10.1016/B978-0-444-63299-9.00011-9

Hannun, R. M., & Razzaq, A. H. A. (2022, March). Air pollution resulted from coal, oil and gas firing in thermal power plants and treatment: A review. [). IOP Publishing.]. *IOP Conference Series. Earth and Environmental Science, 1002*(1), 012008. DOI: 10.1088/1755-1315/1002/1/012008

Hao, Y., Li, Y., Han, X., Wang, T., Yang, R., Wang, P., Xiao, K., Li, W., Lu, H., Fu, J., Wang, Y., Shi, J., Zhang, Q., & Jiang, G. (2019). Air monitoring of polychlorinated biphenyls, polybrominated diphenyl ethers and organochlorine pesticides in West Antarctica during 2011–2017: Concentrations, temporal trends and potential sources. *Environmental Pollution, 249*, 381–389. DOI: 10.1016/j.envpol.2019.03.039

Harrison, R. M. (2006). Sources of air pollution. *Air Quality Guideliees. Air Quality Guidelines*, 9-31.

Hoek, G., Krishnan, R. M., Beelen, R., Peters, A., Ostro, B., Brunekreef, B., & Kaufman, J. D. (2013). Long-term air pollution exposure and cardio-respiratory mortality: A review. *Environmental Health : A Global Access Science Source, 12*(1), 43. DOI: 10.1186/1476-069X-12-43

Huang, Y. L., & Batterman, S. (2000). Selection and evaluation of air pollution exposure indicators based on geographic areas. *The Science of the Total Environment, 253*(1-3), 127–144. DOI: 10.1016/S0048-9697(00)00425-3

Hung, Y. T., & Ashner, I. (2020). Sulfur dioxide emission and mitigation. In *Handbook of environment and waste management: acid rain and greenhouse gas pollution control* (pp. 627-658). DOI: 10.1142/9789811207136_0018

Izah, S. C., Ogwu, M. C., Shahsavani, A., Kermani, M., & Hasanzadeh, V. (2024). Understanding the one health implications of air pollution. In *Air Pollutants in the Context of One Health: Fundamentals, Sources, and Impacts* (pp. 161–185). Springer Nature Switzerland. DOI: 10.1007/698_2024_1129

Kallenborn, R., Hung, H., & Brorström-Lundén, E. (2015). Atmospheric long-range transport of persistent organic pollutants (POPs) into polar regions. [). Elsevier.]. *Comprehensive Analytical Chemistry*, *67*, 411–432. DOI: 10.1016/B978-0-444-63299-9.00013-2

Karri, V., Schuhmacher, M., & Kumar, V. (2016). Heavy metals (Pb, Cd, As and MeHg) as risk factors for cognitive dysfunction: A general review of metal mixture mechanism in brain. *Environmental Toxicology and Pharmacology*, *48*, 203–213. DOI: 10.1016/j.etap.2016.09.016

Kingdon, C. (2023). Air pollution is the largest environmental risk to public health and children are especially vulnerable. *bmj, 381*.

Křeček, J., & Hořická, Z. (2006). Forests, air pollution and water quality: Influencing health in the headwaters of Central Europe's "Black Triangle". *Unasylva*, *57*(224), 46–49.

Kulshrestha, U., & Mishra, M. (2020). A review on long range transport of air pollution in south Asia. *Vayumandal*, *46*(2), 21–30.

Kumar, P., Singh, A. B., Arora, T., Singh, S., & Singh, R. (2023). Critical review on emerging health effects associated with the indoor air quality and its sustainable management. *The Science of the Total Environment*, *872*, 162163. DOI: 10.1016/j.scitotenv.2023.162163

Kyung, S. Y., & Jeong, S. H. (2020). Particulate-matter related respiratory diseases. *Tuberculosis and Respiratory Diseases*, *83*(2), 116. DOI: 10.4046/trd.2019.0025

Lebedev, A. T., Mazur, D. M., Polyakova, O. V., & Hänninen, O. (2014). Snow samples as markers of air pollution in mass spectrometry analysis. In *Environmental Indicators* (pp. 515–541). Springer Netherlands.

Lelieveld, J., Evans, J. S., Fnais, M., Giannadaki, D., & Pozzer, A. (2015). The contribution of outdoor air pollution sources to premature mortality on a global scale. *Nature*, *525*(7569), 367–371. DOI: 10.1038/nature15371

Li, G., Wei, W., Shao, X., Nie, L., Wang, H., Yan, X., & Zhang, R. (2018). A comprehensive classification method for VOC emission sources to tackle air pollution based on VOC species reactivity and emission amounts. *Journal of Environmental Sciences (China)*, *67*, 78–88. DOI: 10.1016/j.jes.2017.08.003

Li, R., Chen, W., Xiu, A., Zhao, H., Zhang, X., Zhang, S., & Tong, D. Q. (2019). A comprehensive inventory of agricultural atmospheric particulate matters (PM10 and PM2. 5) and gaseous pollutants (VOCs, SO2, NH3, CO, NOx and HC) emissions in China. *Ecological Indicators*, *107*, 105609. DOI: 10.1016/j.ecolind.2019.105609

Lin, B., & Zhu, J. (2018). Changes in urban air quality during urbanization in China. *Journal of Cleaner Production*, *188*, 312–321. DOI: 10.1016/j.jclepro.2018.03.293

Lioy, P. J. (2018). *Toxic Air Pollution: A Comprehensive Study Non-Criteria Air Pollutants*. CRC Press. DOI: 10.1201/9781351077347

Liu, Y., Xu, J., Chen, D., Sun, P., & Ma, X. (2019). The association between air pollution and preterm birth and low birth weight in Guangdong, China. *BMC Public Health*, *19*(1), 3. DOI: 10.1186/s12889-018-6307-7

Lu, C., Wang, F., Liu, Q., Deng, M., Yang, X., & Ma, P. (2023). Effect of NO2 exposure on airway inflammation and oxidative stress in asthmatic mice. *Journal of Hazardous Materials*, *457*, 131787. DOI: 10.1016/j.jhazmat.2023.131787

Manisalidis, I., Stavropoulou, E., Stavropoulos, A., & Bezirtzoglou, E. (2020). Environmental and health impacts of air pollution: A review. *Frontiers in Public Health*, *8*, 14. DOI: 10.3389/fpubh.2020.00014

Marelle, L., Thomas, J. L., Raut, J. C., Law, K. S., Jalkanen, J. P., Johansson, L., Roiger, A., Schlager, H., Kim, J., Reiter, A., & Weinzierl, B. (2016). Air quality and radiative impacts of Arctic shipping emissions in the summertime in northern Norway: From the local to the regional scale. *Atmospheric Chemistry and Physics*, *16*(4), 2359–2379. DOI: 10.5194/acp-16-2359-2016

Martinez, G. S., Spadaro, J. V., Chapizanis, D., Kendrovski, V., Kochubovski, M., & Mudu, P. (2018). Health impacts and economic costs of air pollution in the metropolitan area of Skopje. *International Journal of Environmental Research and Public Health*, *15*(4), 626. DOI: 10.3390/ijerph15040626

McDonald, B. C., Dallmann, T. R., Martin, E. W., & Harley, R. A. (2012). Long-term trends in nitrogen oxide emissions from motor vehicles at national, state, and air basin scales. *Journal of Geophysical Research*, *117*(D21), 2012JD018304. DOI: 10.1029/2012JD018304

Ministry of Ecology and Environment. China (2018). China Air Quality Improvement Report. http://www.mee.gov.cn

Nava-Ruiz, C., Méndez-Armenta, M., & Ríos, C. (2012). Lead neurotoxicity: Effects on brain nitric oxide synthase. *Journal of Molecular Histology*, *43*(5), 553–563. DOI: 10.1007/s10735-012-9414-2

Peled, R. (2011). Air pollution exposure: Who is at high risk? *Atmospheric Environment*, *45*(10), 1781–1785. DOI: 10.1016/j.atmosenv.2011.01.001

Pikridas, M., Vrekoussis, M., Sciare, J., Kleanthous, S., Vasiliadou, E., Kizas, C., Savvides, C., & Mihalopoulos, N. (2018). Spatial and temporal (short and long-term) variability of submicron, fine and sub-10 µm particulate matter (PM1, PM2. 5, PM10) in Cyprus. *Atmospheric Environment*, *191*, 79–93. DOI: 10.1016/j.atmosenv.2018.07.048

Popescu, F., & Ionel, I. (2010). Anthropogenic air pollution sources. *Air quality*, 1-22.

Qing, W. A. N. G. (2018). Urbanization and global health: The role of air pollution. *Iranian Journal of Public Health*, *47*(11), 1644.

Rajendran, R., Ragavan, R. P., Al-Sehemi, A. G., Uddin, M. S., Aleya, L., & Mathew, B. (2022). Current understandings and perspectives of petroleum hydrocarbons in Alzheimer's disease and Parkinson's disease: A global concern. *Environmental Science and Pollution Research International*, *29*(8), 10928–10949. DOI: 10.1007/s11356-021-17931-3

Raju, M. P., Bhardwaj, L. K., & Srivastava, A. K. (2026). Temporal trends in AQI and precursor pollutants: A long-term case study of Noida. *Journal of Environmental Science and Health. Part A, Toxic/Hazardous Substances & Environmental Engineering*, •••, 1–19.

Raju, M. P., Safai, P. D., Sonbawne, S. M., Buchunde, P. S., Pandithurai, G., & Dani, K. K. (2020). Black carbon aerosols over a high altitude station, Mahabaleshwar: Radiative forcing and source apportionment. *Atmospheric Pollution Research*, *11*(8), 1408–1417. DOI: 10.1016/j.apr.2020.05.024

Roy, R., & D'Angiulli, A. (2024). Air pollution and neurological diseases, current state highlights. *Frontiers in Neuroscience*, *18*, 1351721. DOI: 10.3389/fnins.2024.1351721

Saxena, P., & Sonwani, S. (2019). Criteria air pollutants and their impact on environmental health: Vol. 1. *No. 1*. Springer Singapore.

Sharma, S. B., Jain, S., Khirwadkar, P., & Kulkarni, S. (2013). The effects of air pollution on the environment and human health. *Indian Journal of Research in Pharmacy and Biotechnology*, *1*(3), 391.

Shindell, D., & Faluvegi, G. (2009). Climate response to regional radiative forcing during the twentieth century. *Nature Geoscience*, *2*(4), 294–300. DOI: 10.1038/ngeo473

Singh, P., & Yadav, D. (2021). Link between air pollution and global climate change. In *Global climate change* (pp. 79–108). Elsevier. DOI: 10.1016/B978-0-12-822928-6.00009-5

Sivarethinamohan, R., Sujatha, S., Priya, S., Gafoor, A., & Rahman, Z. (2021). Impact of air pollution in health and socio-economic aspects: Review on future approach. *Materials Today: Proceedings*, *37*, 2725–2729. DOI: 10.1016/j.matpr.2020.08.540

Smeets, J. (1982). Air quality limit and guide values for sulphur dioxide and suspended particulates—A European community directive. *Environmental Monitoring and Assessment*, *1*(4), 373–382. DOI: 10.1007/BF00403837

Smith, W. H. (2012). *Air pollution and forests: interactions between air contaminants and forest ecosystems*. Springer Science & Business Media.

Sram, R. J., Binkova, B., Dostal, M., Merkerova-Dostalova, M., Libalova, H., Milcova, A., Rossner, P.Jr, Rossnerova, A., Schmuczerova, J., Svecova, V., Topinka, J., & Votavova, H. (2013). Health impact of air pollution to children. *International Journal of Hygiene and Environmental Health*, *216*(5), 533–540. DOI: 10.1016/j.ijheh.2012.12.001

Sram, R. J., Veleminsky, M., Veleminsky, M., & Stejskalová, J. (2017). The impact of air pollution to central nervous system in children and adults. *Neuroendocrinology Letters*, *38*(6), 389–396.

Stieb, D. M., Chen, L., Eshoul, M., & Judek, S. (2012). Ambient air pollution, birth weight and preterm birth: A systematic review and meta-analysis. *Environmental Research*, *117*, 100–111. DOI: 10.1016/j.envres.2012.05.007

Tang, J., McNabola, A., & Misstear, B. (2020). The potential impacts of different traffic management strategies on air pollution and public health for a more sustainable city: A modelling case study from Dublin, Ireland. *Sustainable Cities and Society*, *60*, 102229. DOI: 10.1016/j.scs.2020.102229

Tavella, R. A., da Silva Júnior, F. M. R., Santos, M. A., Miraglia, S. G. E. K., & Pereira Filho, R. D. (2025). A review of air pollution from petroleum refining and petrochemical industrial complexes: Sources, key pollutants, health impacts, and challenges. *ChemEngineering*, *9*(1), 13. DOI: 10.3390/chemengineering9010013

Te Brake, W. H. (1975). Air pollution and fuel crises in preindustrial London, 1250–1650. *Technology and Culture*, *16*(3), 337–359. DOI: 10.2307/3103030

Tesfaye, S., Hamba, N., Gerbi, A., & Neger, Z. (2020). Oxidative stress and carcinogenic effect of formaldehyde exposure: Systematic review & analysis. *Endocrinology & Metabolic Syndrome*, *9*(6), 319–330.

Thakur, S., Bhardwaj, L. K., Rath, P., Sharma, Y., & Kumar, A. (2026). Relationship between Persistent Organic Pollutants (POPs) and Antarctic Environment. In *Xenobiotics in Environment* (p. 15). CRC Press. DOI: 10.1201/9781003521754-3

Thompson, J. E. (2018). Airborne particulate matter: Human exposure and health effects. *Journal of Occupational and Environmental Medicine*, *60*(5), 392–423. DOI: 10.1097/JOM.0000000000001277

Tsocheva, I., Scales, J., Dove, R., Chavda, J., Kalsi, H., Wood, H. E., Colligan, G., Cross, L., Newby, C., Hall, A., Keating, M., Sartori, L., Moon, J., Thomson, A., Tomini, F., Murray, A., Hamad, W., Tijm, S., Hirst, A., & Griffiths, C. J. (2023). Investigating the impact of London's ultra low emission zone on children's health: Children's health in London and Luton (CHILL) protocol for a prospective parallel cohort study. *BMC Pediatrics*, *23*(1), 556. DOI: 10.1186/s12887-023-04384-5

Unsal, M. H. (2025). *Dust Pollution as an Indicator of Indoor Environment Quality* (Doctoral dissertation, Vilniaus universitetas.).

Wang, J., Wu, Q., Liu, J., Yang, H., Yin, M., Chen, S., Guo, P., Ren, J., Luo, X., Linghu, W., & Huang, Q. (2019). Vehicle emission and atmospheric pollution in China: Problems, progress, and prospects. *PeerJ*, *7*, e6932. DOI: 10.7717/peerj.6932

Wang, S., Zhang, J., Zhang, Y., Wang, L., Sun, Z., & Wang, H. (2023). Review on source profiles of volatile organic compounds (VOCs) in typical industries in China. *Atmosphere (Basel)*, *14*(5), 878. DOI: 10.3390/atmos14050878

Wang, T., Xue, L., Feng, Z., Dai, J., Zhang, Y., & Tan, Y. (2022). Ground-level ozone pollution in China: A synthesis of recent findings on influencing factors and impacts. *Environmental Research Letters*, *17*(6), 063003. DOI: 10.1088/1748-9326/ac69fe

Weber, R., Watson, A., Forter, M., & Oliaei, F. (2011). Persistent organic pollutants and landfills-a review of past experiences and future challenges. *Waste Management & Research*, *29*(1), 107–121. DOI: 10.1177/0734242X10390730

Wikoff, D., Fitzgerald, L., & Birnbaum, L. (2012). Persistent organic pollutants: an overview. *Dioxins and health: including other persistent organic pollutants and endocrine disruptors*, 1-35.

World Health Organization. (2021). *WHO global air quality guidelines: particulate matter (PM2. 5 and PM10), ozone, nitrogen dioxide, sulfur dioxide and carbon monoxide*. World Health Organization.

Wu, J. Z., Ge, D. D., Zhou, L. F., Hou, L. Y., Zhou, Y., & Li, Q. Y. (2018). Effects of particulate matter on allergic respiratory diseases. *Chronic Diseases and Translational Medicine*, *4*(2), 95–102. DOI: 10.1016/j.cdtm.2018.04.001

Yadav, P., Dhupper, R., Singh, S. D., & Singh, B. (2019). Crop adaptation to air pollution I. Effect of particulate and SO2 pollution on growth, yield attributes and sulphur nutrition of wheat, barley and chickpea. *Indian Journal of Agricultural Research*, *53*(3), 303–308. DOI: 10.18805/IJARe.A-5273

Yadava, R. N., & Bhatt, V. (2021). Carbon monoxide: Risk assessment, environmental, and health hazard. In *Hazardous Gases* (pp. 83–96). Academic Press. DOI: 10.1016/B978-0-323-89857-7.00030-X

Yue, S., Wang, Y., Wang, J., & Chen, J. (2017). Relationships between lung cancer incidences and air pollutants. *Technology and Health Care*, *25*(1, suppl), 411–422. DOI: 10.3233/THC-171344

Zhang, X., Han, L., Wei, H., Tan, X., Zhou, W., Li, W., & Qian, Y. (2022). Linking urbanization and air quality together: A review and a perspective on the future sustainable urban development. *Journal of Cleaner Production*, *346*, 130988. DOI: 10.1016/j.jclepro.2022.130988

Chapter 5

AI-Driven Solutions for Monitoring and Managing Groundwater Pollution

Venkata Kiran Kumar Ravi
http://orcid.org/0000-0003-1628-4936
Siddhartha Academy of Higher Education, India

Sourabh Garg
http://orcid.org/0000-0003-4892-5726
School of Engineering and Technology, Manav Rachna International Institute of Research and Studies, India

Hari Krishan Malhotra
Manav Rachna International Institute of Research and Studies, India

Nibedita Das
Tecnia Institute of Advanced Studies, India

Baburao Gaddala
http://orcid.org/0000-0003-4190-8325
School of Studies of Engineering and Technology, Guru Ghasidas Vishwavidyalaya, India

Sureshkumar Markanday Subramanian
ACE Engineering College, India

V. Bhoopathy
http://orcid.org/0000-0003-2175-6328
Sree Rama Engineering College, India

ABSTRACT

Ecosystems, human health, and long-term sustainability are all badly threatened by polluted groundwater. ML and AI offer promising new approaches to this critical problem. In order to effectively monitor, anticipate, and manage groundwater contamination, this chapter investigates the use of AI and ML. Among the main uses discussed are decision-support frameworks for policy-making and resource allocation, predictive modeling for the spread of contaminants, and real-time monitoring of water quality. This chapter delves into the topic of how to enhance

DOI: 10.4018/979-8-3373-4422-5.ch005

data collecting and get valuable insights by combining AI-driven solutions with IoT devices. Additionally, we delve into topics like data scarcity, model reliability, and the ethical implications of AI applications. Through analysis of past and future cases, this chapter demonstrates how AI and ML have the ability to revolutionize methods for controlling groundwater pollution and fostering sustainable management of water resources.

1. INTRODUCTION

Located beneath the surface of the Earth, groundwater is a vital resource for human consumption, agricultural usage, and industrial processes. Water for more than two billion people on Earth comes primarily from underground sources. This resource is absolutely necessary for life on Earth and plays a pivotal role in the hydrological cycle. According to Zhang et al. (2024) say that ecosystems, surface water, and soil dynamically interact with groundwater systems, making them intrinsically complex. The increasing need for freshwater, prompted by population expansion and environmental decline, and requires a thorough comprehension of the groundwater ecosystem and its susceptibility to pollution. This article analyses groundwater systems, pollution sources, and their effects on humans and the environment. The intricate relationship among the various causes of groundwater contamination and the numerous researches demonstrating these impacts has rendered this a significant global issue.

Substances including chemical products, untreated sewage, agricultural runoff, and industrial effluents contaminate groundwater. Human health and ecosystems that rely on clean water are put at risk when contaminants drastically change groundwater quality. Yáo et al. (2022) discovered that individuals from economically and socially disadvantaged backgrounds are at a heightened risk of health issues stemming from the use of groundwater contaminated with heavy metals, nitrates, and parasites. Ecosystems and freshwater supplies are further jeopardised by the infiltration of contaminants, perhaps resulting in enduring environmental issues. Effectively mitigating groundwater contamination necessitates interdisciplinary research and cooperation among communities, governments, and scientists. This research will analyse several examples and scientific data to clarify the present condition of contamination of groundwater and assess the effectiveness of various management options (Das et al., 2020). Management of groundwater contamination frequently involves regular sampling and laboratory analysis. These methods are effective to some extent, but they cannot yield predictions or real-time insights. In order to fill this void, this chapter explores how AI and ML can provide powerful resources for intelligent and proactive groundwater management.

2. TYPES OF GROUND WATER POLLUTION

As a result of the industrial revolution and people's failure to understand the link between chemicals and environmental degradation, groundwater is vulnerable to contamination from a variety of sources. Contamination of groundwater poses a significant obstacle to its localisation and detection(Zamil Al-Sudani, 2019). Sewagetanks, urban sewage canals, and pipelines are only a few examples of the real-life examples of undiscovered subsurface pollution, as are leakages from chemical distribution infrastructure and petrochemicals. Over the past hundred years, one of the most challenging issues related to pollution and management has been the storage of products from mining and industrial complexes underground or in other unregulated ways, which allowed the contamination of land.

2.1 Heavy Metals

Historically, there has been significant interest in the effects of heavy metals resulting from the introduction of sewage waste to land use and originating from dumps. Other sources of heavy metals comprise mining and smelting, fossil fuel combustion, metalworking, electrical and chemical industries, military battles and exercises, in addition to the initiation of athletic activities. In recent years, the acknowledgement of the possible harmful effects and enduring contaminants of heavy metal pollutants has concentrated on the safeguarding of soil and groundwater.

2.2 Contaminated Land

Contaminated soil poses a risk of groundwater pollution if pollutants may migrate into the water body and maintain potentially hazardous concentrations. The repercussions of pollution resulting from human actions, both historically and currently have caused and will persist in causing significant groundwater contamination.

2.3 Nitrate Contamination of Groundwater

The primary inorganic contaminant of the primary problem with groundwater is the presence of nitrate ions, commonly located in aquifers near rural and suburban areas.. While pristine groundwater often contains nitrate nitrogen concentrations below 2 ppm, the presence of nitrate in groundwater mostly stems from four sources:

- Soil cultivation.
- Utilization of nitrogen fertilizers, inorganic substances, and animal dung.
- The atmosphere deposition.

- People waste in sewage systems.

The generation of manure accounts for at least 7 million tonnes of the 12 million tonnes of nitrogen used as fertiliser in American agriculture each year. Since removing nitrate from well water is a costly ordeal, most of the time the oxidised nitrate that results from nitrogen reduction in the soil makes its method to the soil, where it solubilises in water and is attenuated.In the interest of public health, water that is highly nitrate-contaminated is often not served to humans.

2.4 Landfill

The ongoing discussion over dilute and disperse and containment landfill design tactics is crucial to understanding the possible effects of landfills on groundwater quality. To understand how landfill leach ate affects groundwater, one must be aware of the components of the leachate. Waste type, landfill architecture and operations, analytical processes, and timing are the main factors that determine the composition of leachate. The majority of the references provided are about mixed household waste, which is biodegradable, has a low initial density, and is vulnerable to long-term consolidation settlement because of its organic portion. On the other hand, single-use disposal facilities usually only treat one or two types of trash it disposing of specific industrial solid wastes. The mineral deposits, coal scrap metal, quarrying fines, and Pulverized Fuel Ash (PFA) from coals-fired power plants are all examples of mining sector wastes that can be found at these locations. The topic of mono-disposal land filling was not extensively covered. An engineered landfill, or dumpsite, is a major source of pollution that threatens groundwater resources.

2.5 Microbiological Contaminants

Human or animal excrement can lead to microbial contamination of groundwater. Sewage may harbour several pathogens, including protozoa, viruses, and bacteria. Waterborne contaminants can present a considerable threat to human health. Microbiological pollutants can infiltrate the subsurface ecosystem through leaky sewage, digging sinks, septic tanks, soaking liquids, mines for removal, garbage dumps, or terrestrial discharge as manure. The introduction of liquid waste and the contamination of groundwater from surface pollutants will substantially affect the flow and degree of pollution in the groundwater. Investigative surveys and groundwater contamination episodes have concentrated on bacterial and, to a lesser degree, virus contamination in their analyses of microbiological pollutant transport patterns within the aquifer environment. The term refuse refers to non-biodegradable, non-floating material that has been generated by human activities and is considered both home and industrial solid waste. Many disastrous consequences, including pollution, environmental dangers, and unsightliness, have resulted from the mismanagement of solid materials. The possible contamination of leachate by garbage is another cause of groundwater pollution.

2.6 Contaminated Land

The potential for pollutants from polluted land to travel to aquatic environments and maintain potentially dangerous quantities in groundwater is a real concern. Serious contamination of groundwater has resulted from and will continue to result from pollution caused by human activities in the past and present.

3. ROLE OF GROUND WATER IN PUBLIC HEALTH

3.1 Sustainable Water Management

To ensure the long-term viability of groundwater supplies, it is important to manage extraction rates in harmony with natural recharge(M. Shaikh & Birajdar, 2024b).Ensuring a consistent supply of potable Ensuring water availability for present and future generations is a vital goal of sustainable groundwater management.

3.2 Accessibility and Consistency

It compared to surface water bodies; groundwater sources are typically better able to withstand changes in weather and seasons. Groundwater is a reliable and essential resource, providing water year-round, regardless of meteorological conditions or surface water deficits. Wells and boreholes provide access to groundwater, allowing its use in various geographical and topographical conditions (Cuthbert et al., 2019).

3.3 Reliable Drinking Water and Providing Safe

Groundwater is vital for safeguarding public health, offering an ongoing supply of clean and potable water to consume for populations globally. In contrast to surface water sources, which are vulnerable to pollution and seasonal fluctuations, well managed groundwater is often safeguarded from pollutants and provides a dependable source of superior quality water. This uniformity is crucial for fulfilling the fundamental drinking water requirements of urban as well as rural residents (Li et al., 2022).

3.4 Mitigating Waterborne Diseases

Surface water bodies are typically more vulnerable to pathogen contamination than well maintained groundwater sources. Groundwater is intrinsically more dependable for use than surface water because of the natural filtration process that transpires as water infiltrates soil and rock strata. This attribute is crucial for enhancing public health and diminishing the incidence of waterborne infections (Andrade et al., 2018).

3.5 Global Reliance on Groundwater

Groundwater is essential for farming, manufacturing, and domestic use. Groundwater often constitutes the main supply of drinkable water for residences, especially in areas that are semi-arid or arid. A significant number of folks depend on water sources and boreholes linked to aquifers to fulfil their Daily hydration needs. Agriculture consumes substantial quantities of water and mostly relies on it for irrigation. The use of groundwater resources improves agricultural output, especially in areas with limited surface water availability (M. A. Shaikh, 2023).Groundwater is extensively utilised in various industrial operations, such as manufacturing and cooling. Access to groundwater is essential for the stability and sustainability of industrial processes, hence underpinning economic activity and employment.

Groundwater is vital for supplying clean and dependable drinking water, impacting global public health. Effective management strategies are essential for safeguarding groundwater, a vital resource that supports the welfare of populations worldwide.

4. KEY CHALLENGES IN GROUNDWATER POLLUTION MANAGEMENT

The management of groundwater and associated groundwater resources are coordinated through integrated management (Jakeman et al., 2016). In order to achieve the three pillars of sustainable outcomes economic balance, social welfare, and environmental protection it has considered the interaction of non-groundwater policies across geography and time (Ross, 2018).One component of integrated groundwater management is integrated assessment, which is also a meta-discipline(Syafiuddin et al., 2020). A scientific meta-discipline is essential for the implementation of integrated groundwater management, as it supplies critical information to tackle management challenges and promotes social learning, hence enhancing decision-making processes. Groundwater management issues with public policy pertaining to long-term environmental management can be discovered by an integrated evaluation.

It is impossible to address individual problems with groundwater management because of how interrelated. As an example, conventional gravity irrigation systems would not be able to replenish groundwater supplies to their full potential. Unlike in India, the agency in charge of controlling groundwater has not upgraded this irrigation infrastructure. Overexploitation of groundwater resources occurs because Indian communities only consume groundwater once and do not reuse or recycle it for other purposes, like agriculture. Concerns have arisen as a result of policy interventions; for example, a policy that was meant to address groundwater management has ended up interfering with other policies pertaining to groundwater activities.

Significant shifts in crop yields and heightened rivalry among regional agricultural sectors have resulted from the persistent execution of pumping limitations aimed at reducing groundwater resource consumption (Jadeja et al., 2018). Isolating and fixing one issue with groundwater management can unintentionally lead to the emergence of new or worsening issues with other groundwater systems. Resolving the issue of inadequate groundwater management requires a collaborative evaluation and treatment of the relevant policy segments, including those dealing incorporating the effects of climate adaptation, water supply, industrial and management of urban pollutants, environmental factors, agricultural methodologies, and land planning and development.

The Malaysian government is creating the Malaysian Climate Change Adaptation .Index to evaluate the preparedness of each state for the probable impacts of climate change. The index would utilize measures of vulnerability and preparation across various sectors, including water resources, the water industry, the environment, governance, and socio-economic factors, among others. The development and assessment have been carried out by the Ministry of Energy and Natural Resources. The purpose of this index is to compare and follow the progress of all states by providing a scatter plot of vulnerability against readiness.

Human context elements, including cultural, social, economic, and political contexts, are integral to integrated groundwater management. It comes to managing groundwater resources, the population also known as the human setting has become an additional obstacle. The demand for groundwater resources is rising in tandem with the population. For example, managing groundwater systems or resources is becoming more difficult due to the fast population growth in China, which is leading to overexploitation of these resources (Coulon et al., 2016). Anthropogenic activities, such as building and farming, have increased their use of groundwater resources, which has led to the pollution of 16% of China's soil and contamination of 83% of the soil, as well as the formation of groundwater contamination.

5. AI AND ML IN GROUNDWATER MANAGEMENT

The aforementioned problems have novel answers in the form AI and ML technology. Their uses in managing and controlling groundwater pollution are described in the sections that follow.

Figure 1. Potential applications of AI in GWM

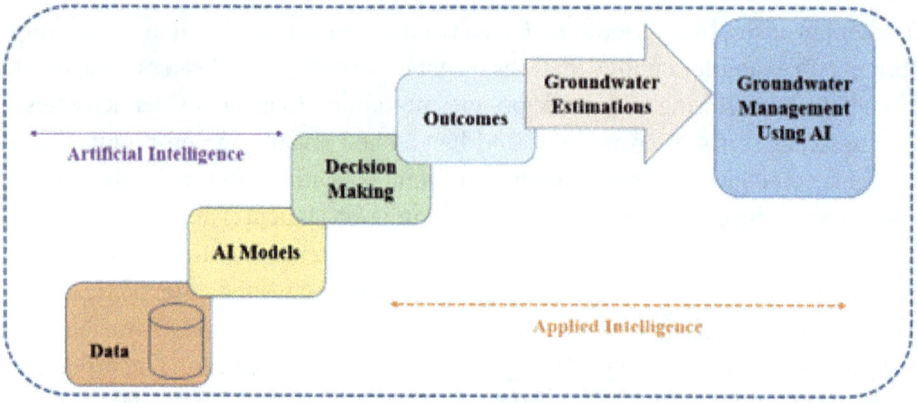

Figure 1 shown in the various uses of ML in managing groundwater.

5.1 Predictive Modeling

About a third of the world's water comes from underground sources, making it the most important freshwater resource. The ease and accessibility of groundwater resources is inversely proportional to socioeconomic development(Kombo et al., 2020). Most people believe that this is due to human actions and the consequences of climate change. However, hydraulic qualities, the quantity of evaporate transpiration, and other natural processes can cause water table levels to change seasonally. Groundwater has received scant attention despite its great potential to contribute to sub-Saharan Africa's socio-economic development. Concerning Rwanda's groundwater resources, very little is known. An approach to better managing groundwater resources is to develop more accurate systems for predicting the changing levels of groundwater throughout the year (Peng et al., 2017).

Tools for better understanding extremely complicated natural systems have been developed as a consequence of advancements in computer modelling, processing power, and data processing.

There has been a lot of research on how ML techniques can be applied in hydrology(Shortridge et al., 2015). The usefulness of ML techniques in groundwater research has been demonstrated. An essential field of study is the application of ML to the problem of hydrological modelling. Compared to physical models, these ones work better for investigations into water resources (Kenda et al., 2018).

Compared to other models, ML is superior in terms of simplicity and accuracy (Sahoo, 2017).The most popular techniques for predicting groundwater resources include Support Vector Machines (SVMs),Artificial Neural Networks (ANNs), and feature similarity-based methods like K-Nearest Neighbour (KNNs). It is also possible to use nearest neighbour with samples that are noisy.In its study, the ASCE Task force committee investigated the use of ANN in water resource modelling in a broad sense

Furthermore, in their evaluation of ANN, SVM, ELM, and Genetic Programming (GP), (Natarajan & Sudheer, 2020) also considered how well each method simulated groundwater levels. Their results imply that compared to GP, SVM, and ANN, ELM obtains better precision. The use of SVMs in hydrology is extensively discussed in. In order to estimate variations in the water table heights using temperature and precipitation data, recent study has looked at how well a Non-linear Auto-Regressive network with exogenous input (NARX) works. Although NARX was found to perform well in seasonal predictions of groundwater depths for various aquifer types by (Wunsch et al., 2018), SVM was recommended to forecast daily levels rather than NARX.

5.2 Real-Time Monitoring

One of the most important sources of potable water for humans is groundwater. For this reason, future sustainability depends on accurate groundwater resource measurement, monitoring, and management. Despite this, the aquifer's groundwater quantity is insufficient, land subsidence is a problem, saltwater intrusion is expensive, overexploitation is rampant, and water contamination is both unsanitary and of poor quality (Yin & Tsai, 2018).Another worldwide pressing concern is the long-term viability of groundwater supplies and their management(Aderemi et al., 2021). Intelligent water systems, advanced healthcare, innovative agribusiness, sophisticated climate control, and numerous other domain have gradually adopted the idea of the IoT over the years (Premsankar et al., 2018). Connected sensors, mobile phones, utility gadgets, and industrial components are what the Internet of Things (IoT) is all about. These items are linked through networks that can analyse data. Thus, groundwater resource management has discovered a use for smart sensors and pressure transducers, with the help of the conventional IOT architecture, which allows for the continuous capture, analysis, and processing of data (Yin & Tsai, 2019)

One common method of implementing the IOT is to use the internet to transfer data created by smart devices to a distant cloud. Conventional internet architecture for the IOT has grown inefficient for processing and analysis due to massive data transfers from IoT devices to the cloud. This is because doing all of these calculations puts an enormous strain on the network that is connected to cloud computing(Raju & Varma, 2017). The cloud's inefficiency and inability to scale to handle massive amounts of real-time IoT data owing to communication latency and network bandwidth has been highlighted in the majority of published works. In addition, identifying regions with reduced groundwater resource amount has been made much easier with the help of IoT in groundwater level management and groundwater data-driven model building(Liu et al., 2019). Nevertheless, geodetic methods' local regional assessments are rendered useless by the low resolution and indirect nature of the sensing equipment installed in the observation wells. To address the demands of groundwater resource management models for efficient and scalable computing, advancements in IoT technology, geophysical modelling, and data-assimilation are required.

5.3 Contaminant Source Identification

The problem of groundwater contamination is getting worse and needs immediate response from the environmental community(Voigt, 1989). It is unfortunate that water treatment is often required to make water drinkable. Instead of investing in costly treatment systems, sustainable management practices could preserve groundwater quality (Katsanou & Karapanagioti, 2017). Understanding the source of a pollutant and how it is released over time is crucial for preventing its spread in an aquifer. Both the (Gómez-Hernández & Xu, 2022)evaluations emphasize how much interest there has been in tracing the origin of pollutants throughout the last forty years. One example of an inverse problem in hydrology is the task of identifying the source of a pollutant using just a small number of concentration readings.

According to (Barati Moghaddam et al., 2021)there are three primary areas where inverse methods can be used to quantitatively resolve the issue of contaminant source identification: optimization-based, stochastic-based, and mathematics-based. Several approaches have been proposed in the literature to address this type of problem. Mathematical approaches use numerical or analytical techniques to directly tackle an inverse source problem. Regularization and stabilizing techniques are two popular approaches to stability issues. In stochastic methods, the estimation parameters are treated as random variables, and the problem is framed within a stochastic framework. Methods based on optimization combine simulation and optimization techniques.

For all inputs (starting and boundary conditions), the model for simulation finds solutions to the transport and flow equations. The next step is to use an optimization technique to reduce the discrepancy between the actual and simulated data.

5.4 Optimization of Remediation Strategies

This approach has been used in several publications, such some researchers have turned to so-called global optimization (GO) approaches due to the fact that the optimization problem of the pumping strategy for the removal of contaminated plumes is extremely nonlinear and nonconvex(Yang, 2011). Other fields have made good use of GO methods since the 1960s; optimization of industrial design is one such example. The availability of GO dramatically expanded with the advent of high-speed workstations and personal computers. Genetic algorithms or (GA) gained immense popularity in the late 1980s as a means for numerous scholars and practitioners in the field of civil engineering to employ a straightforward and efficient optimization method.

The goal of global optimization is to discover the optimal solution to optimization problems with constraints, even though these issues may have multiple local optimal solutions. For issues where other optimization methods fail owing to the presence of numerous extreme and/or analytically defined functions provide a unique opportunity for the GO methodology to shine. A GO problem including box limitations might be framed as discover an optimizer y^* that could

$$h^* = h(y^*) = \min_{y \in Y} h(y) \tag{1}$$

Within the finite interval (box) region of the m-dimensional Euclidean space, the objective function $h(y)$ is defined as

$$Y = [y \in \mathbb{R}^n : \; b \le y \le d \, (for\ each\ dimension)] \tag{2}$$

A high-value penalty function beyond the defined restrictions can be introduced to this constrained optimization problem, transforming it into an unconstrained optimization problem. It determining an optimizer's precise value becomes a challenge in practice, it talk about its estimate and, by extension, the minimum estimate.

5.4.1 Genetic Algorithms

The concept of simulating a search process of natural evolution is the basis of Genetic Algorithms (GA), which are a subset of Evolutionary Algorithms (EA). However, these models are simplistic representations of biological reality. Using concepts from genetics and biology, EA searches are steered with a random element to sidestep local minima. At each iteration, for instance, a random sample is used to choose pairs of parent people (points) who are then chosen based on their fitness (function value). These pairs then recombine to create new offspring. For the following generation, only the finest of these are chosen. Another possibility is that the progeny may mutate, or move about in space at random. Parents who take good care of themselves are more likely to have healthy, fit offspring. Three distinct types of EAs have been created:

- Evolution for Strategies (ES)
- Evolution for Programming (EP)
- Genetics for Algorithms (GA)

Over the past ten years, GAs have become increasingly popular as an all-purpose optimization technique. A Canonica GA was utilized in this study (Gómez-Hernández & Xu, 2022) the best-performing GA was selected for this study after comparing multiple versions and sets of parameters (e.g., mutation rate, crossover frequency, population size, etc.). These versions included ones with the tournament and fitness rank selections. Fitness rank selection, one-point crossover, bit mutation, 15-bit variable coding, and retention of the best points (strings) are all features of this GA variation arrived at thus far (elitism). In order to avoid unnecessary re-evaluations, the presence of repeating strings is checked for in every generation.

5.4.2 Adaptive Cluster Covering (ACCO)

A practical amalgamation of well acknowledged concepts of reduction, clustering, and covering, this method (Solomatine, 1998) is created. To begin the global search, groupings of points in the search space that are close to each other are found; these groups represent the most promising sub domains, and within each of these groups, rectangular sub domains are linked. Next, a predetermined number of points is selected at random from each sub domain. After that, at every step, it check how the goal function is doing. Covering is iterated many times, with each iteration shrinking the subdomain and moving it closer to concentrations of locations with the smallest function values. Several local searches are initiated from the (best) spots obtained in the ACCO phase in the ACCOL variation of the ACCO algorithm.

5.5 Decision Support Systems (DSS)

In the era of information technology, DSS have been created to simplify and improve the accuracy of decision making. The goal of creating these systems is to guide and assist decision-making groups in finding the best options. Decisions about water usage and resources, particularly groundwater, have been made easier with the invention of various support systems within the past 20 years(Aliyari et al., 2018). The database, model base, and decision-making basis are the three main components of a DSS. Technically, a primary program (graphical user interference) makes use of these three components.

DSS is a tool that can streamline planning and management. The inability to choose, create, and use tools like DSS is due to the lack of clarity surrounding objectives, methods, impressive and impressed components, and the overall management pattern. Some have raised the issue of the indicators used to assess management options as a top priority. Indicators like these are defined according to things like conditions, purposes, programmable elements, data availability, task precision, and so on. It is important to take into account the following items.

Figure 2. The fundamental components of DSS

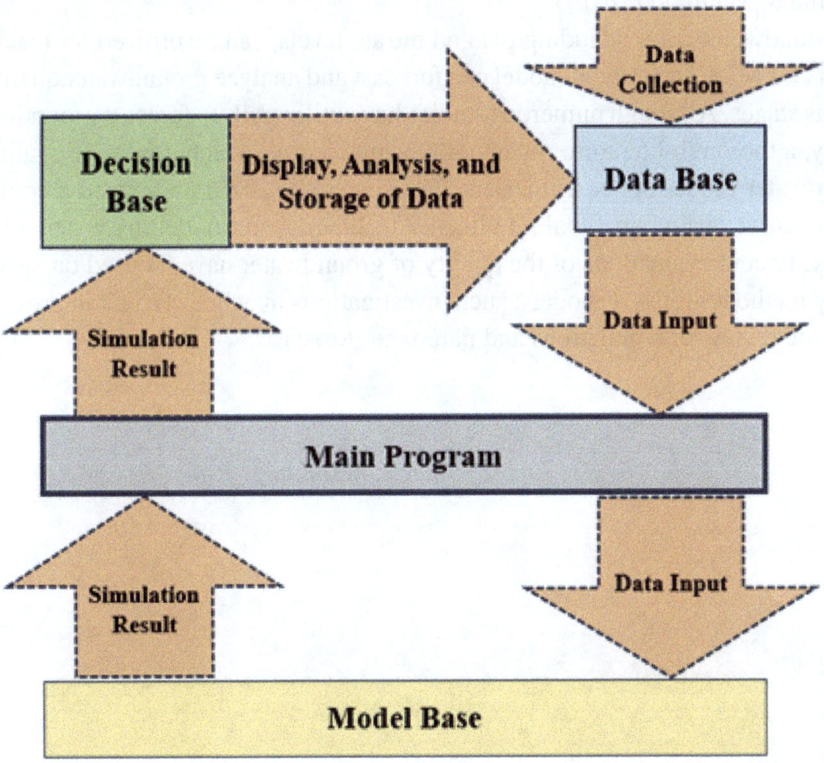

Figure 2 shows the primary program location, the basic components of DSS, and how to use this system to get the best conclusion.

6. RECENT DEVELOPMENTS AND LATEST RESEARCH

Machine learning is an efficacious method for generating predictive models from data. Machine learning models, a branch of artificial intelligence, employ inductive reasoning to assess and formulate rules from data, instead than depending on a fixed set of formulas (Haggerty et al., 2023). These methodologies can identify trends in data and forecast the quality of water indices in the supply of freshwater, including groundwater as well as surface water. Machine learning may obtain features of high dimension with nonlinear relationships. Modern processing hardware and architecture (Reuther et al., 2020) have improved the effectiveness of Machine Learning algorithms (MLAs), particularly deep learning (DL) (Emmert-Streib et al., 2020; Shrestha & Mahmood, 2019).

Groundwater data, including pH and nitrate levels, can be utilized for teaching models for ML . The trained model can forecast and analyze groundwater quality at various stages. Although numerous studies have utilize ML to forecast groundwater quality, a thorough literature review of machine learning techniques for quality of groundwater modelling is still absent. Tiyasha et al. (2020) performed a comprehensive survey using artificial intelligence techniques for modelling water surface quality. Recent evaluations of the quality of groundwater have focused on specific survey methodologies or models; such investigations may be deficient in providing a holistic review of simulations and parameter forecasts.

Figure 3. Machine learning algorithms for model groundwater quality

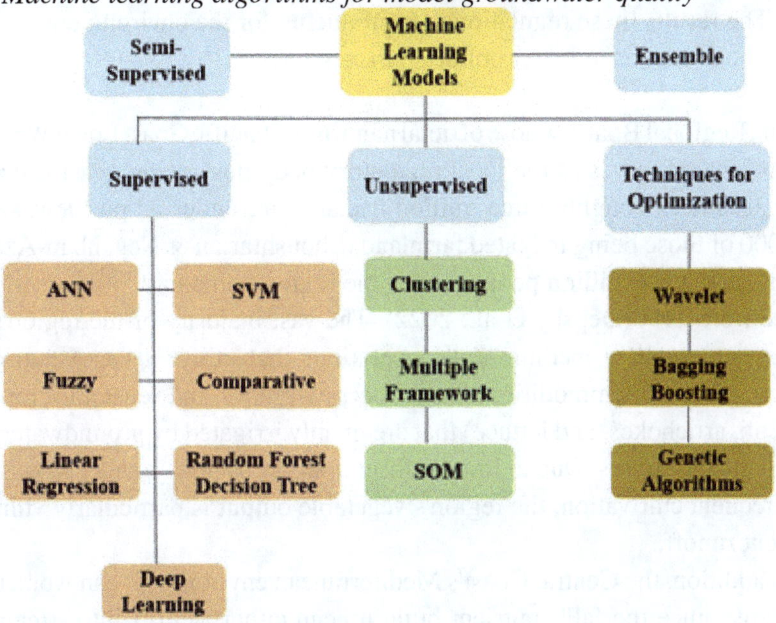

Figure 3 shows how the ML models utilized to model groundwater quality were organized in this study. With each learning type comes a list of model categories these include supervised, unsupervised, semi-supervised, and ensemble learning.

ML model training entails iteratively optimizing the model's parameters or finding a closed-form methodology to ascertain the optimal parameters for the predictive model. To make ML models more efficient and accurate, many different approaches have been suggested (Sze et al., 2017).

7. CASE STUDIES

7.1 AI for Nitrate Pollution Control in Agricultural Regions

The absence of research on the processes via which water agencies develop and execute NPS pollution policies within their respective jurisdictions is a significant knowledge gap it comes to regulating NPS pollution runoff from agricultural sources(DOWD et al., 2008). Nonpoint source contamination in agriculture has been regulated by local water agencies for a long time. But there are no evaluations of in North American studies are

- The policymaking and implementation backgrounds of these regulations
- The results these regulations are producing for the environment
- The obstacles to better pollution management

The Regional Board, whose official name is the Pacific Coast Local Water Quality Control Board serves as the local regulatory body possessing quasi-executive and judicial duties. Roughly three million hectares are under its purview, with about 176,000 of those being irrigated farmland (Shoushtarian & Negahban-Azar, 2020). An estimated 1.4 million people called the region home in 2006, a nearly 20% increase from 1990(Depsky et al., 2022). The vast majority of the region's farmers operate on small to medium-scale operations that utilise a lot of nutrients. The highest-grossing commodities are strawberries, grapes, and vegetable crops (such as broccoli, artichokes, and lettuce) that are mainly irrigated by groundwater on farms less than 20 hectares. Due to low fertiliser uptake efficiency, short growing cycles, and frequent cultivation, the region's vegetable output is particularly vulnerable to nutrient runoff.

In addition, the Central Coast's Mediterranean environment can worsen nutrient discharge since the fall's nutrient build up can either run off into streams during early rain events or leak below the root zone. Sediments, nutrients, and pesticide toxicity are three areas where water quality has been found to exceed requirements, all because of these climatic and agro ecological factors. Lastly, nitrate contamination of groundwater has been a problem along the Central Coast for many years due to agricultural practices. No matter how careful one is with fertiliser, using these polluted waters for irrigation could make the problem worse for the surface water. Because many species of particular concern rely on the agricultural areas in the area as their primary habitat. The Central Coast is home to one varieties of vulnerable Coho Salmon, three subspecies of endangered or threatened Steelhead, and the final extant generation of the Californian Sea Otter. Additionally, farmers may encounter challenges related to the endangered red-legged frog (Rana aurora). The Bay of Monterey National Marine Sanctuary, among the most extensive marine sanctuaries worldwide, receives water from multiple watersheds. Elkhorn Slough, a vital remnant tidal wetlands in the country of America and a National Estuarine Research Reserve declared by NOAA, is located on California's Central Coast. The southern section encompasses the Morro Bay National Estuarine Research Reserve and large salt marsh ecosystems.

7.2 Real-Time Groundwater Monitoring in Urban Areas

It chose Bengaluru and Singapore as our case studies because of their rapid digital transformation. By embracing and utilising digital technologies for water management, Singapore has come a long way in achieving water security and sustainability(Banerjee et al., 2022). The city of Bengaluru, India's Silicon Valley, has begun its digitalisation journey with the aim of bettering water management and governance. To fully grasp the obstacles and difficulties in attaining an effective water management system, it is necessary to compare the two cities' adoption and application of digital technology.

7.2.1 Singapore

With a total territory of 714 square kilometres and a population of 5.2 million, the Southeast Asian city-state of Singapore is truly remarkable. Despite the country's limited catchment area, it receives a considerable amount of rainfall each year about 2,500 mm. Due to its lack of an underground water source or natural aquifer, Singapore is confronting the formidable problem of freshwater shortage at a time its population is expanding at an alarming rate.

Singapore surmounted water supply limitations and attained sustainable, cost-effective water management through an integrated and innovative strategy; this, coupled with rigorous planning and over four decades of diligent effort, enabled the nation to achieve water self-sufficiency. The national water agency of Singapore is the Public Utilities Board, tasked with water gathering, manufacturing, distribution, and reclamation. Prior to 2001, PUB oversaw the management of potable water, electricity, and gas. Consequently, the Department of the Ministry of the Environment delegated the responsibility for sewage and drain to PUB as well. At present, all inhabitants of the city have access to uninterrupted, high-quality piped water. In Singapore, daily water consumption totals 430 million gallons, with 45% designated for household use and 55% employed by non-domestic sectors. The water is derived from one of the Four National Taps: desalinated water, premium reclaimed water (NEWater), imported water, or local water. The PUB predicts that, due to population and economic expansion, water consumption might rise fourfold by 2060, with households representing 30% of this requirement and the remaining 70% derived from other sources.

7.2.2 Bangalore

The Indian city of Bengaluru is home to 12.9 million people and spans around 800 square km; it serves as the capital of the Karnataka state. The city has a semi-arid climate and receives an average of 900 mm of rainfall per year, with the majority of that falling during the summer monsoon. It is located on a hilltop at a height of 900 m above msl. Water consumption increased at a rapid pace due to the growth of public and private sectors, as well as the resulting influx of city dwellers. The

BWSSB, or Bangalore Water Supply and Sewerage Board, was formed in 1964 to supply water to the city.

One of the main rivers in southern India, the Cauvery, was harnessed by BWSSB through the Cauvery Water Supply Scheme (CWSS). Four phases have been developed over the years, and today a 100 km route brings roughly 1,450 million litres of water every day. Groundwater is heavily used, particularly in the city's periphery where BWSSB has not yet installed piped water, because water extracted from Cauvery is insufficient to meet the city's water demand (Sekhar et al., 2017). Bengaluru has a lot of problems, like the fast increase in water demand and a big percentage (over 30%) of water loss from leaks and illegal consumption (Paul, 2019). In most cases, there is a lack of systematic management of storm water, and the city does not consider it a water supply.

8. RATIONALE OF THE STUDY

The Water Quality Initiative's overarching goal is to simplify complex water quality data so that researchers and the general public may make better use of it(Agrawal et al., 2021). While some methods have shown success in several nations, notably India the corresponding calculations can be very laborious and complex. Therefore, the procedure is inherently prone to drawing unforeseen errors (Sánchez et al., 2007). Consequently, the following are the constraints on WQI computations:

- Time-consuming
- Protracted procedure,
- Intricate process,
- Various equations are employed for Water Quality Index estimates, resulting in discrepancies.

Given the preceding, it may be clear that the WQI does not have a standardised methodology. In order to overcome these issues, a small number of researchers have put up a non-physical method that uses AI and ML to accurately forecast WQI (Leong et al., 2021). An AI-based model can quickly generate a WQI value after successful training by removing the need to calculate the sub-indexes. Nonlinear structures, complex trend calculation, managing massive datasets with data of varying dimensions, and insensitivity to missing data are only a few of the advantages of AI algorithms that are gaining more and more attention. Data collection and analysis methods and precision have a significant impact on ML-AI algorithms' forecasting capabilities. Thanks to the ever-increasing power of computers, scientists are now able to use a wide variety of ML-AI model configurations.

Many methods have been successfully used to forecast water quality all over the globe, including ANNs(Isiyaka et al., 2019) adaptive neuro-fuzzy inference systems (Karim et al., 2020), and SVMs. There is an increasing need for new approaches to managing groundwater contamination, and this project intends to provide that demand. Conventional approaches are ineffective in preventing contamination and are frequently reactive. A new approach to proactive and predictive management is possible with the help of AI and ML, which can lessen the financial and ecological burden of groundwater contamination. More effective solutions for pollution management and resource sustainability can be implemented by stakeholders by harnessing these technologies.

9. MATERIALS AND METHODS

9.1 Data Collection

A wide variety of sources, such as public databases, IoT sensors, and remote sensing platforms, contributed to the data set. We looked at things like geographical data, water quality indexes, and pollution sources.

9.2 AI and ML Techniques

It stated that the complete dataset, including GWQI and its predictors, was split into two parts: train and test. In order to create ML algorithms, we utilised the train set, and then we validated them using the test set. In the pre-processing phase, the dataset was oversampled and normalised(Sahour et al., 2023). In cases when the range of predictors varies, normalisation is particularly useful for improving the accuracy of ML systems. As an added bonus, it makes the dataset more consistent in terms of how the models handle and use the data. In oversampling, certain classes are given a greater percentage of the observations in the dataset compared to their actual proportion in the population.

9.2.1 K-Nearest Neighbors (KNN)

The KNN training samples used to classify objects make up the input to the KNN classier. Grades of their closest neighbours receive them. By averaging the values of the k nearest neighbours, the KNN classifier produces its output. KNNs perform local approximations of functions and wait to compute anything until the function is evaluated. Compared to the distant ones, the KNN can significantly improve the average.

9.2.2 Artificial Neural Network (ANN)

This is carried out by a network of nodes that imitate the actions of neurones in living things; these neurones communicate with one another to facilitate learning in ANNs(Sheik et al., 2024). This study employed a, Multilayer Perceptron (MLP), ANN that was trained using a back propagation method. The three layers of an ANN are the input, hidden, and output layers. The ANN approach can identify potential nonlinear linkages by creating actual links between input variables and their related goals.

Figure 4. A step-by-step guide for using the NN algorithm

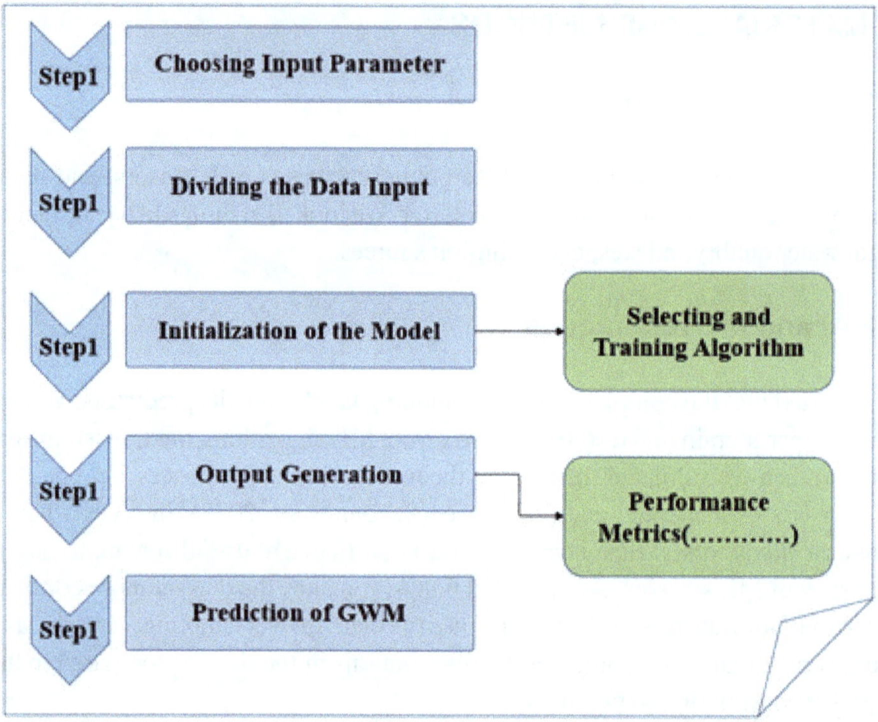

Figure 4 shown in a detailed process for running the Neural Networks (NN) algorithm.

9.2.3 XGBoost (XGB)

In this study, a GWQI class was predicted from the collection of predictors using the XGBoost (XGB) classification method. By boosting or strengthening an already-effective model, gradient boosting allows for the combination of multiple weak models into a more robust one. Thanks to the XGB's scalable gradient-boosting implementations, boosted tree techniques can be computed more quickly and efficiently. When XGB generates trees, it does so in parallel rather than sequentially. We use partial sums of gradient values to assess the quality of separations in every conceivable split in the training set. This model's efficiency in class prediction is a direct outcome of gradient boosting.

9.2.4 Support Vector Machine (SVM)

Among the many algorithms used to solve classification problems, SVM stands out. In N-dimensional spaces, SVMs are employed for point classification using hyper planes. The size of the hyper plane is proportional to the number of features. If there are just two input features, the hyper plane is a simple line; if there are three, the hyperplane becomes a two-dimensional plane.

9.2.5 Gaussian Classification Model (GCM)

An infinite-dimensional extension of a multivariate Gaussian distribution is called a Gaussian Classification Model (GCM). A multivariate distribution is applied to the data produced by a Gaussian process. As an example of a multivariate Gaussian distribution, consider a set of observations, where each observation represents a single point. It is possible to reverse-engineer a GCM to fit this data set in this case.

9.2.6 Random Forest (RF)

The RF is a supervised classification algorithm that uses a randomly built set of decision trees. The combination of bagging and tree randomisation makes committee-based predictions more accurate than individual trees. The decision tree is constructed using many samples, with the bulk of the data being used for classification. Bootstrap aggregating or bagging is used to train tree learners on random forests.

9.3 Model Validation

Using the confusion matrix, accuracy, precision, recall, and F1-score, which are common metrics for binary classification tasks, we assessed the model's performance (Afzal, 1990). All things considered, these metrics give a good picture of how well the algorithm can identify phishing attempts.

Accuracy: It many URLs were successfully categorised relative to all URLs.

$$Accuracy = \frac{VM + VO}{VM + VO + HM + HO} \quad (3)$$

Recall: Whether the model can detect malicious URLs

$$Recall = \frac{VM}{VM + HO} \quad (4)$$

Precision: The model's accuracy in detecting valid URLs

$$Precision = \frac{VM}{VM + HM} \quad (5)$$

F1-Score: Equivalent to the harmonic mean of recall and precision, it gives a fair evaluation of the model's performance.

$$F1Score = \frac{2.Precison.Recall}{Precision + Recall} \quad (6)$$

VM: Validity of URLs accurately predicted.
VO: URLs identified as potentially malicious
HM: URLs that are wrongly assumed to be valid
HO: URLs that are mistakenly believed to be malicious

These metrics evaluate the model's accuracy in URL classification, taking into account both true and false positives and negatives, in order to distinguish between authentic and phishing URLs. Results from the experiments, such as the analysis of features, the schedule for learning rates, and the progress made during training, are display data in a way that reveals how the model is being trained.

Model accuracy and resilience were evaluated using cross-validation procedures and performance indicators like R-squared values, Root Mean Square Error (RMSE), and Mean Absolute Error (MAE).

10. RESULT AND DISCUSSION

Results from an evaluation of the Gaussian Classification Model (GCM) with an Attention Mechanism (GCM) model have shed light on its efficacy in detecting phishing attempts.

Figure 5. Confusion matrix for GCM model

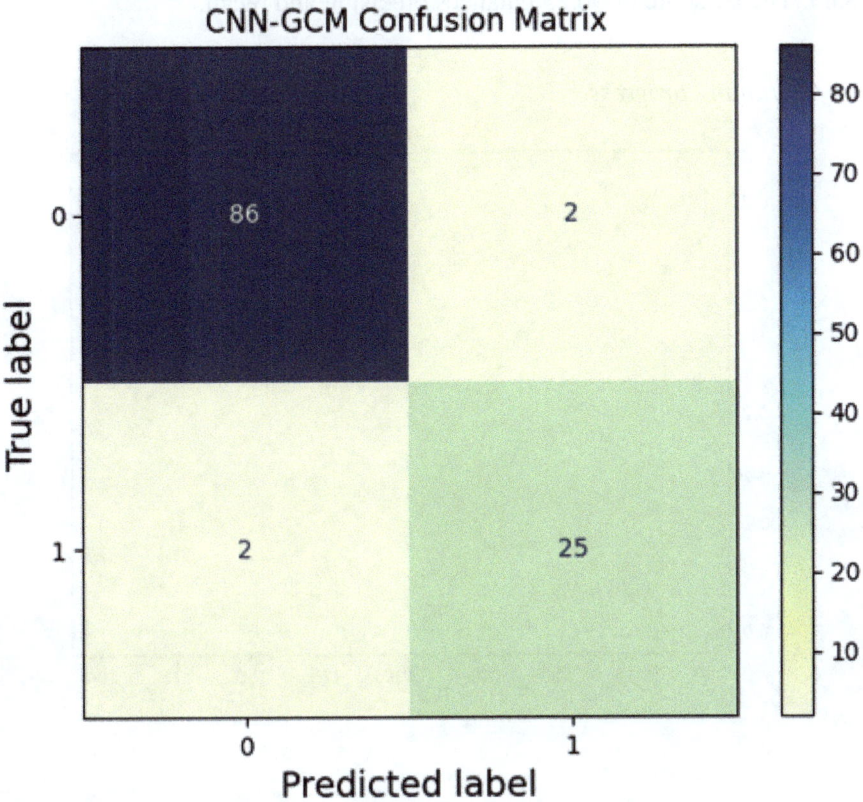

Figure 5 shown in the top-performing models' recall, accuracy, and precision are compared to a baseline model that is currently in use.

Table 1. Performance metrics of the proposed model

Metric	Metric	Value
True Positive	Accuracy	**90.09**
True Negative	Recall	**88.83**
False Positive	Precision	**90.09**
False Negative	F1-Score	**90.23**

Table 1 shown in provides detailed information the model's precision score of 90.09 demonstrates its ability to reduce false positives, increasing its reliability in detecting phishing threats. The model accurately detects phishing URLs with a recall of 88.83. The F1 score of 90.23 balances precision and recall.

Figure 6. Training progress

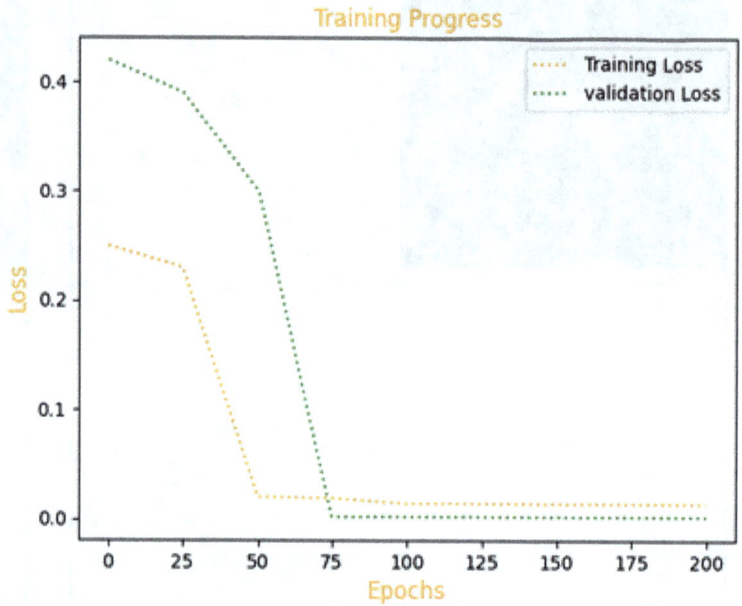

Figure 6 shown in a helpful insight into the model's convergence and warnings of over fitting or under fitting were provided by plotting the training and validation losses over epochs.

Figure 7. Learning rate schedule

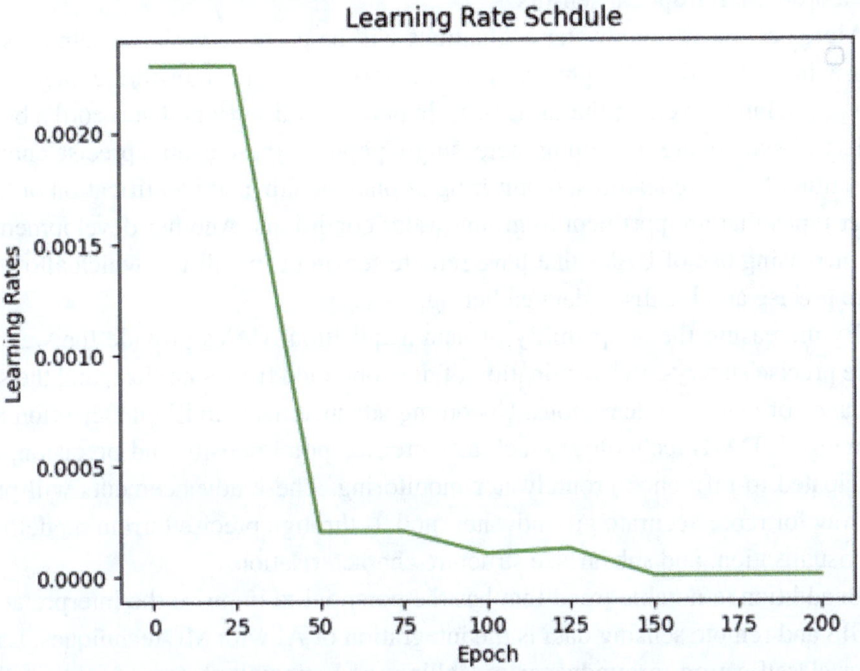

As shown in Figure 7, the Learning Rate Schedule graph illustrates how the rate is initially set at a high level to facilitate rapid convergence and then gradually decreases for weight refinement. The learning rate scheduler modifies the learning rate when neural network is being train.

11. EMERGING TECHNOLOGIES

With improved spatial and temporal resolutions, groundwater-related parameter monitoring will be more precise thanks to new developments in satellite sensor technologies, which are ripe for a revolution in the field(M. Shaikh & Birajdar, 2024a). Groundwater mapping, surface and subsurface feature change detection, and more precise land cover classification are all areas that stand to benefit from the anticipated enhancement in data quality. The growing use of Synthetic Aperture Radar (SAR) technology has been a game-changer. SAR can now monitor in all kinds of weather, day or night, and it can even work through cloud cover, which was a problem for optical sensors.

More accurate groundwater evaluations will be possible with the help of SAR data, which should shed light on ground deformation, soil moisture content, and changes in land cover. At the same time, hyper spectral sensors that record a broad variety of spectra are becoming increasingly popular; these enable precise characterisation of surface features, monitoring of plant health, and identification of land cover types that are pertinent to groundwater conditions. Another development is the increasing use of UAVs that have remote sensing capabilities, which allow for more precise and localised data gathering.

By increasing the adaptability of data acquisition, UAVs provide the way for more precise surveys, the verification of data obtained from satellites, and the surveillance of inaccessible regions. Upcoming advancements in Light Detection and Ranging (LiDAR) technology, such as increased point density and precision, are anticipated to influence groundwater monitoring. These advancements will pave the way for more accurate groundwater models through precise terrain modelling, 3D visualisation, and subsurface structure characterisation.

In addition, a notable trend that has the potential to improve the interpretation of GIS and remote sensing data is the integration of AI with ML techniques. Land cover classification, groundwater modelling, and anomalous event identification are all areas that could benefit from the precision that ML and AI offer. In addition, IOT sensors are being used more and more to monitor groundwater-related data in real-time.

When integrated with GIS, it facilitates the prompt identification of changes. Groundwater monitoring projects are becoming collaborative, accessible, and scalable due to the rising utilization of cloud computing and large-scale data analytics for the efficient storage, processing, and analysis of extensive geographic datasets. A prominent trend in the dynamic digital landscape is the exploration of Virtual Reality (VR) and Augmented Reality (AR) applications for visualizing and engaging with three-dimensional spatial data. Augmented and virtual reality technologies are anticipated to enhance the comprehension and distribution of groundwater-related data, resulting in improved public engagement and decision-making. The dissemination of remote sensing and GIS information is essential for enabling standardized procedures, bolstering research and monitoring initiatives globally, and is underscored by a focus on international collaboration and open data initiatives.

12. CONCLUSION

Some of the most basic need for human survival and civilisation include an adequate supply of clean water, both in terms of volume and quality(Agrawal et al., 2021). In fact, groundwater utilisation for home, industrial, and irrigation uses has multiplied over the last few decades due to the phenomenal development in both the global population and scientific and technological know-how. The availability and quality of groundwater have been affected by factors such as fast urbanisation, overexploitation, and improper waste disposal. The use of pesticides and agricultural chemicals has become more common due to fast urbanisation and overpopulation, and this has led to their leaching and mixing into the subsurface. The use of AI and ML presents a significant improvement over previous methods for mitigating and managing groundwater contamination. The effectiveness and sustainability of groundwater management procedures are enhanced by these tools, which enable predictive modelling, real-time monitoring, and optimised repair. To assure their appropriate implementation, however, ethical and regulatory considerations must be given considerable weight.

REFERENCES

Aderemi, B. A., Olwal, T. O., Ndambuki, J. M., & Rwanga, S. S. (2021). A Review of Groundwater Management Models with a Focus on IoT-Based Systems. *Sustainability (Basel), 14*(1), 148. DOI: 10.3390/su14010148

Afzal, M. (1990). Machine intelligence applications in agriculture. *AMA. Agricultural Mechanization in Asia, Africa and Latin America, 21*(2).

Agrawal, P., Sinha, A., Kumar, S., Agarwal, A., Banerjee, A., Villuri, V. G. K., Annavarapu, C. S. R., Dwivedi, R., Dera, V. V. R., Sinha, J., & Pasupuleti, S. (2021). Exploring Artificial Intelligence Techniques for Groundwater Quality Assessment. *Water (Basel), 13*(9), 1172. DOI: 10.3390/w13091172

Aliyari, H., Kholghi, M., Zahedi, S., & Momeni, M. (2018). Providing decision support system in groundwater resources management for the purpose of sustainable development. *Journal of Water Supply: Research & Technology - Aqua, 67*(5), jws2018130. Advance online publication. DOI: 10.2166/aqua.2018.130

Andrade, L., O'Dwyer, J., O'Neill, E., & Hynds, P. (2018). Surface water flooding, groundwater contamination, and enteric disease in developed countries: A scoping review of connections and consequences. *Environmental Pollution, 236*, 540–549. DOI: 10.1016/j.envpol.2018.01.104

Banerjee, C., Bhaduri, A., & Saraswat, C. (2022). Digitalization in Urban Water Governance: Case Study of Bengaluru and Singapore. *Frontiers in Environmental Science, 10*(March), 1–12. DOI: 10.3389/fenvs.2022.816824

Barati Moghaddam, M., Mazaheri, M., & Mohammad Vali Samani, J. (2021). Inverse modeling of contaminant transport for pollution source identification in surface and groundwaters: a review. *Groundwater for Sustainable Development, 15*(November 2020), 100651. DOI: 10.1016/j.gsd.2021.100651

Coulon, F., Jones, K., Li, H., Hu, Q., Gao, J., Li, F., Chen, M., Zhu, Y.-G., Liu, R., Liu, M., Canning, K., Harries, N., Bardos, P., Nathanail, P., Sweeney, R., Middleton, D., Charnley, M., Randall, J., Richell, M., & Pollard, S. (2016). China's soil and groundwater management challenges: Lessons from the UK's experience and opportunities for China. *Environment International, 91*, 1–12. DOI: 10.1016/j.envint.2016.02.023

Cuthbert, M. O., Gleeson, T., Moosdorf, N., Befus, K. M., Schneider, A., Hartmann, J., & Lehner, B. (2019). Global patterns and dynamics of climate–groundwater interactions. *Nature Climate Change, 9*(2), 137–141. DOI: 10.1038/s41558-018-0386-4

Das, A., Kumar, R., Patel, S. S., Saha, M. C., & Guha, D. (2020). Source apportionment of potentially toxic elements in street dust of a coal mining area in Chhattisgarh, India, using multivariate and lead isotopic ratio analysis. *Environmental Monitoring and Assessment*, *192*(6), 396. DOI: 10.1007/s10661-020-08321-2

Depsky, N. J., Cushing, L., & Morello-Frosch, R. (2022). High-resolution gridded estimates of population sociodemographics from the 2020 census in California. *PLoS One*, *17*(7), e0270746. DOI: 10.1371/journal.pone.0270746

Dowd, B. M., Press, D., & Huertos, M. L.DOWD. (2008). Agricultural nonpoint source water pollution policy: The case of California's Central Coast. *Agriculture, Ecosystems & Environment*, *128*(3), 151–161. DOI: 10.1016/j.agee.2008.05.014

Emmert-Streib, F., Yang, Z., Feng, H., Tripathi, S., & Dehmer, M. (2020). An Introductory Review of Deep Learning for Prediction Models With Big Data. *Frontiers in Artificial Intelligence*, *3*(February), 1–23. DOI: 10.3389/frai.2020.00004

Gómez-Hernández, J. J., & Xu, T. (2022). Contaminant Source Identification in Aquifers: A Critical View. *Mathematical Geosciences*, *54*(2), 437–458. DOI: 10.1007/s11004-021-09976-4

Haggerty, R., Sun, J., Yu, H., & Li, Y. (2023). Application of machine learning in groundwater quality modeling - A comprehensive review. *Water Research*, *233*(402), 119745. DOI: 10.1016/j.watres.2023.119745

Isiyaka, H. A., Mustapha, A., Juahir, H., & Phil-Eze, P. (2019). Water quality modelling using artificial neural network and multivariate statistical techniques. *Modeling Earth Systems and Environment*, *5*(2), 583–593. DOI: 10.1007/s40808-018-0551-9

Jadeja, Y., Maheshwari, B., Packham, R., Bohra, H., Purohit, R., Thaker, B., Dillon, P., Oza, S., Dave, S., Soni, P., Dashora, Y., Dashora, R., Shah, T., Gorsiya, J., Katara, P., Ward, J., Kookana, R., Singh, P. K., Chinnasamy, P., & Chew, M. (2018). Managing aquifer recharge and sustaining groundwater use: Developing a capacity building program for creating local groundwater champions. *Sustainable Water Resources Management*, *4*(2), 317–329. DOI: 10.1007/s40899-018-0228-6

Jakeman, A. J., Barreteau, O., Hunt, R. J., Rinaudo, J.-D., & Ross, A. (2016). Erratum to: Integrated Groundwater Management. In *Integrated Groundwater Management* (pp. E1–E1). Springer International Publishing. DOI: 10.1007/978-3-319-23576-9_29

Karim, S. A. A., Kadir, E. A., & Nasution, A. H. (2020). Optimization Based Model Using Fuzzy and Other Statistical Techniques Towards Environmental Sustainability. In Karim, S. A. A., Kadir, E. A., & Nasution, A. H. (Eds.), *Optimization Based Model Using Fuzzy and Other Statistical Techniques Towards Environmental Sustainability*. Springer Singapore., DOI: 10.1007/978-981-15-2655-8

Katsanou, K., & Karapanagioti, H. K. (2017). Surface Water and Groundwater Sources for Drinking Water. In *Handbook of Environmental Chemistry* (Vol. 67, pp. 1–19). Issue December., DOI: 10.1007/698_2017_140

Kenda, K., Čerin, M., Bogataj, M., Senožetnik, M., Klemen, K., Pergar, P., Laspidou, C., & Mladenić, D. (2018). Groundwater Modeling with Machine Learning Techniques: Ljubljana polje Aquifer. *EWa*, *S3*, 697. DOI: 10.3390/proceedings2110697

Kombo, O. H., Kumaran, S., Sheikh, Y. H., Bovim, A., & Jayavel, K. (2020). Long-term groundwater level prediction model based on hybrid KNN-RF technique. *Hydrology*, *7*(3), 1–24. DOI: 10.3390/hydrology7030059

Leong, W. C., Bahadori, A., Zhang, J., & Ahmad, Z. (2021). Prediction of water quality index (WQI) using support vector machine (SVM) and least square-support vector machine (LS-SVM). *International Journal of River Basin Management*, *19*(2), 149–156. DOI: 10.1080/15715124.2019.1628030

Li, P., Wu, J., & Shukla, S. (2022). Achieving the One Health Goal: Highlighting Groundwater Quality and Public Health. *Water (Basel)*, *14*(21), 3540. DOI: 10.3390/w14213540

Liu, Y., Yang, C., Jiang, L., Xie, S., & Zhang, Y. (2019). Intelligent Edge Computing for IoT-Based Energy Management in Smart Cities. *IEEE Network*, *33*(2), 111–117. DOI: 10.1109/MNET.2019.1800254

Natarajan, N., & Sudheer, C. (2020). Groundwater level forecasting using soft computing techniques. *Neural Computing & Applications*, *32*(12), 7691–7708. DOI: 10.1007/s00521-019-04234-5

Paul, R. (2019). Urban metabolism of Bangalore City: A water mass balance analysis. *Journal of Industrial Ecology*, *22*, 1–23.

Peng, T., Zhou, J., Zhang, C., & Fu, W. (2017). Streamflow Forecasting Using Empirical Wavelet Transform and Artificial Neural Networks. *Water (Basel)*, *9*(6), 406. DOI: 10.3390/w9060406

Premsankar, G., Di Francesco, M., & Taleb, T. (2018). Edge Computing for the Internet of Things: A Case Study. *IEEE Internet of Things Journal*, *5*(2), 1275–1284. DOI: 10.1109/JIOT.2018.2805263

Raju, K. R. S. R., & Varma, G. H. K. (2017). Knowledge Based Real Time Monitoring System for Aquaculture Using IoT. *2017 IEEE 7th International Advance Computing Conference (IACC)*, 318–321. DOI: 10.1109/IACC.2017.0075

Reuther, A., Michaleas, P., Jones, M., Gadepally, V., Samsi, S., & Kepner, J. (2020). Survey of Machine Learning Accelerators. *2020 IEEE High Performance Extreme Computing Conference (HPEC), September*, 1–12. DOI: 10.1109/HPEC43674.2020.9286149

Ross, A. (2018). Speeding the transition towards integrated groundwater and surface water management in Australia. *Journal of Hydrology (Amsterdam)*, *567*, e1–e10. DOI: 10.1016/j.jhydrol.2017.01.037

Sahoo, S. (2017). Machine learning algorithms for modeling groundwater level changes in agricultural regions of the US. *Water Resources Research*, *5*(3), 2–2. DOI: 10.1111/j.1752-1688.1969.tb04897.x

Sahour, S., Khanbeyki, M., Gholami, V., Sahour, H., Kahvazade, I., & Karimi, H. (2023). Evaluation of machine learning algorithms for groundwater quality modeling. *Environmental Science and Pollution Research International*, *30*(16), 46004–46021. DOI: 10.1007/s11356-023-25596-3

Sánchez, E., Colmenarejo, M. F., Vicente, J., Rubio, A., García, M. G., Travieso, L., & Borja, R. (2007). Use of the water quality index and dissolved oxygen deficit as simple indicators of watersheds pollution. *Ecological Indicators*, *7*(2), 315–328. DOI: 10.1016/j.ecolind.2006.02.005

Sekhar, M., Tomer, S., Thiyaku, S., Giriraj, P., Murthy, S., & Mehta, V. (2017). Groundwater Level Dynamics in Bengaluru City, India. *Sustainability (Basel)*, *10*(1), 26. DOI: 10.3390/su10010026

Shaikh, M., & Birajdar, F. (2024a). Advancements in Remote Sensing and GIS for Sustainable Groundwater Monitoring: Applications, Challenges, and Future Directions. *International Journal of Research in Engineering, Science and Management,* 7(3), 16–24. https://journal.ijresm.com/index.php/ijresm/article/view/2955

Shaikh, M., & Birajdar, F. (2024b). Groundwater and Public Health: Exploring the Connections and Challenges. *International Journal of Innovative Science and Research Technology*, *9*(2), 1351–1361. DOI: 10.5281/zenodo.10730864

Shaikh, M. A. (2023). Groundwater management and sustainable farming practices: a socioeconomic analysis of their interplay in rural agriculture-a case study of Solapur, Maharashtra. *International Journal for Innovative Science Research Trends and Innovation,* 2023. DOI: 10.5281/zenodo.8402903

Sheik, A. G., Kumar, A., Sharanya, A. G., Amabati, S. R., Bux, F., & Kumari, S. (2024). Machine learning-based monitoring and design of managed aquifer rechargers for sustainable groundwater management: Scope and challenges. *Environmental Science and Pollution Research International, 0123456789*(59), 31572–31605. Advance online publication. DOI: 10.1007/s11356-024-35529-3

Shortridge, J. E., Guikema, S. D., & Zaitchik, B. F. (2015). Empirical streamflow simulation for water resource management in data-scarce seasonal watersheds. In *Hydrology and Earth System Sciences Discussions* (Vol. 12, Issue 10, pp. 11083–11127). DOI: 10.5194/hessd-12-11083-2015

Shoushtarian, F., & Negahban-Azar, M. (2020). Worldwide Regulations and Guidelines for Agricultural Water Reuse: A Critical Review. *Water (Basel), 12*(4), 971. DOI: 10.3390/w12040971

Shrestha, A., & Mahmood, A. (2019). Review of Deep Learning Algorithms and Architectures. *IEEE Access : Practical Innovations, Open Solutions, 7*, 53040–53065. DOI: 10.1109/ACCESS.2019.2912200

Solomatine, D. P. (1998). Genetic and other global optimization algorithms - comparison and use in calibration problems. *3rd International Conference on Hydroinformatics*, 1021–1028.

Syafiuddin, A., Boopathy, R., & Hadibarata, T. (2020). Challenges and Solutions for Sustainable Groundwater Usage: Pollution Control and Integrated Management. *Current Pollution Reports, 6*(4), 310–327. DOI: 10.1007/s40726-020-00167-z

Sze, V., Chen, Y.-H., Emer, J., Suleiman, A., & Zhang, Z. (2017). Hardware for machine learning: Challenges and opportunities. *2017 IEEE Custom Integrated Circuits Conference (CICC), 2017-April*, 1–8. DOI: 10.1109/CICC.2017.7993626

Tiyasha, T., Tung, T. M., & Yaseen, Z. M. (2020). A survey on river water quality modelling using artificial intelligence models: 2000–2020. *Journal of Hydrology (Amsterdam), 585*(January), 124670. DOI: 10.1016/j.jhydrol.2020.124670

Voigt, H.-M. (1989). 1. Introduction. In *Evolution and Optimization* (pp. 11–16). De Gruyter., DOI: 10.1515/9783112568620-002

Wunsch, A., Liesch, T., & Broda, S. (2018). Forecasting groundwater levels using nonlinear autoregressive networks with exogenous input (NARX). *Journal of Hydrology (Amsterdam), 567*, 743–758. DOI: 10.1016/j.jhydrol.2018.01.045

Yang, X.-S. (2011). Optimization Algorithms. In *Studies in Computational Intelligence* (Vol. 356, pp. 13–31). Issue December., DOI: 10.1007/978-3-642-20859-1_2

Yao, R., Yan, Y., Wei, C., Luo, M., Xiao, Y., & Zhang, Y. (2022). Hydrochemical Characteristics and Groundwater Quality Assessment Using an Integrated Approach of the PCA, SOM, and Fuzzy c-Means Clustering: A Case Study in the Northern Sichuan Basin. *Frontiers in Environmental Science*, *10*(June), 1–12. DOI: 10.3389/fenvs.2022.907872

Yin, J., & Tsai, F. T. C. (2018). Saltwater scavenging optimization under surrogate uncertainty for a multi-aquifer system. *Journal of Hydrology (Amsterdam)*, *565*(August), 698–710. DOI: 10.1016/j.jhydrol.2018.08.021

Yin, J., & Tsai, F. T.-C. (2019). Steady-State Approximate Freshwater–Saltwater Interface in a Two-Horizontal-Well Scavenging System. *Journal of Hydrologic Engineering*, *24*(10), 06019008. Advance online publication. DOI: 10.1061/(ASCE)HE.1943-5584.0001836

Zamil Al-Sudani, H. I. (2019). A Review on Groundwater Pollution. *International Journal of Recent Engineering Science*, *6*(5), 13–21. DOI: 10.14445/23497157/IJRES-V6I5P103

Zhang, Y., Xie, Z., Hu, Q., & Hao, L. (2024). Groundwater Pollution Control and Groundwater Management. *Water (Basel)*, *16*(23), 3542. DOI: 10.3390/w16233542

Chapter 6
Open Sewers and Hidden Dangers: A Study of Indian Urban Drainage

S. Sindhuja
http://orcid.org/0000-0003-1472-1127
SRM Institute of Science and Technology, India

Amir Ahmad Dar
http://orcid.org/0000-0002-0379-2272
Lovely Professional University, India

Mahamadyaseen Shiggaon
http://orcid.org/0009-0009-1120-4037
Lovely Professional University, India

Bashir Ahmad Dar
Lovely Professional University, India

ABSTRACT

This paper takes a close look at the origins of these drainage problems, from past inequalities to present planning gaps, and considers how they still impact the most disadvantaged communities. Drawing on evidence from cities such as Delhi, Mumbai, and Bengaluru and comparing it with similar problems in Lagos, Nigeria the research demonstrates how open drains spread disease, cause flooding, and lead to environmental degradation. It also spotlights cities that are experimenting with novel solutions, such as sponge parks and permeable pavements, to better deal with water sustainably. Albeit with all these efforts, issues such as inadequate coordination across departments and financial constraints remain as obstacles. The present

DOI: 10.4018/979-8-3373-4422-5.ch006

research presents pragmatic solutions centring on planning, innovation, community participation, and more effective utilization of resources, hoping to assist in enabling safer, cleaner, and stronger urban areas in India.

1. INTRODUCTION

India is undergoing one of the fastest and widest urban transformations on the globe. As per the Census of India (2011), the population living /in towns and cities grew from 286 million in 2001 to 377 million in 2011, and the United Nations (2018) estimates it to be as high as almost 600 million by 2030. This demographic change has placed enormous pressure on the nation's infrastructure, and one of the most pressing and underdeveloped aspects of public infrastructure is the urban drainage system. The prevalence of open drains uncovered stormwater and waste-water channels persists in most cities, especially in informal settlements, where such systems are indicative of infrastructural underdevelopment as well as ongoing public health hazards.

Indian cities have, historically, shown sophisticated drainage planning. The Indus Valley Civilization (2600–1900 BCE), for instance, had some of the earliest urban sanitation systems, ranging from covered drains to water management structures that were hooked up to individual homes. But the loss of the tradition over the centuries and selective building of infrastructure in British colonial times resulted in glaring differences. Colonial town planning selectively introduced closed drains, reserving them for cantonments and administrative sectors while conferring indigenous neighbourhoods with exposed and poorly maintained systems (Chaplin, 2011; Prakash, 2002). Such infrastructural inequities are still present in present-day urban planning, where wealthy neighbourhoods are given preference to informal settlements in sanitation coverage (Bhan, 2019).

Urban drainage in India has been characterized by researchers as not only a technical problem but also one that is deeply political. Chaplin (2011) contends that sanitation captures deep-seated hierarchies of urban planning and governance (Chaplin, 2011), whereas Bhan (2019) and Anand (2017) have demonstrated how open drainage is used to materialize and symbolize the exclusion of marginalized groups (Bhan, 2019; Anand, 2017). In Mumbai, Anand explains how open drains constitute the borders of inclusion and exclusion within the city's infrastructural texture. Evidence from ethnographic research in the informal settlements of Delhi, showing that individuals who dwell alongside open drains are both physically exposed and socially stigmatized (Syal, 2024).

The health effects of open drains have been documented extensively. Some study observed directly correlated with each other, where close proximity of a location with open drains contributes to a rise in water- and diseases including cholera, typhoid, and malaria in Delhi, and other parts of Indian urban centres (Singh, 2008; Sanyour & Chaturvedi, 2025).. Shah & Patel (2021) have even proven that poor air quality contributed by proximity near open drainage raises respiratory problems (Shah & Patel, 2021). Where drains stagnate or overflow routinely, mosquitoes and flies flourish and the air that surrounds them has high concentrations of toxic pathogens and odorous gases.

There are also serious environmental issues at stake. The Central Pollution Control Board (2021, 2022) has repeatedly noted that urban open drains are the primary cause of river water pollution in rivers such as the Yamuna and Ganga. Uncontrolled flows from residential and industrial zones enter surface water bodies, leading to high BOD, coliform, and eutrophication. Su et al. (2023) demonstrated that such drains also discharge large amounts of greenhouse gases, mainly methane, because of anaerobic decay (Su et al., 2023). Accounts by the Tamil Nadu and Maharashtra Pollution Control Boards have pointed towards the same in Chennai and Mumbai.

Climate change is escalating such threats. The Indian Institute of Human Settlements (2022) and the National Disaster Management Authority (2023) point to the increasing unpredictability and severity of rainfall events, which are overwhelming outdated drainage systems. Floods triggered by poor drainage now routinely disrupt economic activity, displace families, and contaminate drinking water supplies. Chennai's 2015 flood disaster and Bengaluru's 2022 waterlogging are stark reminders of the danger. A Comptroller and Auditor General (2022) performance audit showed that mismanagement and delays in the Mithi River Development Project in Mumbai exposed thousands to seasonal flooding.

Indian cities are, in turn, coming up with creative solutions. The Bengaluru Development Authority (2022) launched pilot "sponge park" projects to collect runoff and encourage infiltration. Pune and Surat incorporated permeable paving and flood management systems that are mixed with technology along with green infrastructure. TERI (2022) and IWMI (2022) propose decentralized systems such as rain gardens, soakaways, and wetlands, which can be mixed with existing infrastructure to reduce floods and greywater treatment. Kerala's Mazhapolima rainwater recharge program also demonstrates that community-based approaches can scale well.

In spite of these initiatives, financing and governance deficits persist. CPHEEO estimates that more than 2.3 lakh crore are required to upgradation of India's drainage and sanitation infrastructure. Many Urban Local Bodies (ULBs) are however perennially underfunded, short of staff, and lack consolidated responsibilities among departments (Chaplin, 2011). The Finance Commission (2021) and NIUA (2022) have emphasized the requirement for stronger fiscal decentralization, grants linked to performance, and incentives for public-private partnerships.

This paper responds to the multi-dimensional crisis of India's urban drainage. It investigates the historical, political, and environmental contexts of the issue with specific focus on health and equity issues. Based on local research, policy analyses, and overseas comparisons particularly with the Nigerian cities confronting similar drainage issues the paper suggests context-specific, sustainable, and equitable drainage solutions for a more resilient urban environment (Fayomi et al., 2024). The Figure 1 shows the open drains in India.

Figure 1. The open drains which impact the public health

While the public health consequences of open drainage systems—such as water-borne, vector-borne, and respiratory diseases—are widely documented across Indian cities. Instead, it situates health outcomes within the broader structural, historical, and governance-related determinants that shape infrastructural inequality (Ramchunder et al., 2009).

1.1 Objectives of the Study

This study aims to:

1. Analyze the historical and socio-political background of the open drainage systems in Indian cities.
2. Evaluate the environmental and health effects of open drainage to the population.
3. Assess institutional and financial obstacles in city drainage management.
4. Make comparison between Indian cities and selected world cases to draw policy findings.
5. Introduce combined, climate-resilient, and fair drainage reform policies.

Although the poor drainage has well-known consequences in terms of health of the population, this chapter does not simply list the disease outcomes. Instead, it puts drainage in a more global context of infrastructural inequality, environmental pressure and climate insecurity. The elaborate empirical relationships between open drainage exposure and disease prevalence are assembled subsequently in Section 5 (Findings) and the introduction is based on structural drivers of the subject, not a repetitive description of epidemiology.

1.2 Research Questions

This study is guided by the following research questions:

1. How have historical and socio-political processes shaped contemporary open drainage systems in Indian cities?
2. What are the environmental and public health implications of open drainage exposure?
3. What institutional and financial barriers hinder sustainable drainage reform?
4. How can climate-resilient and technology-enabled solutions improve urban drainage governance?

2. LITERATURE REVIEW

2.1 Unravelling the Causes and Currents of India's Open Drainage Troubles

India's urban drainage trouble is an old one to one built into the very fabric of history, disease, policy incompetence, and a general lack of care for the environment. Even as cities expand faster than ever, current infrastructure has not been able to match this pace. An increasing number of publications by scholars, official agencies, and international organisations illuminates the how and why of open drainage continuing to pose a challenge in Indian cities. This review draws on major themes in that literature to set out the challenges and the way forward.

2.2 Historical Legacies and Unequal Infrastructure

India's narrative of drainage infrastructure starts centuries back. Archaeological evidence from the Indus Valley Civilization especially in Mohenjo-Daro shows highly sophisticated systems with covered drains linked to houses and public areas (Jansen, 1989). But in later history, especially under British colonial rule, drainage planning became extremely uneven. Researchers such as Chaplin (2011) have demonstrated how infrastructure was focused in elite areas such as cantonments and European quarters, with Indian areas being left with minimal or no drainage facilities whatsoever (Prakash, 2002).

This imbalance did not cease with independence. Studies by Prakash (2002) and Kundu (2011) indicate that India's post-colonial urban development continued to ignore informal settlements and low-income neighbourhoods, solidifying a gap between modern sewerage and open drainage along economic and social class (Kundu, 2011; Prakash, 2002). Now, this infrastructure disparity is evident in cities from the point of gated communities with underground sewers to adjacent slums with open, clogged drains passing by doorways.

2.3 Environmental Impacts and Public Health Risks

The findings of different studies confirm that open drains are a serious threat to the health of the population. Individuals living near open drainage in Delhi had much greater incidences of diseases like diarrhea, typhoid, and hepatitis (Singh, 2008) also found the same issue in other Indian cities and associated standing water in the drains with diseases transmitted by mosquitoes, including dengue and malaria. Research by Shah and Patel (2021) adds another layer and claims higher gastroenteritis in slums of Mumbai with improper drainage (Shah and Patel, 2021). The environmental cost is also very grim. The reports by the Central Pollution Control Board (2021, 2022) demonstrate the pollution of rivers like Yamuna and Ganga with deadly amounts of coliform bacteria and organic waste due to untreated sewage in open drains. Such polluted water often gets into the ground in broken or unlined drains, which contaminate the quality of ground waters. Other studies like the one by Su et al. (2023) have also expressed alarm at emissions, concluding that broken drains with rotting organic matter release toxic gases like methane and nitrous oxide, which contribute to the production of urban greenhouse gases on a background basis (Su et al., 2023).

2.4 Policy, Governance, and Institutional Barriers

Regardless of this knowledge, there have been disjointed policy responses. According to NIUA (2022) and CPHEEO (2022), scant coordination with minimal overlap is common in a single municipality, where drainage, stormwater, and sewerage services are often handled by separate departments (Chaplin, 2011). Such a two-fold role creates delays in the project, poor management, and loopholes in accountability. Funding is also a major bottleneck. The 15th Finance Commission has indicated that most Indian municipalities fail to collect user charges that are specifically on drainage and many of them do not budget adequately on maintaining the systems. Unable to upgrade infrastructure, Urban Local Bodies (ULBs) do not have the revenue base that might support them. According to researchers like Mishra and Mohanty (2018), cities should be better decentralized to allow them more control of their revenues to than re-invest in services like drainage (Mishra and Mohanty, 2018).

2.5 Innovations and Global Lessons

However, it is moving towards the right direction. Bengaluru has tried out nature-based interventions such as sponge parks which absorb rainwater and reduce water runoff. The permeable pavements of Pune and the sensor and live information-based smart flood warning systems of Surat are technological solutions to the traditional drainage. Combined constructed wetlands, green rooftops, and bioswales in the ground level that are encouraged by reports of TERI (2022) and IWMI (2022) encourage the harvesting of stormwater at the ground level in an environmentally friendly manner. Other countries like Nigeria also have the same problems outside of India. A study by Fayomi et al. (2024) confirms that poor drainage has led to frequent cases of cholera outbreaks in Lagos and other cities (Fayomi et al., 2024). However, according to the same study, co-designed infrastructure and community-based maintenance have proven to be effective in cases where the governments collaborate with citizens (Anand, 2017).

2.6 The Gaps Ahead

Though the literature is informative, it also indicates that there are some gaps. Majority of studies are city-based, and they fail to address the differences in climatic, topographical and urban structure in India, at regional level. Additionally, few studies incorporate technical interventions with social perspectives such as the way people live and/or learn to live with open drains in their daily lives. Such authors as Anand (2017) make us remember that drainage is not only a physical or engineering problem but also a matter of dignity, access, and inclusion in the city (Anand, 2017).

2.7 City Drainage System and Smart Drainage Systems

Nevertheless, in the recent sources (2023-2025), the urban drainage begins to be put into the perspective of climate resilience. Research highlights the greater importance of blue-green infrastructure, digital hydrology models, IoT-based monitoring device, and AI-based flood forecasting as a tool to improve the adaptive capacity of cities. Smart drainage systems integrate sensor-based water-level monitoring, predictive maintenance systems and GIS-based vulnerability mapping to optimize the real-time response and long-term planning. These technological advances are a shift towards conventional stormwater management and flexible and evidence-based management of urban infrastructure. The new financing methods of urban drainage reform including municipal bonds, climate adaptation funds, blended finance structures, and performance-based infrastructure grants also interest new scholarship. These plans introduce the drainage reform to the frame of sustainable development financing plan, climate adaptation commitments, in particular, to SDG 11 and SDG 13 (Veerappan, 2024).

3. METHODOLOGY

The study employs a mixed-methods design, which involves a qualitative policy analysis, secondary data evaluation, and comparative case studies to investigate the issues and probable solutions related to the topic of open drainage systems in urban India. Its methodology was carefully designed in such a way that it aimed to examine the issue in different dimensions such as historical, infrastructural, environmental, and social and also draw conclusions based on both national and international experiences.

3.1 Research Design and Scope

The research is analytical and descriptive in nature. It tries to explore the underlying causes and the extent of impact of open drainage systems and it tries to discuss existing or proposed solutions. The study is focusing on major Indian cities such as Delhi, Mumbai, Chennai, and Bengaluru that have always had a problem with urban drainage. Comprative data is also observed to enrich the analysis and to bring out the international importance through other rapidly growing urban cities in Nigeria like Lagos and Ibadan which have similar strain in urbanization and drainage failure.

3.2 Data Collection

Representative and reliable sources were used in retrieving data in order to have an in depth understanding:

> FG Reports: Main reports and documents of government were analyzed in such authorities as the Ministry of Housing and Urban Affairs, Central Pollution Control Board, Finance Commission of India, and National Disaster Management Authority among other governmental bodies of different states.

These gave the latest update on overall policy trends, infrastructure condition, and budgetary allocations.

Literature: The relevant perspectives on the development of drainage systems throughout the history, the politics of urban infrastructures, and the health-related concerns of the population due to poor sanitation were provided by academics such as Chaplin, Bhan, Anand, and Shah.

Policy and Technical Reports: Reports on the city-level like Delhi Master Plan 2041, Mithi River audit report, and Smart City progress reports by Pune, Bengaluru, and Chennai were studied to examine the performance and implementation of the drainage-related projects.

International Case Studies: Cross country case studies and international literature of other countries such as Nigeria were compared. The study was borrowed in foreign journals on development around the world and contracts awarded by organizations like the World Bank, which offered certain context of similar city issues and adaptation context-specific elsewhere in the Global South.

3.3 Data Analysis Methods

Two main types of analyses were the ones used.

Thematic Analysis: Common themes of health risks of the people, financial strain, fragmented governance and exposure to climatic conditions were systematically abducted in the literature and policy reports. These themes were then coded to highlight the challenges that were persistent and obstacles that were common in cities.

Comparative Policy Review: An evaluation of Interventions and strategies was conducted in different contexts- in India and other countries- to determine those that have been effective, in what circumstances, and why others have not.

Spatial Insights: Although original GIS mapping is not used in this study, it relies on geo-spatial findings of other institutions like the Indian Institute of Human Settlement and Surat Municipal Corporation who have already mapped the spatial distribution of their drainage-related risks and infrastructure shortages.

3.4 Limitations

This study relies primarily on secondary sources, and therefore the findings can only be limited to access and quality of published data. The absence of primary fieldwork or face to face interviews limits the ability to record on the ground opinions. However, this gap is partially filled with the existing ethnographic works that give extensive narratives and accounts of the concerned population. The method of multi-sourced and cross-checked still permits the robustness of the analysis and grounding within the context regardless of all these limitations.

4. CASE STUDIES

The Indian urban drainage systems are highly diverse in terms of structure, functionality and socio-political setting. Analyzing some examples at city level will help to identify the manner in which the local authorities solve drainage problems, what are the recurrent problems, and which of them are scalable solutions. This part presents four case studies of three examples of Indian cities Delhi, Mumbai, and Bengaluru and one international example where a comparison is made to emphasize global issues in a generic way, Lagos, Nigeria.

4.1 Delhi: Slum Chronic Drainage Issues

As the capital of India, Delhi has been grappling with decades old drainage problems particularly in its poor and informal settlements. Although the majority of richer areas in general possess covered sewerage, the majority of the city have uncovered drains. In a report released in 2020 by the Delhi Jal Board, over 40 percent of city stormwater drains were clogged, and rain and sewage water were not separated (Bhan, 2019).

Figure 2. Drainage access disparity across income groups in Delhi (2020). The figure illustrates a stark inequality, with 90% coverage in high-income wards compared to only 18% in informal settlements, highlighting infrastructure inequity

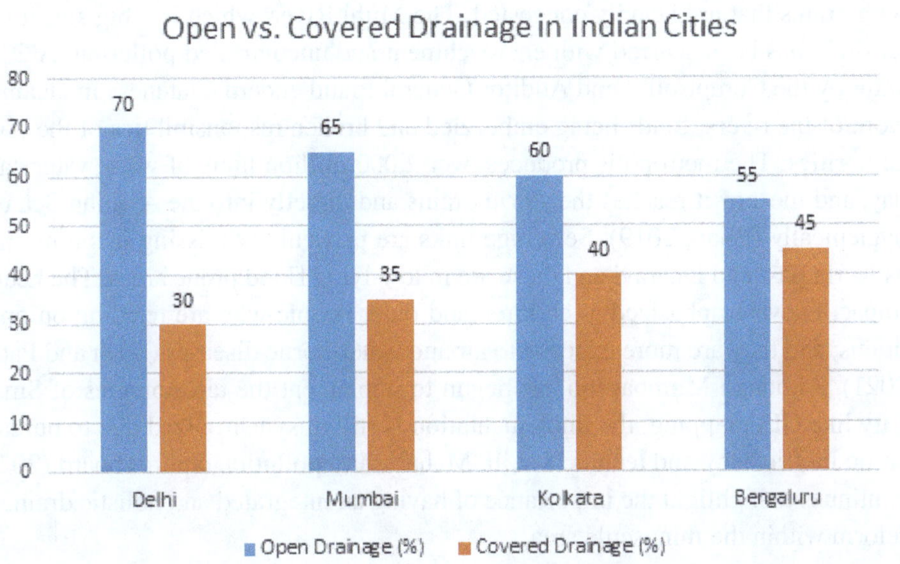

The difference between drainage coverage between income groups 90% in high income districts versus only 18% in slums as illustrated in Figure 2. Regions such as Okhla and Seelampur are likely to experience overflow of monsoon leading to frequent epidemic of waterborne diseases such as malaria and gastroenteritis (Sanyour & Chaturvedi, 2025). Even though efforts have been put to ensure open drains are retrofitted by covering them with concrete, it is not a long-term practice. Moreover, the inability to coordinate different departments is likely to lead to delayed work that is required such as desilting and repair. According to NGO and academic studies, including Syal (2024), to a large percentage of people living in the city, these drainage systems are not only a health risk, but also a direct expression of state apathy and negligence (Syal, 2024).

4.2 Mumbai: Monsoon Flooding, Pollution, and Deficits of Infrastructure.

Mumbai is the financial capital of India and has been identified to experience seasonal flooding. Despite the fact that the city is not well off, it is still marred with drains that are heavily congested. The Mithi River, which is a big storm water drain has been scarred with encroachment and uncontrolled pollution. A 2022 audit by the Comptroller and Auditor General found endemic lateness in cleaning work of the rivers, funds being embezzled and broken responsibilities of the civic authorities. The metropolis produces over 2,000 million liters of wastewater each day, and most of it reaches the storm drains and directly into the Arabian Sea unhygienically (Bhan, 2019). Sewerage links are particularly missing in slums such as Kurla East and Dharavi and these are in low-lying flood prone areas. The health impact is overemphasized to children and older people who are residing on open drains, and they are more exposed to air and water-borne diseases (Shah and Patel, 2021). Although Mumbai too has begun to implement the technologies of Smart City like GIS mapping, the implementation is still unsynchronized due to bureaucratic inefficiency and technical skill. Maharashtra pollution control board (2023) continues to highlight the importance of having an integrated and holistic drainage reform within the minuscule area.

4.3 Bengaluru: Innovating Nature-based Drainage

Gengaluru is one of the Indian urban cities that boast of early sustainable urban drainage promotion. In 2022, Bengaluru has proposed Sponge Park Initiative a project to develop green urban parks as a means to absorb rainwater, reduce surface runoff, and recharge the surrounding aquifers. Besides being able to contain flood water, these polyfunctional parks are also recreational areas that contribute to the increase of liveability of the city. The city of Bengaluru is similarly limited by land-use challenges of the past. The rapid urban development has led to commercial and residential compound encroachment of lakes and natural wetlands and thus limiting the natural drainage of the city. The specialists of the Indian Institute of Human Settlement (2022) warn that although engineered solutions can be effective, there is no way they can uncrowdfund the lost ecological infrastructure. However, HSR Layout and Whitefield pilot projects are positive. Installed bioswale pavements and bioswale installations along the Bellandur Lake have been able to prevent water-logging and promote local biodiversity. The Bengaluru example can be extended to include climatic sensitive design implementation in mainstream infrastructure frameworks (Varshini et al., 2022).

4.4 Lagos, Nigeria: A Parallel in the World with Some Lessons to be Learned.

A contrasting picture is provided by the experience in Lagos, the biggest city in Nigeria, outside of India. Lagos, like most cities of the Global South, has witnessed a very rapid population growth, and the city is overwhelmed with uncontrolled development, which places a heavy load on the city to maintain the drainage systems in good conditions. There are open drains and canals that are filled up especially in slums. According to Fayomi et al. (2024), this water in a standing state in these systems is the direct cause of regular cholera and spreading of diseases that are transmitted by the mosquitoes (Fayomi et al., 2024). It was intensified by development which blocked the storm waterways of nature to cause flash floods during the wet season. All this, however, has brought about community based solutions. The cooperatives of the local residents who have assumed the responsibility of the drainage systems by their own arrangements and clearing blockages have created reliability of the system and increased civic engagement. International NGOs have also guaranteed climatic resilience and wastewater treatment by partially funded pilot restoration of wetlands. The case of Lagos highlights the global concern of the drainage crisis in India. It highlights the role of grassroots involvement, adaptive governance and factors of green planning that could be game changers in the case of Indian cities that face pressure of the same nature.

5. FINDINGS AND ANALYSIS

The research discovers a multifaceted but uniform pattern of infrastructural abandonment, environmental degradation, and health inequalities linked to open drainage in Indian cities. Based on a combined analysis of data, policies, and case examples, some key themes and observations are revealed. These findings not only present an account of the present urban drainage situation, but also a policy intervention agenda for the future.

5.1 Infrastructure Access Inequalities and Drainage Shortfalls

Perhaps the moststark observation is unequal access to drainage infrastructure across socioeconomic groups and places as shown in Figure 3. Wealthy neighbourhood's possess underground sewerage along with appropriately upgraded stormwater drainage, whereas slums and squatter settlements are provided with narrow, open drains. To cite an example, in Bengaluru city, more than 75% drainage coverage is available in affluent wards and fewer than 20% slum localities possess functional drainage.

Figure 3. Compares open vs. covered drains in Delhi, Mumbai, Bengaluru, and Kolkata; highlights infrastructure imbalance

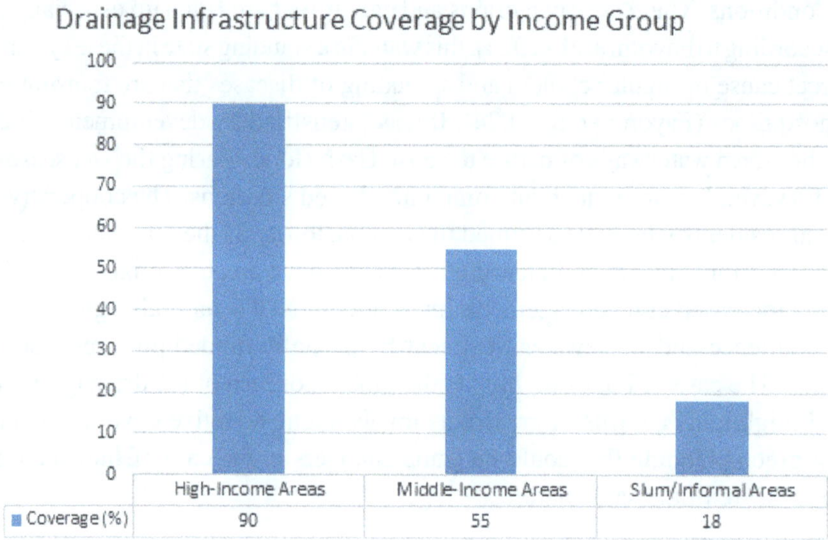

This physical isolation is also systematic. Low-drainage support communities tend to be excluded from urban planning, trapping them in a cycle of neglect and exclusion. These results are consistent with the wider literature on urban injustice and infrastructural inequality, confirming drainage to be both as much of a governance as an engineering problem.

5.2 Public Health Impacts and Disease Burden

The case studies evidently reveal a positive association between open drain exposure and augmented health risks. Open drains in Delhi and Mumbai slum pockets have been linked with water-borne and vector-borne outbreaks of diseases like typhoid, malaria, and cholera (Shah & Patel, 2021). Even unhealthy air around stationary drains has been shown to bear a relationship with respiratory diseases, especially among children and the elderly.

These are typical tendencies. The Nigerian case study illustrates this fact by demonstrating that poor and highly populated communities all over the world suffer from disproportionately high health problems as a result of poor drainage. Apart from that, these conditions not only place a burden on health systems but also reduce production and impose economic burden to vulnerable households.

The results indicate a distinct and consistent correlation of exposure to open drainage system and high probability of public health hazards among the cities under study. Instead of repeating the statistics of the diseases that were presented in the previous sections, this analysis combines structural trends that have connected infrastructural deficit to health susceptibility. The societies that live around open drains, especially in informal settlement areas, get a disproportionate exposure to water-borne and vectors-borne diseases because water and its stagnant water masses and breeding mosquitoes and water-borne diseases and contamination of the surrounding water. Moreover, long-term exposure to foul smells and airborne contamination of decomposing waste also adds to the feeling of respiratory discomfort and low health status in general, particularly in children and the elderly. Notably, such health burdens are not only biological outcomes but they are spatially and socially patterned. The imbalance between the open drainages in the low-income neighbourhoods indicates infrastructural inequity, thus converting the issue of sanitation failure to an environmental justice issue. This tendency is enhanced by the comparative example of Lagos which proves that poor governance of the drainage systematically increases the susceptibility of the marginalized urban population to the diseases. Accordingly, the public health crisis that is linked to the open drains should be addressed as a governance and planning issue as opposed to an exclusively epidemiological issue.

5.3 Environmental and Ecological Stress

Another noteworthy observation is the environmental impact of uncontrolled open drains. Wastewater gets discharged directly into rivers and lakes in cities as a whole without treatment, polluting water and aquatic life perishing. The Central Pollution Control Board reports abnormally high biochemical oxygen demand (BOD) and microbial pollution in most urban water bodies that receive open drainage inflows.

In urban areas such as Mumbai and Chennai, urbanization of natural floodplains and wetlands through hardened construction has cut natural drainage capacities in half. The consequence: regular urban floods even during mid-level rains. Periodic restoration notwithstanding, there is still a serious shortage of integrated ecological thinking in the currently prevailing drainage master plans.

5.4 Institutional Fragmentation and Policy Bottlenecks

The study places a governance failure at the core of the problem. Planning and drainage management usually fall between various departments stormwater, sewerage, solid waste, and land-use planning causing poor coordination and ineffectiveness. Practically all cities do not have one master drainage plan or one unit that has systematic responsibility.

Figure 4. Displays rising investments in drainage by central/state governments, foreign aid, and PPP sources

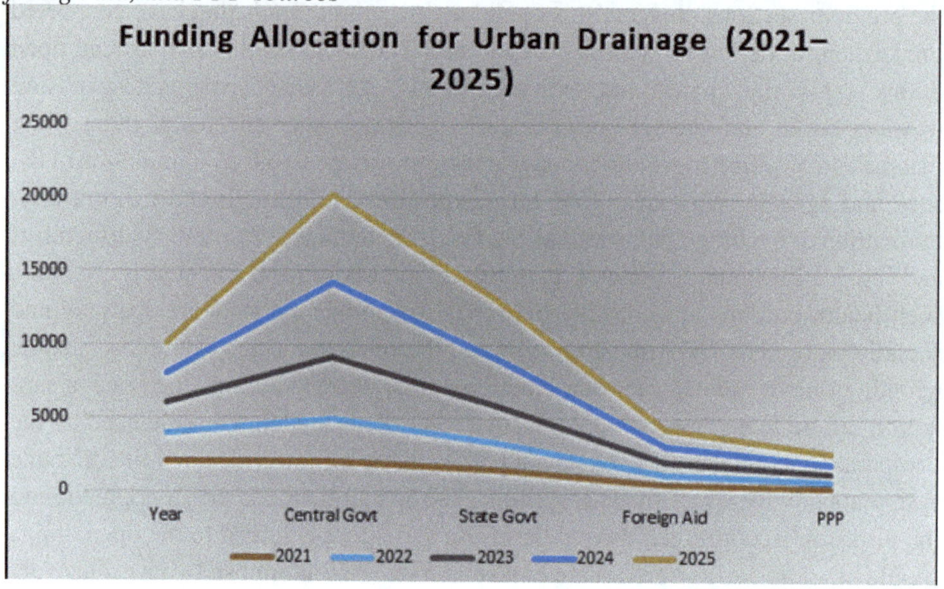

Besides, consistent underfinancing still hinders progress. Urban authorities have rarely distinct drainage maintenance budgets, and even fewer levied dedicated drainage user charges. The funding gap combined with poor technical capacity and red tape significantly delays project implementation as shown in Figure 4.

5.5 Promising Innovations and Localized Solutions

While the system itself has its pitfalls, the research uncovers some good projects too. Sponge parks in Bengaluru, intelligent surveillance systems in Surat, and co-operative drainage for settlements in Lagos show that one can have context-specific and scalable innovation. All such examples have similar characteristics: stakeholder involvement, ecological emphasis, and collaboration among agencies.

But these are usually pilot-scale or the result of exceptional local leadership. Without institutional backing and policy continuity, these innovations will be short-lived and isolated.

6. RECOMMENDATIONS

Addressing the pervasive and chronic problems associated with open drainage systems in the urban areas of India demands measures beyond temporary quick fixes. It demands a whole-of-government, multi-faceted approach that is inclusive, climate-sensitive, and based on long-term thinking. Drawing lessons from this research, the recommendations below are presented to inform more resilient, more equitable, and more sustainable urban drainage infrastructure.

6.1 Develop Integrated Urban Drainage Master Plans

Most Indian cities still do not have a comprehensive plan to deal with drainage, sewerage, and stormwater systems. It is critical that municipalities create Integrated Urban Drainage Master Plans that are integrated with overall urban development and land-use plans.

The plans must:

- Identify and conserve natural drainage lines, wetlands, and low-lying areas
- Provide for inter-departmental coordination, which should involve the Public Works Department (PWD), urban planning bodies, and water boards
- Use geospatial technologies, community feedback, and flood-risk mapping to identify priority vulnerable areas
- To be pertinent, these master plans must be refreshed periodically and incorporated into Smart City programs and municipal master plans.

6.2 Prioritize Decentralized and Nature-Based Solutions

Concrete stormwater systems are costly and usually inappropriate for the ecological and urban processes of Indian cities. Instead, there is increasing evidence for nature-based solutions (NbS) such as:

- Rain gardens, bioswales, and constructed wetlands that filter and manage stormwater

- Pervious pavements, soak pits, and other decentralized options for localized water absorption

These strategies are more biodiverse, lower surface runoff, and recharge groundwater. Effective pilot projects in Bengaluru, Pune, and Surat prove the viability of these systems and offer scalable models replicable in other urban environments, most importantly in informal and underserved settlements.

6.3 Improve Institutional Coordination and Transparency

One of the principal obstacles to drainage reform is institutional fragmentation of responsibilities. In order to overcome this, cities must:

- Create a nodal agency or task force in urban local bodies to deal with drainage as a whole
- Foster data sharing and collaborative planning between sewerage, stormwater, waste management, and road infrastructure agencies
- Put in place real-time performance monitoring, citizen grievance apps, and open dashboards for public accountability

Enhanced coordination will facilitate more effective implementation of projects and minimize delays resulting from bureaucratic overlap.

6.4 Enhance Financing Mechanisms and Foster Sustainability

Modernization of urban drainage structures needs long term and diversified financial support other than the traditional municipal budgetary allocations. Cities need to embrace new financing paradigms that can bring together both state and privately-owned funds and be long-term financially sustainable in light of the magnitude of infrastructure deficits and the growing risks of flooding caused by climate change. A possible process is the issuance of the municipal bonds, especially the green bonds that will be focused on stormwater management and the upgrades of wastewater infrastructure. Urban local bodies rated by credit rating agencies can use bond market as a means of raising preliminary finance to undertake drainage retrofitting, sewer-drain separation and smart monitoring infrastructure. The repayment systems can be backed up by the property tax, drainage service charges, or guarantees by the state to enhance investor confidence. Moreover, drainage reform follows the same priority during climate adaptation and as such, could be able to tap national and international climate adaptation funds. City resilience efforts which mitigate flood risks and increase water security are increasingly supported by development banks and international financing platforms on climate change. Municipalities can open up concessional loans, grants and technical assistance by integrating drainage initiatives and climate action plans as part of SDG 11 (Sustainable Cities) and SDG 13 (Climate Action). The alternative way of distributing the fiscal burden while ramping up infrastructure reform is through blended finance schemes, where the funds of the state and its private investors are combined with multilateral assistance. Well designed Public-Private Partnerships (PPPs) may be used in the integration of technology, maintenance contracts over a long period, delivery of infrastructure based on performance, as long as there are effective accountability mechanisms. Lastly, performance based grants in line with suggestions of the 15 th Finance Commission can be used to encourage quantifiable results. The transfers across governments could be linked to a decrease in waterlogging cases, a better drainage cover in informal settlements, or regular fixes standard. These outcome-based fiscal models move the process of governance away by focusing on expenditure-based budgeting and towards result-based infrastructure management.

Reform of drainage should be supported by stable and diversified funding. Some of the main steps are:

- Implementing user-based drainage charges, or incorporating charges in property tax systems, with proper subsidies for low-income families
- Promoting public-private partnerships (PPPs) for financing major upgrading, especially where monitoring technology or treatment plants are involved
- Receiving climate adaptation finance via institutions like the Asian Development Bank (ADB), World Bank, and other development partners

Also, central and state government funds can be made to conform to performance indicators, e.g., decrease in waterlogging events or increase in wastewater treatment coverage, to promote outcome-oriented investment.

6.5 Promote Community Involvement and Awareness

Drainage problems cannot be solved by infrastructure alone there has to be community outreach. Residents have to be engaged not only as recipients but as owners and partners in the maintenance of the system.

Suggested measures are:

- Establishing local drainage management committees at ward or neighbourhood's level
- Conducting awareness campaigns to deter littering in drains and encourage civic awareness

- Equipping youth groups, resident associations, and NGOs with basic training for drain cleaning and maintenance assistance
- Enabling communities promotes ownership, enhances maintenance at the local level, and strengthens resilience from below.

6.6 Combine Climate Resilience and Forward-Looking Design

Due to the growing variability of rainfall, extreme events, and higher temperatures in the city attributed to climate change, urban drainage systems have to be reformulated. Conventional methods of drainage planning which are premised on past rainfall statistics are no longer adequate. Rather than that, master drainage frameworks should be adopted using climate-responsive hydrological modeling, scenario-based infrastructure sizing, and adaptive capacity planning. Climate adaptable drainage is not just an engineering accommodation but a strategic adjustment strategy which facilitates urban disaster preparedness and sustainability in the long run.

With rainfall variability and urban flooding due to climate change, drainage needs to be planned to adapt, not react. Forward-looking planning must:

- Apply hydrological modelling to select the right drain size, route, and material
- Include blue-green infrastructure such as retention ponds, urban wetlands, and green roofs in planning and retrofitting work

Figure 5. It shows the blockages and roads during rainy season

- The blockage and roads during rainy seasons as shown in Figure 6.
- Link flood management practices with urban resilience systems, moving from a focus on emergency response towards sustained risk mitigation

These practices are particularly vital for coastal and flood-risk cities, where infrastructure is increasingly pushed to the breaking point by the scale of extreme weather events.

Climate-resilient drainage planning must integrate predictive hydrological modeling, rainfall variability projections, and adaptive infrastructure sizing. As extreme precipitation events become more frequent, drainage systems must be designed for future climate scenarios rather than historical rainfall data. Aligning drainage reforms with SDG 11 and SDG 13 strengthens access to climate finance and enhances institutional accountability

Furthermore, integrating blue-green infrastructure such as retention ponds, urban wetlands, green roofs, and permeable corridors enhances natural water absorption while reducing surface runoff. These measures strengthen ecosystem services, improve groundwater recharge, and mitigate urban heat island effects. Embedding drainage reform within broader climate adaptation strategies also enables cities to access global climate finance and resilience-oriented investment programs.

6.7 Innovative Financing Mechanisms for Urban Drainage Reform

The urban drainage reform in India needs a long-term and diversified financing strategy that is no longer based on the annual municipal budgets allocations. Considering the magnitude of the infrastructure shortage and growing exposure to climate-driven risks, cities should use new financial instruments that would be able to mobilize local and foreign funds. Conventional grants cannot serve the ever increasing investment demand to modernise, retrofit and climate-resilient expansion of drainage networks. The use of municipal bonds, especially green municipal bonds that are specifically allocated to sustainable infrastructure development is one of the promising mechanisms. Big urban local bodies (ULBs) may, by bonding, finance stormwater management improvements, sewer-drain separation systems, as well as intelligent monitoring methods. Repayment sources of revenue can be attached to the property taxes, draining user fee, or the guarantees supported by the state. In combination with credit rating reform and better financial transparency in cities, bond financing can allow cities to raise large amounts of upfront capital to renew infrastructure. Moreover, urban drainage reform is in agreement with the climate adaptation priorities and thus it can qualify the climate adaptation funds of national and international sources. Development banks and global climate funds are now more and more supporting the multilateral projects that make cities less prone to floods, enhance water resilience, and combine blue-green infrastructure. Framing drainage investments as climate resilience plans and positioning them in line with Sustainable Development Goals (SDG 11 and SDG 13) will open the door to concessional loans, grants, and technical support to municipalities. The other new approach is the blended finance that involves using the combination of public resources, private finance, and development aid to mitigate the risk and ensure that the private entities are involved. Cities can achieve extensive drainage upgrades with large scale drainage projects through risk sharing mechanisms, viability gap funding, and outcome based financing structures without excessive effects on the municipal budgets. Likewise, properly designed Public-Private Partnerships (PPP) may enable integration of technology, real time monitoring, waste water treatment plants, long term maintenance agreements with the upkeep that regulatory protection and accountability mechanisms are well stipulated. Lastly, the performance-based grants consistent with the recommendations of the 15 th Finance Commission could be a catalyst in enhancing the drainage governance. Intergovernmental fiscal transfers that can be linked to quantifiable results such as a decrease in the incidences of waterlogging, the presence of adequate drainage in the informal settlements, better wastewater treatment ratios, or even regularity in the maintenance cycles can help change the mindset of the municipalities that focus their budgetary resources to expenditures, but not results. These types of funding mechanisms based on incentive promote accountability, transparency and long-term service delivery changes. Together, these financing options offer a route to financially sound, climate-sensitive and technologically advanced city drainage systems. In the absence of these structural fiscal reforms then technical and ecological innovations will not ascend effectively across the rapidly urbanizing cities in India.

6.8 Smart and Technology-Enabled Drainage Governance

Technological modernization plays a critical role in transforming urban drainage from reactive crisis management to predictive infrastructure governance. Smart drainage systems incorporate IoT-based water-level sensors, automated flow regulators, GIS-based vulnerability mapping, and AI-driven flood forecasting tools. These systems enable real-time monitoring of blockages, early warning dissemination, and data-driven maintenance scheduling.

By integrating predictive analytics and digital dashboards within municipal governance platforms, cities can shift from seasonal desilting cycles to continuous performance monitoring. Such systems also improve transparency and accountability by enabling citizen-accessible reporting platforms and grievance redress mechanisms. When combined with institutional reforms and financing innovation, smart drainage technologies can significantly enhance urban resilience and operational efficiency.

7. MANAGEMENT AND POLICY IMPLICATIONS

The findings of this study carry significant implications for urban governance, municipal finance, climate adaptation planning, and infrastructure management. Drainage reform must transition from a reactive engineering function to a strategic governance priority embedded within long-term urban resilience planning.

Indian urban drainage reform demands a radical transformation of fractured and reactive practices in urban drainage management to institutionalized, prospective governance models that instigate resilience, equity and sustainability into daily municipal governance (Bhat et al., 2025). Municipal managers should no longer implement a desilitation cycle with a crisis-like cycle which takes effect annually or prior to the monsoon season but rather implement a predictive model of maintenance systems, which is based on the hydrological modelling, GIS mapping, and real time monitoring technology. It is also important to integrate drainage infrastructure in formal capital budgeting systems so that stormwater systems are viewed not as secondary utilities but as central urban systems with lifecycle costing, depreciation planning, and multi-year investment plans. This method can help municipalities to invest in upgrade, retrofitting and technological modernization and not only in emergency allocations following the occurrence of floods. Institutional reforms are also required to defeat departmental silos that divide the stormwater management, sewerage, land-use planning and solid waste management; organized data-sharing platforms, interoperable digital dashboards, and centralized drainage task forces would go a long way in enhancing coordination and accountability. Moreover, introduction of performance-based infrastructure measurements (i.e. decrease in the number of waterlogging cases, increase in drainage coverage in informal settlements, decrease in the number of pollutants entering urban waterways, and quicker reaction to blockages) can change the evaluation into expenditure-driven auditing into outcome-driven governance. Lastly, the drainage reforms should be aligned strategically with the other national and global development commitments, specifically the SDG 6 (Clean Water and Sanitation), SDG 11 (Sustainable Cities and Communities), SDG 13 (Climate Action), and national and state-level action plans on climate adaptation. Incorporating drainage management as part of climate resilience helps to not only enhance the urban flood preparedness but also open access to climate adaptation capital, green infrastructure funds, and international development collaboration. Essentially, it can be stated that sustainable management of urban drainage requires an overall change in the planning, financing, monitoring, and institutional integration that makes the drainage infrastructure the core of the promotion of the health and environment of the city, and urban stability over long-term, as opposed to a secondary technical service.

CONCLUSION

The prevalence of open drainage systems in India's cities is not just the result of old infrastructure it is a failure of deeper sorts, including urban governance, socio-spatial justice, and environmental management. With increasing city density and the intensification of climate change impacts, the failure of these systems to serve the public good of health and to address storm water issues is becoming a matter of acute urban concern.

This study highlights the manner in which open drains particularly in urban slums are not just unhygienic and hazardous, but directly dangerous. They spread disease, contaminate rivers and groundwater, and worsen flooding. While areas of affluence may enjoy closed drainage and regular upkeep, marginalized communities are left to experience the direct damage of infrastructure neglect. These trends are not random; they represent systemic inequalities based on fractured institutions, constrained municipal capacity, and exclusion of vulnerable populations from planning processes.

Nevertheless, there are grounds for hope. Cities like Bengaluru, Pune, and Surat are starting to adopt innovative, decentralized, and nature-based drainage systems initiatives that merge resilience with sustainability. Globally, cities like Lagos provide insights into the strength of community-driven maintenance, adaptive governance, and low-cost ecological infrastructure.

The future necessitates a rethink of how drainage is understood and valued. It needs to be addressed not as a technical by product, but as a paramount public utility a one that impinges on the health of the city, equality, and the ability to survive climate change. This entails strategic urban planning, investment in blue-green infrastructure, efficient governance, and most of all, true engagement from communities most impacted.

By placing emphasis on the multi-dimensional effects of open drainage systems, assessing both local and international reactions, and providing evidence-based suggestions, this paper seeks to contribute to a more sustainable and inclusive vision for India's cities. Ultimately, the intention is not just to rehabilitate old drains, but to change the systems and attitudes that enabled them to exist in the first place.

REFERENCES

Bhan, G. (2019). Notes on a Southern urban practice. *Environment and Urbanization, 31*(2), 639–654. DOI: 10.1177/0956247818815792

Bhat, A. M., Tripathi, I. M., Mohapatra, P. K., & Bhagat, C. (2025). Reimagining stormwater management: Sustainable drainage pathways for resilient Indian cities. *Environmental Conservation*, •••, 1–7.

Chaplin, S. (2011). *The politics of sanitation in India: Cities, services and the state.* No Title.

CPHEEO. (2022). *Central Public Health & Environmental Engineering Organisations.* Ministry of Housing and Urban Affairs.

Fayomi, G. U., Onyari, E. K., & Jaiyeola, A. T. (2024). Quantitative assessment of open drainage and storm water runoff capacity in Nigeria: Consequences and way forward. *Case Studies in Chemical and Environmental Engineering, 9*, 100766.

Jansen, M. (1989). Water supply and sewage disposal at Mohenjo-Daro. *World Archaeology, 21*(2), 177–192. DOI: 10.1080/00438243.1989.9980100 PMID: 16470995

Kundu, A. (2011). *Trends and Processes of Urbanisation in India. Human Settlements Group.* IIED.

Mishra, A. K., & Mohanty, P. K. (2018). Urban infrastructure financing in India: Applying the benefit and earmarking principles of taxation. *Journal of Social and Economic Development, 20*(1), 110–128. DOI: 10.1007/s40847-018-0059-1

Prakash, G. (2002). The urban turn. *Sarai reader.* Oxford University Press., *2*(7), 2-7.

Ramchunder, S. J., Brown, L. E., & Holden, J. (2009). Environmental effects of drainage, drain-blocking and prescribed vegetation burning in UK upland peatlands. *Progress in Physical Geography, 33*(1), 49–79. DOI: 10.1177/0309133309105245

Sanyour, A., & Chaturvedi, S. (2025). Expert perspectives on climate change and public health in port cities: a case study of Mumbai. In *Urbanization, Climate Change, and Health: Integrating Strategies for Sustainable and Resilient Cities* (pp. 25–47). Springer Nature Switzerland. DOI: 10.1007/978-3-031-99132-5_2

Shah, N., & Patel, R. (2021). Spatial analysis of gastroenteritis prevalence in Mumbai's informal settlements. *Journal of Water, Sanitation, and Hygiene for Development : a Journal of the International Water Association, 11*(3), 456–468.

Singh, R. (2008). *3.3 Wastewater Related Risks and Social Vulnerability: A Case Study of Delhi. Resilience and Social Vulnerability, 121. Anand, N. (2017). Hydraulic city: Water and the infrastructures of citizenship in Mumbai*. Duke University Press.

Su, Q., Dai, H., Xie, S., Yu, X., Lin, Y., Singh, V. P., & Karthikeyan, R. (2023). Water–energy–carbon nexus: Greenhouse gas emissions from integrated urban drainage systems in China. *Environmental Science & Technology*, *57*(5), 2093–2104. DOI: 10.1021/acs.est.2c08583 PMID: 36696288

Syal, S. (2024). The (unprivileged) polluter pays: Conflict of Rights in Delhi's stormwater drain-adjacent 'informal' settlements. *Planning Practice and Research*, *39*(6), 960–993. DOI: 10.1080/02697459.2024.2383824

Varshini, S., Hungund, K., Kundu, S. K., Santhanam, H., & Dhyani, S. (2022). Assessing ecological risks of urban air and water environment to analyse the scenarios for mainstreaming nature-based solutions: a case study of Bengaluru City, India. In *Blue-Green Infrastructure Across Asian Countries: Improving Urban Resilience and Sustainability* (pp. 163–191). Springer Singapore. DOI: 10.1007/978-981-16-7128-9_8

Veerappan, S. (2024). Edge-enabled smart storm water drainage systems: A real-time analytics framework for urban flood management. *Journal of Smart Infrastructure and Environmental Sustainability*, *1*(1), 52–59.

ADDITIONAL READING

Bengaluru Development Authority. (2022). *Sponge Parks Initiative: Impact Assessment Report*. BDA Publications.

Central Pollution Control Board. (2021). *Status of Water Quality in India*. Ministry of Environment, Forest and Climate Change.

Central Pollution Control Board. (2022). *Water Quality Assessment of Major Urban Water Bodies*. Ministry of Environment, Forest and Climate Change.

Finance Commission of India. (2021). *15th Finance Commission Report for 2021–2026*. Government of India.

Indian Institute of Human Settlements. (2022). *Natural Drainage Systems in Indian Cities: Status and Challenges*. IIHS Publications.

International Water Management Institute (IWMI). (2022). Constructed Wetlands for Urban Wastewater Treatment: Performance Assessment of Pilot Projects in South Asian Cities. IWMI Research Report.

National Disaster Management Authority. (2023). *Urban Flooding Incidents in Indian Cities: Analysis and Lessons*. NDMA Publications.

Pune Municipal Corporation. (2023). Permeable Paving Project: Implementation and Impact Assessment. PMC Technical Report.

Surat Municipal Corporation. (2022). *Integrated Flood Management System: Technical Specifications and Performance Analysis*. SMC Publications.

The Energy and Resources Institute. (2022). Cost-Benefit Analysis of Sustainable Urban Drainage Systems in Indian Cities. TERI Research Paper.

United Nations. (2018). *World Urbanization Prospects: The 2018 Revision*. UN Department of Economic and Social Affairs.

Chapter 7
Optimizing Urban Mobility Through Adaptive Routing and Intelligent Signal Systems:
Smart Routes, Smarter Cities

Sitesh Kumar Singh
http://orcid.org/0000-0002-7108-0808
College of Engineering, National University of Science and Technology, Oman

Jaishri Gothania
http://orcid.org/0000-0002-2656-642X
Lingaya's Vidyapeeth, India

Rituraj Jain
http://orcid.org/0000-0002-5532-1245
Marwadi University, India

Abby Balkis Gorayah
http://orcid.org/0009-0006-7326-2249
College of Engineering, National University of Science and Technology, Oman

ABSTRACT

This chapter will discuss how urban mobility can change with adaptive routing and intelligent signal systems (ISS), as it is an efficient way to improve traffic flow and emissions and optimise the journeys of people commuting within smart cities. By

DOI: 10.4018/979-8-3373-4422-5.ch007

using technologies, including AI, IoT, GIS, and V2X communication, such systems deliver real-time, data-based traffic control. The chapter analyses the global trends of implementation, assesses the constraints of legacy systems, and explores the issue of socio-spatial equity, cybersecurity threats, and forms of governance. It combines interdisciplinary strategies such as socio-technical systems, multi-level governance, and innovation diffusion to postulate a strategic model of inclusive efficiency and futuristic urban mobility.

INTRODUCTION

Urbanization and Mobility Stress

The rate at which cities are growing across the world is explosive and the world urban population is escalating at alarming rates up to the year 2050. This mushrooming of urbanization comes with huge traffic planning and transport infrastructure management problems. In most urban areas, the conventional traffic control methods based on fixed-time signal control systems and fixed routes protocol are becoming critically inadequate to meet the current transportation needs. These include congestion, overly long travel delays, increased emissions, and commuting dissatisfaction which are the effects of ill-conditioned transportation networks being unable to adapt on a real-time basis.

Traditional traffic networks use an already programmed logic, not updated to the live traffic flows, loss of service or commuting patterns. Their inflexibility does not reflect on the dynamically changing traffic conditions thus bringing about inefficiencies that reflect economic productivity, in addition to playing a role in environmental performance. With growing residential densities comes an amplified effect of outdated systems or systems that lack flexible efficiencies of mobility infrastructures, and it is therefore time and critical to switch to smarter and more flexible mobility systems.

Objective

This chapter seeks to investigate the utilization of adaptive routing and intelligent traffic signal systems that will enhance the mobility in cities by overcoming the limitations of the traditional static traffic management approaches. It attempts to discuss how real-time data, predictive analytics, artificial intelligence and connected technologies can be used to achieve dynamic route optimization and responsive signal control but stresses the need to connect routing platforms with traffic signal infrastructure to improve network-wide efficiency. The chapter also assesses environmental, economic and societal benefits of such systems, such as reduced congestion, lower emissions, and improved travel reliability, and identifies some of the important implementation challenges and research opportunities going forward in the advancement of smart urban transportation.

The Shift to Smart Urban Mobility

The smart city paradigm has anchored a new territorial vision of governments since urban services are revisited through data-driven governance, which is placed at the center of future-ready cities. This transformation has currently placed mobility, an important city activity, on the hot seat. Rather than building more physical infrastructure, cities are increasing looking at digital intelligence to maximize mobility by analyzing data in real-time, automating routines, and predictive modelling. To have a fundamental idea regarding the concept of smart urban mobility, Figure 1 has been given below.

Figure 1. Conceptual representation of smart urban mobility

Good examples of this are the adaptive routing and intelligent signal systems (ISS). Adaptive routing makes future navigation possible based on real-time circulation states, where vehicles are directed to avoid clogged zones and cut on travel time. ISS on the other hand changes the traffic signal phases automatically following the input of sensors based on the current use of the road. The two methods rely on new technologies, including Artificial Intelligence (AI), the Internet of Things (IoT), and Geographic Information Systems (GIS), allowing previously impossible levels of control and despite their flexibility.

As is highlighted by Singh and Jain (2025), the transition to intelligent transport system implies an essential move from a fixed system-of-the-systems infrastructures toward dynamic, cyber-physical system functioning in which the components may collaborate freely to ensure the best possible flow and safety.

Technological Enablers: AI, IoT, and GIS

The combination of three technologies AI, IoT and GIS makes the digital core of smart urban mobility as shown in Figure 2. Pastugh AI, traffic systems get to make predictive analysis whereby past and present data are used to predict congestion or delays. Continuous data sources that can be rendered by IoT technologies, such as vehicle sensors, traffic cameras, and connected mobile devices, can be considered as sources of data input to traffic control algorithms. GIS has a spatial structure to analyze visualize and simulate highway conditions and optimize routes to clear traffic.

Figure 2. The fusion of AI, IoT, and GIS technologies of smart urban mobility

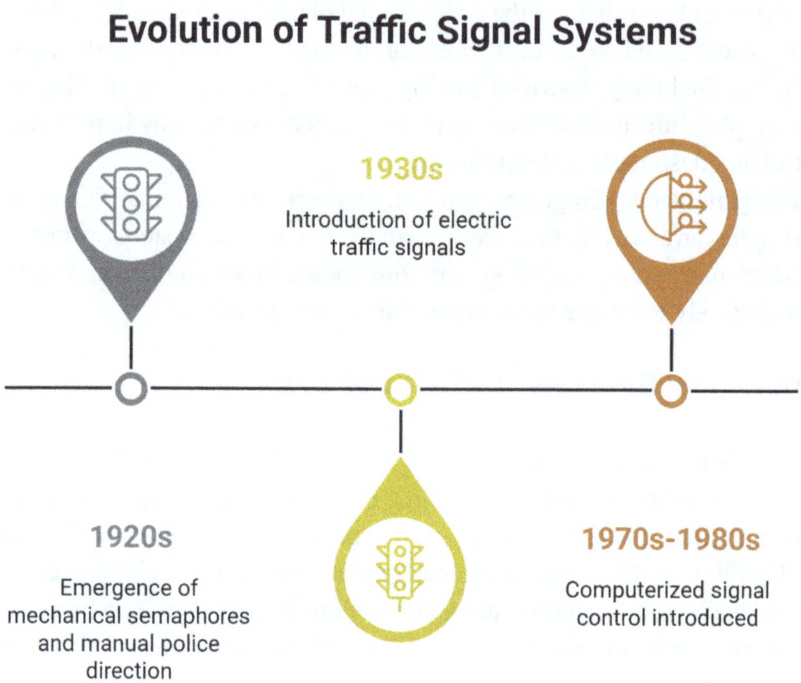

Such technologies are actively being implemented in the cities who attempt to build an intelligent transport ecosystem. Research by Ali et al. (2020) has been conducted on the use of AI and IoT in optimizing traffic signal timing and their energy use and waiting times were significantly reduced. Their work demonstrates how data fed decentralized traffic signal controllers could operate autonomously on adjusting the phases and decrease idle time (up to 35 per cent). The additional benefits of GIS integration to these systems include the ability of the traffic planners to visualize the high-congestion corridors in maps and the ability to simulate the various intervention strategies within various geographic contexts.

These digitally advanced systems are not merely a way to simply move traffic, but they contribute to sustainable initiatives by reducing carbon footprint, and enhanced travel efficiency especially when there is high demand.

Alternatives to Infrastructure Expansion

In the past the problem of urban traffic congestion had been solved by expanding the capacity of the road through infrastructural increase. Nevertheless, such a practice is becoming more criticized due to the lack of its sustainability as a solution to the problem of overloading. Duranton and Turner (2011) showed that higher infrastructure construction can generate more demand which eventually causes the same level of congestion as before thus causing a phenomenon called induced traffic.

On the other hand, adaptive routing and ISS are more economical in terms of utilization of available infrastructure. As an example, Adaptive Signal Control Technology (ASCT) uses deep reinforcement learning models that have demonstrated such promise in the real world. Genders and Razavi (2016) presented a signal controller based on a deep Q-network, which reduced delays at an intersection by more than 30%. By making use of real-time inputs, these systems can optimize changes in signals according to the needs of the signal and not at some fixed time intervals and thus can respond to the changing traffic in an adaptive manner.

Smart mobility systems would offer a viable way out to cities surrounded by space challenges with a climate imperative resulting in the eliminations of cost, space, disruption of environmental crash courses of physical infrastructure expansion.

Global Cases and Implementation Insights

Intelligent mobility solutions are currently popular, but implementation is still in different stages across different countries due to differences in technological maturity as well as governance and socio-economic realities. Such cities, where the data management and cross-sectoral collaboration system is integrated, are more likely to achieve success in implementing adaptive traffic solutions.

Yang et al. (2020) analyzed smart bus networks and smart traffic systems based on AI surveillance, V2I communication, and servers in Hangzhou, China, where the practical consequences of applying the technology benefits were in a reduced length of commute and better emissions. Notably, the study insists that the key to the long-term sustainability of such innovations lies in institutional sponsorship via the use of policy systems and leadership between the governments and the private sector.

The introduction of intelligent systems is, however, under great constraint. These are the compatibility with legacy infrastructure, large investment requirements, regulatory lag and privacy concerns regarding data. The solution is holistic planning using both technical innovation and policy- making as well as stakeholder inclusion.

PROBLEM STATEMENT

Urban areas across the world are facing unprecedented population growth, vehicle ownership growth, and increased travel demand. As cities have grown, transportation networks are finding it difficult to keep up with growing congestion, variable travel times, rising fuel use and poor air quality. Traditional traffic management systems - mostly relying on static signal timing, fixed assumptions about traffic routes, and limited access to real time data are no longer adequate to cope with dynamic and complex aspects of modern urban mobility. This mismatch in increasing demand for travel and old-fashioned traffic control strategies has led to considerable socio-economic and environmental problems that jeopardize the sustainability and livability of cities. One of the basic problems involves the inefficiency of current routing and traffic signal systems. Conventional mechanisms for navigation and traffic control are often working in isolation from one another and are based on historical data or pre-programmed plans of timing rather than the real time traffic conditions. As a result, they do not respond effectively to sudden disruptions such as accidents, roadworks and weather events or temporary surges in demand. This results in poor traffic distribution because some corridors can be very congested while alternative routes are not used. This lack of coordination between routing platforms and traffic signal systems is further increasing the congestion by causing bottlenecks that spread throughout the network. Another important concern is increasing variability and unpredictability of urban travel. Daily traffic patterns are no longer consistent with the onset of flexible work schedules, ride hailing, e-commerce delivery, and multimodal mobility options. Static traffic signal plans are unable to adapt to these varying patterns, leading to excessive delays on intersections, increasing stop and go traffic, and low throughput on the network. Intersections are known to be major sources of delay and emissions as inefficient signal timing for the intersection causes unnecessary idling and inefficient movement of vehicles. Without adaptive and intelligent control, intersections still exist as isolated islands of control as opposed to integrated parts of a responding urban mobility system. Environmental and public health issues also add to the urgency of this problem. Prolonged congestion contributes to climate change and urban health risks because it increases greenhouse gases and fuel consumption and pollution. Cities are under growing demand for achieving sustainability goals and mitigating the emission of transportation, but existing traffic management strategies offer little help for environmentally optimized traffic routing or signal coordination. The lack of intelligent systems that are focused on smoother traffic flow and reduced idling outputs detracts from wider efforts to ensure sustainability and smarten cities. Additionally, the accelerated development of data sources, including connected vehicles, mobile devices, sensors and intelligent transportation systems, has opened an opportunity that is not yet being taken advantage of. While enormous amounts of mobility data are produced every day, many cities are not equipped with joined-up frameworks that can convert mobility data into real-time traffic management strategies that can be implemented immediately. Fragmented data platforms and limited interoperability mean that cities are not taking advantage of all the possibilities of advanced analytics, machine learning, and artificial intelligence to optimize traffic. Therefore, there is an urgent need to develop adaptive routing and intelligent signal systems that can dynamically react to real-time traffic conditions, integrate diverse sources of data and network-wide traffic management strategies. By closing the gap between routing decisions and signal control, cities can be on their way to smarter, more efficient, and more sustainable transportation networks. Addressing this problem is fundamental to decrease congestion, increase travel reliability, cut emissions, and ultimately improve the quality of life in the rapidly growing urban environment.

LITERATURE REVIEW

Introduction to Traditional Traffic Management

Deterministic, infrastructure-oriented models of traffic control that are based on the infrastructure have historically underpinned urban transport systems. Some of the most long-standing aspects of these models include fixed-time traffic signal systems as well as pre-planned routing algorithms. These were transformational in the early and mid-20th century when cities were less dense, and the traffic seems more predictable. They, however, now fail more and more to effectively address the complex, dynamic and high-density traffic environments that are common in contemporary cities.

There are fixed time signals, which are based on cyclical programming regardless of real traffic conditions. On the same note, route choices in early routing systems rely on preset routes optimized based on past traffic averages and not in real-time. This type of solution, all sizes fit all, is reactive in nature and cannot respond to the sudden influx of changes or situations like accidents and weather disturbances or a sudden influx of vehicles. With urban expansion exceeding the flexibility of old systems, traffic congestion, ecological harm and commutation aggravation have aggravated.

The Historical Evolution of Traffic Signals

Some of the first structures of traffic management in cities were developed in the 1920's where there was mechanical semaphore and manual police control on main streets. The grandchild of this were electric traffic signals that were developed in the 1930's and increasingly standardized in the middle of the 20th century. Averaged daily traffic counts were used to set fixed-cycle signals to control the intersections, and this mechanism was the default one.

By the 1970's and 1980's, signal control was computerized, which allowed behavior to be a little responsive in respect to fixed detectors installed in the roads. Nevertheless, they were still based on centralized rule-based logic and could not be too dynamic in reacting to the changes in traffic patterns. Such systems, as N. Sharma and Sahu (2018) observed, even in the case of being linked to traffic sensors, were operating within strict guidelines that could not optimize their performances in real-time. Figure 3 is a pictorial form of the history of evolution of traffic signals.

The costs of these legacy systems, though low, and designing them to scale up reasonably easily, the philosophy underpinning them was handling congestion with inbuilt rules, as opposed to flexibility. The problem with this legacy now is in the fact that many developing cities are trying to lay smart infrastructure upon obsolete structures.

Figure 3. The historical evolution of traffic signals

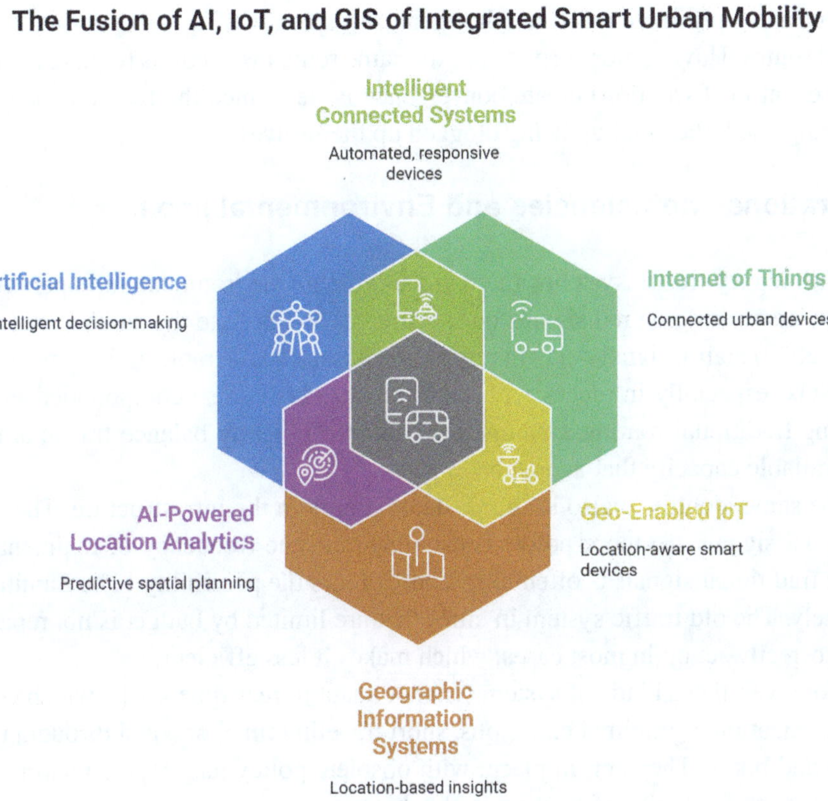

Scalability and Adaptability Constraints

Scalability is one of the fundamental shortcomings of traditional traffic systems. With the increase in the size of cities and the number of intersections, and the multiplicity of routing options, the static signal systems become very cumbersome to control. Recalibration of fixed-time signals must be performed on a large network over a considerable time; they need a manual adjustment time-to-time which is usually done using an old survey on traffic patterns.

Kohler and Strehler (2012) noted that fixed-time strategies lead to more postponing at times of variable demand and do not adjust to incidents, construction areas, or even rushes. In addition, they cannot give precedence to facilitated vehicles, community transportation or put-in jeopardy users like pedestrians and cyclists without adding expensive supplementary equipment.

Likewise, rule-based routing algorithms are inflexible. These systems cannot anticipate congestion in real time; they generally spread traffic over inferior or congested routes. Having more drivers use the same route instructions (e.g., in the case of an event or disruption) create convergence issues; since the recommendations are being used, they end up being clogged up themselves.

Operational Inefficiencies and Environmental Impacts

With legacy traffic systems, there are significant environment and operational costs. Cars waiting on red signals burn resources and pollute the air. As Scandella et al. (2022) claim, signal delay in city areas causes considerable emissions of CO_2 and NOx, especially in places with high traffic. This is also compounded by the existing traditional routing systems that cannot efficiently balance traffic among the available capacity that adds more VKT.

The same identity is also in the waste of energy in the infrastructure. The conventional signals are more power consuming, and the frequency of maintenance of the traditional signals is often long since it lacks the possibility to be monitored remotely. The old traffic system in cities that are limited by budget is not repaired or incorrectly set up in most cases, which makes it less efficient.

Moreover, these kinds of systems cannot adapt to real-time goals, which could be optimization of minimal emissions, short traveling time, or equal throughput of buses and bikes. They act, in place, with obsolete policy targets like maximizing the quantity of vehicles of any sort with whatever purpose.

Urban Planning Misalignment

The other major criticism of the traditional traffic systems is their lack of correspondence with the present-day city planning objectives. Back and forward 20th century traffic planning and 21st century mobility is completely different in their concern and improvement in terms of throughput of the automobile visibility with its vicious circle of the traffic planning of the automobile usage, but the 21st century mobility is more sustainable, equitable and multimodal usage concentration. Transit signal priority (TSP), bicycle phase detection and pedestrian adaptive crossing are not natively supported by fixed-time systems.

Traditional tools of planning do not harmonize transportation with land use, and they do not capture the extent to which neighborhoods and socioeconomic classes differ regarding the patterns of mobility. It results in institutional injustice in which, in the framework of the optimized flows, central business districts enjoy short distances with the subsequent delays and inaccessible peripheral zones.

Conventional networks also promote automobile dependency where little attention is paid to the design of these active means of transportation. As an example, pedestrian wait time is not often in signal phases calculation what results in the insecurity of crossings and reduced walkability.

Legacy System Lock-In and Cost of Replacement

Traditional systems have inherent limitations but exist in many cities due to infrastructure lock-in removal and replacement of existing legacy signal controllers, inter-section re-cabling and reprogramming whole networks is costly in capital and labor. Deployment of adaptive systems is stifled by the lack of funding, government red tape and decentralized government in most cities.

Recent research by Arroyo-Arroyo and Frame (2021) pointed out that, despite savings being readily available, the process of coordinating city departments, utilities, and transport authorities is usually non-existent, slowing down the process of smart upgrade. In addition to this, legacy systems might not work with the new platforms unless they begin exorbitant retrofitting exercises to the transformation agenda.

Its price is not only monetary. The move to adaptive systems needs capable human resources, a high level of data infrastructure, and the need to reform regulations. Most cities fail to fulfill these requirements resulting in disparity in the quality of planning and modernization of the traffic corridors.

Global Examples of Legacy Friction

Other cities such as Cairo, Manila, and Lagos depict how hard it is to shift into new ways. Traffic is very high in these cities in which the infrastructure is fragmented and outdated. The fixed-time signals may have poor synchronization, and they will not respond to fluctuation in demand, and they tend to fail in case of power outages. This has resulted in cases whereby traffic is usually regulated using unofficial means by police officers or unsung members.

On the other hand, cities that have been able to modernize successfully like Tokyo and Stockholm managed to do so by adopting the capability of adaptation in the course of time. Such is the case in Tokyo, as a gradual shift was carried out to move away to a fixed time system, towards a semi-actuated one, and ultimately fully adaptive system, which had been aided by constant data gathered and multi-decade planning.

This shows modernization is not a thing of binaries transition but a phased change that must support previous realities, institutional preparedness and behavioral patterns of the locality.

Reframing the Problem

There has been an emerging agreement among researchers and practitioners that legacy systems not only happen to be obsolete, but also incompatible with the future of mobility. Intelligent transport needs to be adaptive, context sensitive and have the abilities of machine learning, none of which can be handled by rule based static logic.

Furthermore, introduction of connected and autonomous vehicles (CAVs), on-demand, and multi-modal mobility requires a layer of coordination that can synthesize different data flow into consistent and real-time decisions. Inefficiencies present in the traditional systems are no longer the only limitations because they are obstacles to innovation and sustainability.

The inability of the system to respond to the changing circumstances and complexities in traffic is alleged by many scholars to be the real problem of traffic management not the congestion. So, change cannot be simply about hardware or software, but it is about architecture of your system, and philosophy of planning.

METHODOLOGY

The methodology of this chapter preparation is based on the review of various literatures from global Scopus indexed sources such as Science direct, MDPI, etc. from the published articles 2020 onwards. Total of 122 relevant literatures were studied, out of which 33 which best fits with this study were cited for preparation of this chapter. This chapter presents a detailed literature study on the various methods, case studies and challenges of urban mobility optimization through adaptive routing and intelligent signal systems.

EMERGENCE OF ADAPTIVE ROUTING STRATEGIES

Introduction to Adaptive Routing

The cities used to be regulated by strict route assignments and fixed signal patterns. Nevertheless, such approaches fail to cover the cost-efficient transportation needs in an environment in which mobility requirement has become dynamic and volatile. With cities becoming highly dynamic and information-centric systems, the static routing techniques are being replaced by dynamic routing strategies in routing systems in which the routing infrastructure is updated in real-time in response to traffic conditions, user-feeds, sensors and so on.

Adaptive route provides real-time and decentralized decision-making, which will give freedom to change vehicular paths dynamically to reduce travel time, fuel consumption, and congestion. With mobile app data in hand, coupled with GPS data, traffic sensors, and crowd-sourced user data, adaptive systems devolve the controls on routing away from centralized control to autonomous networks operating according to algorithms. By doing so, they make the paradigm of transport in the cities contextually aware and self-optimal.

Enabling Technologies: GPS, IoT, and Crowdsourced Platforms

On the center of adaptive routing lie real time technologies: Global Positioning System (GPS), Internet of Things (IoT) sensors and user input driven data as illustrated by Figure 4 and is capable of integration. GPS-enabled smart phones help mobile navigation services such as Google Maps, Waze, and Apple Maps to keep track of vehicle speeds and traffic jam. Such apps collect the data of the users in real time to advise the best routes and the dynamic adjustments according to changing circumstances.

Figure 4. Adaptive routing technologies integration

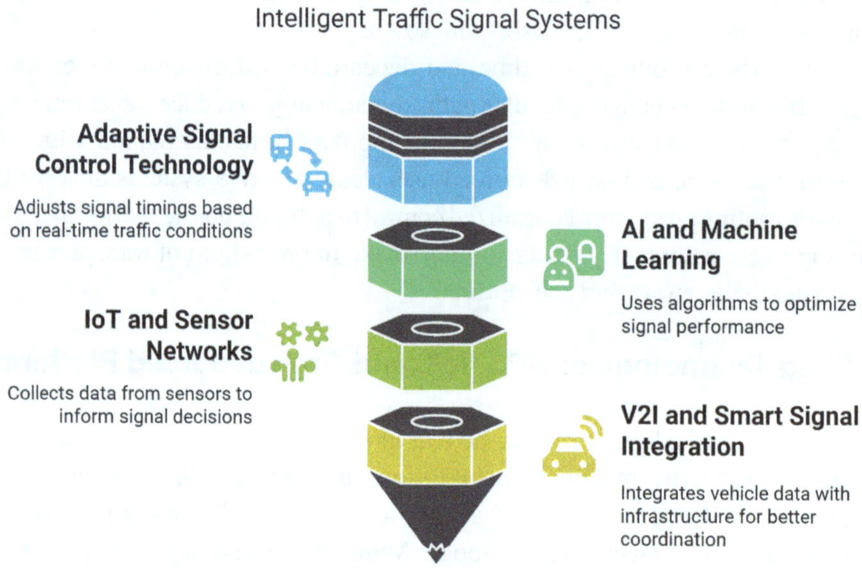

Relative to urban traffic congestion in crowded corridors, Karthika et al. (2025) asserted that with such GPS based routing systems, the throughput of the urban traffic congestion has increased tremendously. Their research revealed that those commuters who utilized the dynamic routing applications had lower commuting time than their counterpart's game to static routes especially during rush hours or road along verses.

Crowdsourcing improves the performance of these systems in accuracy and responsiveness. There are apps like Waze, which allow you to report accidents, road hazards, police sights, or other constructions. This information is then used as input into adaptive routing algorithms, which change directions not only of the reporting driver but thousands of other drivers in the area.

The combination of the three innovations results in a highly responsive routing environment that is significantly more resistant to temporary disruptions than older systems by combining forces of GPS, the Internet of Things, and crowdsourced intelligence.

Decentralized Routing and Vehicle-Based Decision Making

Decentralization of decision-making is one of the revolutionary changes in adaptive routing. Routing decisions in traditional models are located centrally to the traffic control centers. On the contrary, adaptive routing brings routing intelligence to vehicles and edge devices, which are semi-autonomous based on real-time input.

An improvement in edge computing makes this decentralization possible and helps vehicles and mobile devices to keep data locally without relying on a central server. Trigka and Dritsas (2025) illustrated how the routing architecture that will support faster decision-making, low latency, and fault-tolerant navigation may be achieved through edge-assisted routing architectures, particularly in high-mobility areas (highway layout or intersections).

Vehicles do not only consume data but also act as participants in the total traffic ecosystem in this paradigm. The movement, speed, and delay of each user help the system gain a real time view of traffic. Therefore, self-organization occurs in the decentralized adaptive routing systems so that external interventions are limited.

Case Study: Singapore's Intelligent Routing Ecosystem

Singapore is an urban mobility global best practice. The Intelligent Transport System (ITS) of the city combines the information of the GPS, the sensor networks, traffic cameras, and public transport connections to carry out the traffic measurements and manage traffic of the whole island. Expressway Monitoring and Advisory System (EMAS) and Parking guidance system operate concurrently with the GPS-enhanced routing applications to spread traffic as evenly as possible and keep out hot spots in the traffic.

In a study conducted by Kumar (2024), such adaptive systems enhanced average travel speed through the network of expressways by 12% and cut the amount of CO_2 emissions by 18% relative to the conventional approach. Moreover, Singapore developed MyTransport.SG app that enables citizens to obtain dynamic routing recommendations in real-time considering the changes in traffic and multimodal transit, which demonstrates that the cooperation between the governmental and user-based systems is effective.

There are three pillars underlining the success of the adaptive routing strategy in Singapore (1) stable data infrastructure, (2) the centralized access to transport data, and (3) the readiness of the population to use intelligent navigation software. It is an effective model that can be applied in other cities that want to shift focus in their traffic management of being reactive to proactive.

Challenges in Adaptive Routing Implementation

Although an adaptive routing has advantages, there are technical, ethical and infrastructural challenges that are posed by the adaptive routing. The question of reliability and redundancy of the data is one of the most important. There may be noise or bias in case of excessively relying on user-reported or mobile-sourced data. As an example, directing many users to a side street with the help of an app is likely to generate unintentional traffic jams or safety hazards in residential neighborhoods.

Another ethical issue is data privacy and surveillance, when user tracking at the real-time is used to make routing decisions. In a warning issued by Alam (2024), unless strong data governance regulations were put in place, intelligent routing systems could undermine personal privacy or introduce the unintended algorithmic discrimination of serving the needs of more affluent neighborhoods in route optimization to the detriment of others.

Another bottleneck is infrastructure. Adaptive routing is based on dense sensor networks and consistent mobile connectivity, both of which are not attained in most developing cities. In addition, the old road systems might not allow diversification of the routing such that the adaptive algorithms are weakened.

Lastly, there is human behavior and trust that is a crucial element. Users cannot be trusted to follow recommendable paths, especially when such directions seem to be illogical or in some other ways inconvenient. Trust in adaptive systems needs transparency, reliability and alignment to user expectations.

Future Prospects: Integration with Autonomous Vehicles and V2X

The emergence of the Connected and Autonomous Vehicles (CAVs) and the Vehicle-to-Everything (V2X) communication is directly related to the adaptive routing of the future. Such systems will be able to communicate in real-time with other cars, traffic side infrastructure, and traffic management systems to minimize routes like never before.

According to Singh et al. (2025), all of that will change once adaptive routing and autonomous navigation encounter. Traffic control is going to change. The cars will not only take decisions based on their determination but also on the network intelligence. This integration might lead to significant decrease in delays, accidents, and impact on the environment since traffic flows would be synchronized in the city-wide level.

Moreover, routing through blockchain is under investigation so that full transparency and trust can be introduced to data exchanges between vehicles and platforms to reassure about the possibility of data being edited or monopolized by tech providers.

INTELLIGENT TRAFFIC SIGNAL SYSTEMS

Introduction to Intelligent Traffic Signal Systems

The nervous system of urban mobility is an urban traffic signal system. These systems are traditionally constructed using static control logic however, this is increasingly being changed due to smart technologies to make them adaptive and intelligent platforms. Intelligent Traffic Signal Systems (ITS) is a technology that uses Artificial Intelligence (AI), Internet of Things (IoT), Vehicle-to-Infrastructure (V2I) communication, as illustrated in Figure 5 to set up dynamic traffic light timings to accommodate a real-time traffic flow, the number of vehicles on the road, pedestrian traffic, and emergency events.

In contrast to the traditions of fixed-cycle traffic signals, ITS platforms are responsive to the prevailing situation, improving the amount of traffic passing and reducing the number of idling periods and risks of accidents. With their move to Smart City systems, ITS technologies become the foundation in the endeavor to reach the sustainability, efficiency, and resilience levels of urban mobility.

Figure 5. Intelligent traffic signal system

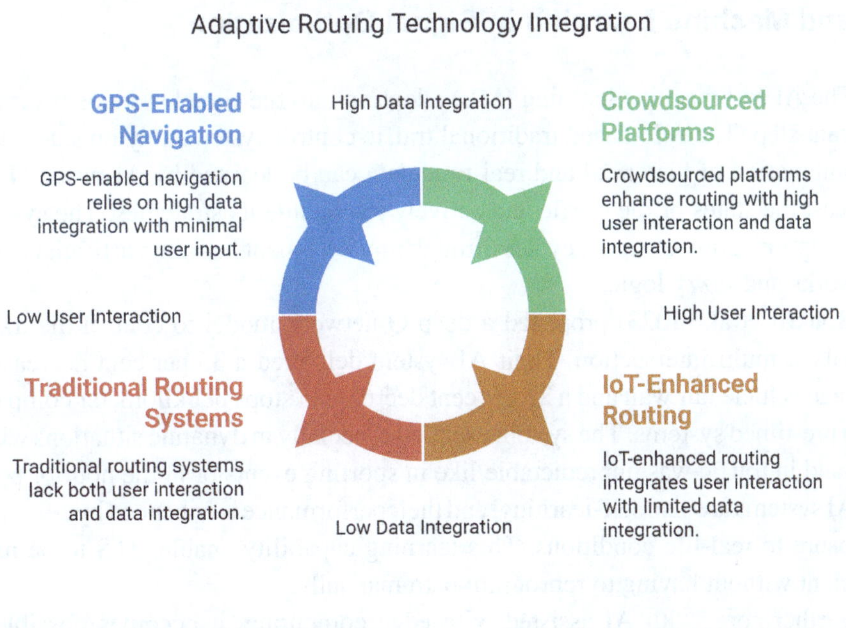

Adaptive Signal Control Technology (ASCT)

Adaptive Signal Control Technology (ASCT) at the center of modern ITS implementation is the alteration of the signal timing and timing of the signal that is according to the sensations of the traffic to the intersections. The traditional systems are coded in a way that depends on the past traffic and historic data is unchanging and possibly useless. Conversely, ASCT will gather real-time measurements made by sensors and will operate with them via algorithms and optimize signals according to the circumstances.

The ASCT systems are found to minimize the delay of vehicles at the intersection by 20-30% even in busy conditions (Van Kooten et al., 2017). Their investigation of the implementation of ASCT deployment in the Netherlands showed that contiguous signal plans deployed with 510-second update cycles showed significant gains in intersection throughput and travel-time reliability.

Notably, ASCT can be implemented as locally isolated intersections to city-wide coordinated corridors. Although signals operate independently, it can be adopted to meet the larger urban traffic policies. ASCT, in combination with incident detection systems, can also re-priorities, e.g. by establishing green corridors (allowing traffic under green) on priority routes in case of emergency or broadcasting to drivers via a navigation unit to pull over.

AI and Machine Learning in Signal Optimization

The AI and machine learning (ML) adoption into the signal systems is the significant step. The rule-based traditional traffic control systems rely on static data; a combination of historical and real-time data can be learned by AI-based ITS to forecast the states of the traffic and actively reschedule its strategies. The systems apply optimization of signal cycles through reinforcement learning, artificial neural networks and fuzzy logic.

Bouktif et al. (2023) proposed a deep Q-network model to control the traffic signals at multi-intersection. Their AI system delivered a 35 per cent decrease in average vehicle fair wait and a 22 per cent decrease in stops per kilometer compared with pre-timed systems. The system excelled especially in dynamic situations where demand in traffic was unpredictable like in sporting events or in the hour of peak.

AI systems are also self-learning, and their performance is enhanced by continued exposure to real-life conditions. This learning capability enables ITS to be more efficient without having to reprogram path manually.

Furthermore, with AI assisted with edge computing, it becomes possible to achieve distributed signal-level intelligence, where every intersection will be able to take real time, local decisions, but contribute to an overall network strategy.

Internet of Things (IoT) and Sensor Networks

IoT technology delivers physical sensing and communication infrastructures of ITS. Smart traffic lights have links with inductive loop detectors, video cameras, Bluetooth/Wi-Fi sensors and infrared pedestrian detectors. Such devices gather localized traffic information vehicle speed, length of queue, presence of pedestrians and relay them to the traffic management systems.

Lei et al. (2007) finds that the use of IoT-based sensors in city corridors enhances the throughput of the vehicles in combination with adaptive traffic signal platforms. In a study conducted in Shanghai they demonstrated that intersections with multi-sensor nodes backed with vehicle classifiers worked better than the legacy system in controlling congestion particularly in emergency peaks.

To supplement, IoT can do long-distance monitoring and servicing of signal devices. Using traffic agencies will allow the status of equipment to malfunction in real time, ensuring that downtimes are as minimal as possible to enhance service reliability. IoT systems can go so far as to adjust signals to weather data that slows down cycles when it is raining heavily or lengthens pedestrian phases in a heat wave on the concept of climate resilience.

It can also fuse data across transport modes, i.e. bicycles, buses, and pedestrians, so that signal systems meet not only multimodal needs, but they can give priority to vulnerable road users, too.

Vehicle-to-Infrastructure (V2I) and Smart Signal Integration

The future applications in ITS will be the Vehicle-to-Infrastructure (V2I) communication where the cars share information with the traffic signals, other cars on the road and centralized systems. V2I enabled vehicles can predict the change of signals through the Dedicated short-range communication (DSRC) or 5G cellular networks, optimize their speed profile and coordinate with other traffic to facilitate traffic flow.

Jiang et al. (2022) proved a V2I-enhanced signal control framework for Beijing that led to a 25 percent decrease in the fuel consumption of vehicles and an 18 percent decrease in the emission of CO 2. Vehicles in the testbed controlled the speed to reduce stops and starts giving the high eco-driving behavior and delay in the intersection.

These systems enhance safety as well. Put another way, when a pedestrian is sensed crossing later in a phase, V2I-connected signals can slacken the phase or signal within reach vehicles to adjust or stop beforehand. This mutual traffic environment is transformative regarding imputed vehicle intelligence to cooperative trafficking units.

Case Study: London's SCOOT System

The city of London has been an early user of ITS with its SCOOT (Split Cycle Offset Optimization Technique) system, which is a real time adaptive strategy of managing traffic signals. SCOOT controls the signal timings at coordinated corridors using sensors integrated into roads to monitor vehicle flows. The system is incorporated in the wider approach of Transport for London to manage its traffic, such as priorities and buses, and incident management.

As stated by Singh et al. (2025), SCOOT system resulted in an average decrease in vehicle delay of 12 percent in essential arterial routes and increased journey time certainty of the public transport services. The case explains how traditionally built cities can successfully modernize their signal apparatus via gradual installation of cheap systems.

GIS AND SPATIAL ANALYTICS IN TRAFFIC OPTIMIZATION

Introduction to GIS in Urban Mobility

The tools in urban transportation planning were made pivotal in traffic control and planning by the Geographic Information Systems (GIS). The systems help the cities break out of stale maps and move into dynamic and spatially aware decision-making models. Integration of geospatial information, real time traffic and predictive analytics gives GIS based platforms an edge in their ability to give a holistic view of the mobility patterns within an urban environment which could be used by planners to determine bottlenecks, simulation of traffic conditions, and optimization of the transport systems.

Smart city GIS was developed during a time when visualizing and decision support systems were needed in the transportation sector (road management) throughout the process of fitting capacity to demand, predicting congestion, and modeling the effects of new policies or alterations in infrastructure. The capability of observing traffic as a geographic, fluid process controlled by the land use, time of the day, and weather conditions, as well as the socio-economic factors: this is a special opportunity that is offered by spatial analytics.

Core Functions of GIS in Traffic Analysis

GIS platforms provide large extent of functions dedicated to the field of traffic management as illustrated in Figure 6 such as:

- **Congestion Mapping:** GIS looks over the locations of congestion using the information *provided by loop detectors, mobile applications, and road camera sources.*
- **Network Analysis:** GIS facilitates the shortest path routing of travel, service area mapping and accessibility modelling to various user populations (e.g. public transport travelers, pedestrians).
- **Temporal-Spatial Analysis:** The ways people behave with traffic varies depending on time; GIS helps record such changes so that planners can vary their signal timing or route layout at different points in time.
- **Predictive Modeling:** GIS platforms embed machine learning models that will forecast the future level of congestion based on historical and real congestion levels.

Figure 6. Core functions of GIS in traffic analysis

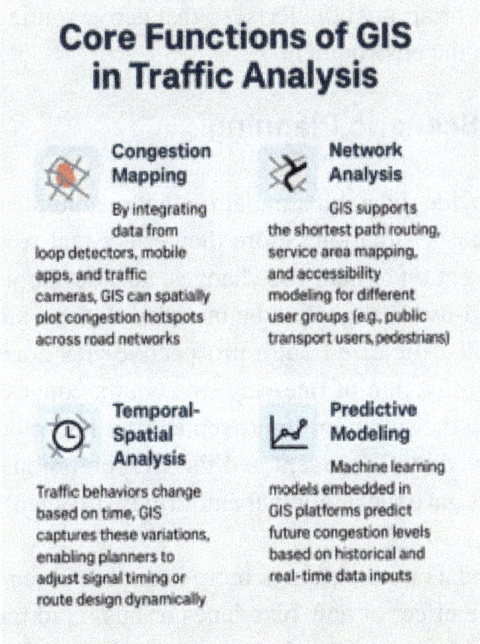

Salazar-Carrillo et al. (2021) observe that spatial modeling tools can help cities take their traffic management efforts into a proactive planning phase wherein traffic congestion can be predicted and thus preemptively addressed prior to its occurrence using GIS environments.

GIS-Based Predictive Traffic Modeling

Predictive modeling is one of the most influential tasks of GIS related to traffic optimization. This procedure presupposes the performance of historical traffic data and predicting the conditions under different settings, including population growth, road closure, or intervention policies. With predictive GIS tools, it is possible to forecast the dates of traveling, course of actions, and mode change, and modifications of infrastructure and traffic controls can be done beforehand.

Shi et al. (2008) created a neural network model that involves GIS technology to predict the morning peak congestion in Shanghai. They found that their findings showed 92 percent accurate predictions, which gave the traffic controllers the ability to lengthen signal cycles and release route advisories prior to actual deteriorations. The promptness of such long-range vision is priceless in fast developing urban settings where not every situation allows infrastructure to keep up with the growth.

Predictive GIS modeling is also useful in climate-sensitive traffic management, where the modelling simulates the traffic movement in advance in the case of extreme weather events or air quality alerts, so that active routing or scheduling may reduce the impact on the environment.

Simulation and Scenario Planning

GIS tools can provide dynamic simulation that enable planners to perform the simulation of the what-if situation before they effect real word changes. Various types of policy intervention or land-use changes, new developments, infrastructure improvements or land-use changes can be included in these simulations.

Abuh and John (2023) carried out a prospective GIS analysis to simulate the effects which the introduction of one-way streets in a congested business area in Abuja, Nigeria has on the situation. Improvements to their model were the average vehicle speed improved by 23 percent and the CO_2 emissions had reduced by 19 percent. These results gave the municipal authorities sufficient reasons to go ahead with the intervention.

Even the multimodal coordination is more and more being tested in simulation scenarios, such as the effect of new bike lanes or bus rapid transit (BRT) integrations, or pedestrianizing streets. GIS platforms combine data sets of various agencies to facilitate truly comprehensive planning and operational designs that cut across modal silos.

Spatial Decision Support Systems (SDSS)

It is common to find modern GIS systems being nested within Spatial Decision Support Systems (SDSS) interactive systems that are used by planners and policy-makers to simulate urban mobility policies based on spatial reasoning. These systems integrate transportation models, GIS data layers and socio-economic indicators in making investment decisions and regulatory planning.

One of the major characteristics of SDSS is that it can weigh out several factors, e.g., travel time, pollution exposure, and accident risk to produce best judgments made in conflict goals. An example could be that a policy which aims at decongesting an area could be reconsidered through the prism of equity whereby the policy is applied to low-income areas more so than areas in other income brackets.

Another benefit is that SDSS tools enable real-time stakeholder interaction, since graphical dashboards and map-based interfaces can be used to enhance the availability of complicated information to the audience, during community workshops and consultations.

Real-World Case Study: Stockholm

Stockholm has become one of the main examples of GIS-based traffic optimization. The city has incorporated the data collected by GPS-enabled taxis; vehicles of mass transit means as well as the permanently installed sensors into a central GIS system that help to monitor and model the traffic patterns in real-time. With predictive analytics, the city is also able to simulate traffic conditions of the 15-minutes intervals 3 hours in the future.

Franklin and Karlstrom (2009) attribute the fact that Paris has reduced mean trip travel delay by 18 percent to its predictive traffic modeling system that has made it possible to modify signal settings in advance (control roadworks, dummy marathons etc.).

Along with this collaboration on environmental information the platform also includes GIS (GIS) to impose restricted vehicle travel zones on a zonal basis during smog alert. Such dynamic zoning, which can be facilitated only with GIS-based systems, will enable cities to coordinate traffic control in line with public health and sustainability goals.

Challenges in GIS Integration

Even though the GIS holds the promise of major changes in traffic optimization, there are some issues of integration that cities have:

- **Data Silos:** The transportation, land use, healthcare and emergency management are typically in isolated databases that complicate intermodal GIS modeling.
- **Technical Skills Deficit:** More advanced GIS modeling only comes through expert training, and this might not be present in many municipal agencies.
- **Data Quality & Coverage:** Base maps in certain areas are old or incomplete and traffic data sources (e g sensors or mobile traces) might be non-existent.
- **Privacy Issues:** GIS systems have become more adept at acting with anonymized location information through mobile apps, which causes worries regarding surveillance and information misuse.
- **Licensing and Infrastructure Cost:** The high-end GIS applications and data servers are a high-cost refurbishment and maintenance burden.

In response, many cities are turning to open-source GIS platforms (e.g. QGIS) and building data collaboratives which share information across agencies and sectors.

Future Directions: GIS in AI-Enabled Smart Cities

Hyper-personalized traffic optimization is on its way to moving into the forefront through the integration of GIS with AI, cloud computing, and big data analytics. Individual behavior can now be modeled in systems, short-term incidences of congestion can be predicted, and the traffic management can also be aligned to take a larger perspective that could direct other aspects of urban mobility strategy such as climate resilience, walkability, and safety.

Also, 3D GIS systems and digital twin systems are developing as the next generation of planning tools, allowing virtual simulations of cities to be developed in which planners can perform simulations of reconfigurations of roads, changing signal timings, or even the response in the event of a disaster.

Such innovations represent a transition in GIS which is used as a support tool to GIS that would serve as a strategic control facility of mobility planning that is integrated on all levels of governance and is ready to address both immediate and future-ready requirements.

URBAN CHALLENGES AND EQUITY CONSIDERATIONS

Introduction: Centering Equity and Resilience in Smart Urban Mobility

The introduction of intelligent signal and adaptive routing is one of the revolutionary feats in traffic management. Nonetheless, these advances create an urgent issue of spatial equity and infrastructure inequalities in addition to cybersecurity strength. On the one hand, smart mobility technologies may lead to increased efficiency; on the other hand, they will entrench current inequalities unless they are regulated within the framework of inclusive, data and context-sensitive policies. The section critically analyses the urban conflicting issues that also come with intelligent mobility taking into consideration equitable access, infrastructural preparedness and the vulnerability of digital mobility infrastructures.

Socio-Spatial Inequity in Smart Mobility Deployment

The smart mobility systems are based on the conditions of the sensor-enriched, digitally connected environment that is lacking in disadvantaged communities. When adaptive systems focus on high-data zones, deprived neighbors that lack connectivity or investment in infrastructure may be left out of the benefits of decongestion of traffic, time-saving or multimodal cross-connectivity.

According to Dodel and Hernandez (2024), adaptive traffic systems in various European cities did not benefit middle- and low-income households and commuters who relied on public transport systems, as their basic level of service did not improve at all due to the adaptive traffic system implementations. This aspect is known as mobility-based digital redlining is the result of prejudiced algorithmic development and asymmetrical infrastructural investment.

Singh and Saraswat (2019) in an Indian context placed the significance of the level of service (LOS) forecasting in semi-urban cities. Their research proved the sharp differences in traffic capacity and traffic volume between the urban cores and the fringe areas, and this aspect should be considered in designing the signal systems such that traffic corridors with a high density of vehicles should not be favored by the design compared to lower-density, mixed-used areas.

Infrastructure Disparities and Implementation Challenges

Successful dense sensor networks, communications infrastructure, and high-fidelity computing will be needed to implement effective adaptive traffic control. A significant number of cities (mostly Global South ones) do not have the basic infrastructure to enable real-time sharing of data and cloud-connected optimization of signals.

In a study of more than four cities that was conducted by Abdelati (2024), part of the municipalities that took the survey reported infrastructure deficit as a major obstacle to the implementation of intelligent signal system. The urban peripheries and informal settlements were unearthed as having no fiber-optic service, edge devices, and calibrated sensor inputs, which makes the adaptive traffic control an impractical alternative or a useless service in such locations.

In addition, the study observed that off-the-shelf solutions that remain in their offshore form frequently become the default type of public procurement where low-capacity regions, instead of lowering hierarchies in the cities, reaffirm them.

Cybersecurity Vulnerabilities in Data-Driven Traffic Systems

With cloud computing, edge processing, and sharing real-time data becoming more and more essential components of smart mobility platforms, they will become a significant target of cyberattacks. The attack surface involves traffic sensors, V2I interfaces as well as traffic management central servers.

Aluko (2024) says that intelligent traffic systems have multiple vulnerabilities to false data injection, denial-of-service (DoS) attack, and man-in-the-middle. vehicle-to-infrastructure communication breaches. Such attacks may impact the signal coordination, bring hysteric routing and even cause accidents in key intersections.

Particular attention should be given to those who do not use encrypted third-party data feeds, including crowdsourcing navigation input by third-party apps. Hijacking the feeds may confuse adaptive algorithms, which make inefficient or hazardous control decisions. Aluko (2024) postulates that cyber-resiliency procedures such as encrypted V2X communications, anomaly detection and multi-level data validation should be brought on board to prevent real-time manipulation.

Governance and Policy Gaps

Although there is a significant challenge of technical limitations, governance gaps tend to define the success of smart mobility in fair delivery or failure. The implementation of adaptive systems without the structure of observing their effects on equality, ecological fairness or quality of services has been used in many jurisdictions.

Golub et al. (2019) pointed out in their review of smart mobility pilot programs in Portland that partnerships between the government and the private sector usually did not have equity measurements, making them effectively and insufficiently cover the areas of low-income households. The authors emphasized the need of binding contracts that commit the technology vendors to achieve inclusive deployment, data sharing, and participatory assessment.

Moreover, its bi-partisan control over transport, ICT, and public works can lead to silo environments, making cybersecurity responses less effective as well as creating a lack of citizen trust.

Strategic Recommendations for Inclusive Implementation

Social equity needs to be built into every stratum of intelligent traffic system design to balance the presence of technological innovation in the city. There is a strategic pathway:

- **Equity-Integrated System Design:** GIS-based demand modeling and service level analysis should be implemented to provide adaptive signal systems that respond fairly to all users of the road as well as to neighbors.
- **Open Data and Transparency Requirements:** An adaptive signal vendor must post renditions of anonymized performance data disaggregated by geography, socioeconomic profile, and travel mode.
- **Cybersecurity by Design:** Crystallize secure communication standards, anomaly detection, and live security inspections into all the V2X and ITS systems.
- **Community Co-Design:** Incorporate the historically excluded communities into the design, calibration and testing of adaptive traffic systems to support relevance and trust.
- **Inclusive Procurement and Regulation:** Social and space equity benefits should be incorporated in the terms of tender issued by government and the tenders should also be regulated.

FUTURE DIRECTIONS AND POLICY IMPLICATIONS

Framing Smart Mobility Through Theoretical Lenses

The stepped-up introduction to adaptive routing and intelligent traffic system symbolizes an intellectual progression in addition to a paradigm shift in city attitude toward the governing of mobility. These systems operate in socio-technical ecosystems where elements of technology, institutional structure, urban infrastructure and human behavior interact. The social-technical systems theory puts the co-evolution of these areas into the spotlight and states that technological development on its own cannot deliver sustainable results without adjacency to organizational, political, and societal forms (Kopackova & Libalova, 2017).

In this regard, smart mobility governance should be embedded with system thinking, equity and participatory ethics. With the current vitality about the urban mobility of tomorrow, it is not an issue of solitary gains in technology as technology is not a hierarchy by itself, rather, it is a unanimous shift of many layers of governance and stakeholder entails.

Multi-Level Governance for Smart Mobility Integration

Smart mobility technologies are V2X, adaptive signaling, and predictive route systems, and they take place at various governance levels, local, metropolitan, national, and international. A theory that helps comprehend how these cross-cutting powers should collaborate to have an effective execution and control is multi-level governance theory (Bache, 2012).

There is a need to have policy coherence in transport, telecommunications, environment and digital governance. As an example, adaptive routing systems might be implemented by municipal organizations whose data governance may be subject to the national cybersecurity legislation. Cities should henceforth devise inter-agency coordination platforms such as cross sector mobility task forces and coordinated urban ICT plans.

Systems Thinking and Adaptive Policymaking

Systems thinking Policy makers need to adopt the concept of system thinking where urban mobility becomes an aspect of a bigger system that changes with time. Based on the arguments of Weberg et al. (2025), the smart mobility policy must be iterative and cope through periods of experimentation, stabilization, disruption, and reorganization.

This is more of adaptive policy making that regulatory policies can react in real-time information, travel trends, or other issues that may arise on the minds of citizens. As an example, regulatory sandboxes can enable cities to experiment with mobility technologies and apply them to the entire city only after the testing.

Innovation Diffusion and Scaling

The implementation of smart mobility is in the path suggested by the Diffusion of Innovations theory as proposed by Rogers (as explicated by Bokolo in 2023). Technology is usually tested in early adopting innovation-friendly districts or technologically metropolitan municipalities. Their popularity, noticeability, and policy infrastructures affect most people, creating the drive towards wider institutional usage.

Policies to aid the diffusion of innovation are:

- Performance-based funding as a reward of measurable outcomes.
- Impose open data policies to promote ecosystem collaborations.
- Inter-city learning knowledge transfer platforms.

These allow one to move between local trials and region wide or national deployment both ensuring a scalability and contextual adaptability.

Ethical Governance, Trust, and Public Participation

Algorithms and risks associated with bias, surveillance, and digital exclusion become opportunities where the transport sector has been experiencing escalating datafication. Ethical governance (Floridi, 2018) in digital societies defines three imperatives, i.e., transparency, inclusivity, and accountability.

To be in line with this, the cities ought to:

- Require the route and signal optimization algorithm transparency.
- Conduct equity assessment of the performance of the ITS by race and neighborhood.
- Enable participatory governance across the design, deployment, and evaluation activities by the citizen.

It is very important to establish trust among people. Resistance or unfair outcomes can be met in systems that are even technically heightened without transparency and participation.

Financial and Institutional Readiness

Solid and effective financing models are needed in smart mobility transitions. Deployments may be based on public-private partnerships (PPPs), mobility taxes and using digital infrastructure bonds. Nevertheless, social equity and performance should be some of the conditions of financing so as not to perpetuate inequalities.

The institutions should also be ready in terms of technical expertise, clarity of the regulation and inter-departed processes. Long-term policy resilience entails capacity-creating functions and institutional innovation programs.

Strategic Policy Toolbox for Smart Mobility Futures

Drawing from the theoretical perspectives above, cities can operate smart mobility governance through the following integrated toolbox as presented in Table 1:

Table 1. Strategic policy for the future of smart mobility

Policy Lever	Theoretical Basis	Implementation Strategy
Interoperability & Standards	Multi-Level Governance	Enforce open standards for ITS data and communication
Institutional Coordination	Socio-Technical Systems	Cross-sectoral planning and ICT-mobility integration
Adaptive Regulation	Systems Thinking	Regulatory sandboxes, pilot programs with feedback
Innovation Scaling	Diffusion of Innovations	Knowledge-sharing networks, performance-based scaling
Public Trust & Ethics	Ethical Governance	Transparency mandates, participatory co-design
Resilient Financing	Institutional Economics	PPPs with equity clauses, impact-linked funding

CONCLUSION AND STRATEGIC RECOMMENDATIONS

Conclusion

In the following strongly interdisciplinary chapter, the urban mobility of the past (static traffic systems) is contrasted with the current developments (adaptive routing, intelligent signals as well as geo-spatial analytics). We integrated socio-technical theory, systems thinking, innovation diffusion and ethical mode of governance with the emphasis that anything meaningful requires integrated governance of policy, social equity, and robust institutional structures to bring change in comparison with mere engineering.

Patronized technologies such as IoT, edge computing, V2I and AI are turning into the foundations of urban mobility, which it is possible to achieve with the adaptive, inclusive, secure, and public-values-based governance of technology.

Strategic Recommendations

To guide cities in navigating this complex transformation, the following multi-dimensional strategic roadmap is proposed:

Integrated Systems Governance

- Build Mobility Planning Platforms: Coordination points to plan, implement and evaluate smart mobility facilities are needed (transport / ICT / planning bodies)
- Iterative Adaptive Policies: adopt Iterative Adaptive Policies: Implement regulatory sandboxes and horizon scanning to facilitate intelligent, step wise ramping of innovations.

Technological and Data Infrastructure

- Standardized Open Mobility Frameworks:
- Back Pilot-to-Scale Pathways: New technologies must be advanced into regime-level use through performance-based funding beyond the niche experimentality.

Equity and Public Value

- Commission Equity Impact Assessments: employ the utilization of GIS and SDSS to design and invent adaptive solutions based on underserved areas.
- Enact Algorithmic Transparency: make certain vendors disclose optimization and routing standards, making them answerable and guaranteeing credibility.

Cyber Resilience and Trust

- Security-by-Design: Develop a common approach to encrypted communications, threat awareness and forensic audit across all applications of ITS.
- Increase Transparency to the Public: Post the results of their security performance and incident management plans through civic dashboards.

Funding and Sustainability

- Contract Outcome-based PPPs: Design contracts, so that they incorporate social and environmental performance KPIs, along with technical deployment.
- Charge High captured values: Generate over time development to pay the infrastructure of smart mobility to guarantee the interest of the generation.

Future Research and Policy Agenda

Intelligent Transport Systems (ITS) is in a state of rapid growth; the estimated annual growth rate is 25%. This development also reflects the transformative power of ITS, as well as the existence of open research gaps in some areas, like ethics, digital equity, and resilience, which still need to be studied by scholars and incorporated into the policies.

Furthermore, the system architects ought to think of embracing the solid system engineering frameworks, including the morphological matrix, as exhibited in roadmaps of adaptive signal control. The information received during bibliometric and network studies of the research devoted to ITS, as it can be observed in such works as Heliyon (2021) and those which provide details on the idea of achieve the readiness of the transport systems on the future based on the surveys, as can be found with the help of several journals, bring useful points of view on significant tendencies and points that could be discussed in further studies.

REFERENCES

Abdelati, M. H. (2024). Smart Traffic Management for Sustainable development in Cities: Enhancing safety and efficiency. International Journal of Advanced Engineering and Business Science, 5(1), 0. DOI: 10.21608/ijaebs.2024.242361.1088

Abuh, P. O., & John, J. A. (2023). Geographic analysis of traffic congestion in FCT-Abuja, Nigeria. *Journal of Sustainability and Environmental Management*, *2*(1), 26–32. DOI: 10.3126/josem.v2i1.53114

Alam, T. (2024). Data privacy and security in autonomous connected vehicles in smart city environment. *Big Data and Cognitive Computing*, *8*(9), 95. DOI: 10.3390/bdcc8090095

Ali, M., Devi, G. L., & Neelapu, R. (2020). Intelligent Traffic Signal control System using machine learning techniques. In Lecture notes in electrical engineering (pp. 611–619). DOI: 10.1007/978-981-15-3828-5_63

Aluko, N. S. E.Sunday Ebenezer Aluko. (2024). Cybersecurity and defense in intelligent transportation systems. *World Journal of Advanced Engineering Technology and Sciences*, *13*(1), 871–879. DOI: 10.30574/wjaets.2024.13.1.0469

Arroyo-Arroyo, F., & Frame, G. (2021). Traffic Management in African cities: The way forward. In *SSATP*. SSATP.

Bache, I. (2012). Multi-Level governance in the European Union. In *Oxford University Press eBooks*. DOI: 10.1093/oxfordhb/9780199560530.013.0044

Bokolo, A. J. (2023). Examining the adoption of Sustainable eMobility-Sharing in Smart Communities: Diffusion of Innovation Theory Perspective. *Smart Cities*, *6*(4), 2057–2080. DOI: 10.3390/smartcities6040095

Bouktif, S., Cheniki, A., Ouni, A., & El-Sayed, H. (2023). Deep reinforcement learning for traffic signal control with consistent state and reward design approach. *Knowledge-Based Systems*, *267*, 110440. DOI: 10.1016/j.knosys.2023.110440

Dodel, M., & Hernandez, D. (2024). Smarter but more unequal transport? How socioeconomic and digital inequalities hinder adoption of mobility apps in the Global South. *Travel Behaviour & Society*, *38*, 100911. DOI: 10.1016/j.tbs.2024.100911

Duranton, G., & Turner, M. A. (2011). The fundamental law of road congestion: Evidence from US cities. *The American Economic Review*, *101*(6), 2616–2652. DOI: 10.1257/aer.101.6.2616

Floridi, L. (2018). Soft ethics and the governance of the digital. *Philosophy & Technology, 31*(1), 1–8. DOI: 10.1007/s13347-018-0303-9 PMID: 30322997

Franklin, J. P., & Karlstrom, A. (2009). Travel time reliability for Stockholm Roadways. *Transportation Research Record: Journal of the Transportation Research Board, 2134*(1), 106–113. DOI: 10.3141/2134-13

Genders, W., & Razavi, S. (2016). Using a deep reinforcement learning agent for traffic signal control. *arXiv preprint arXiv:1611.01142.* /arXiv.1611.01142DOI: 10.48550

Golub, A., Satterfield, V., Serritella, M., Singh, J., & Phillips, S. (2019). Assessing the barriers to equity in smart mobility systems: A case study of Portland, Oregon. *Case Studies on Transport Policy*, 7(4), 689–697. DOI: 10.1016/j.cstp.2019.10.002

Heliyon. (2021). https://www.cell.com/heliyon/issue?pii=S2405-8440(25)X0004-9

Jiang, Z., Yu, D., Luan, S., Zhou, H., & Meng, F. (2022). Integrating traffic signal optimization with vehicle microscopic control to reduce energy consumption in a connected and automated vehicles environment. *Journal of Cleaner Production, 371*, 133694. DOI: 10.1016/j.jclepro.2022.133694

Karthika, I., & Pushpalatha, D., D, I. H., S, D., K, M., & S, H. (2025). Real time adaptive routing for efficient urban mobility. In 2025 International Conference on Visual Analytics and Data Visualization (ICVADV) (pp. 1052–1058). IEEE Xplore. DOI: 10.1109/ICVADV63329.2025.10961486

Köhler, E., & Strehler, M. (2012). Combining static and dynamic models for traffic signal optimization inherent load-dependent travel times in a cyclically time-expanded network model. *Procedia: Social and Behavioral Sciences, 54*, 1125–1134. DOI: 10.1016/j.sbspro.2012.09.827

Kopackova, H., & Libalova, P. (2017). Smart city concept as socio-technical system. *2017 International Conference on Information and Digital Technologies (IDT)*, 198–205. DOI: 10.1109/DT.2017.8024297

Kumar, S. (2024, December 18). Singapore Smart Traffic System is Redefining Urban Mobility. IntelliStride. Retrieved June 23, 2025, from https://www.intellistride.com/blog/how-singapores-smart-traffic-system-is-redefining-urban-mobility/

Lei, W., Qi, K., Hui, X., & Qidi, W. (2007). Traffic Intelligent Optimization and Local Traffic-Flow Control Inside Shanghai-EXPO-Area. In *2007 IEEE International Conference on Networking, Sensing and Control* (pp. 856–861). IEEE Institute of Electrical and Electronics Engineers. DOI: 10.1109/ICNSC.2007.372893

Salazar-Carrillo, J., Torres-Ruiz, M., Davis, C. A.Jr, Quintero, R., Moreno-Ibarra, M., & Guzmán, G. (2021). Traffic Congestion Analysis Based on a Web-GIS and Data Mining of Traffic Events from Twitter. *Sensors (Basel)*, *21*(9), 2964. DOI: 10.3390/s21092964 PMID: 33922627

Scandella, M., Ghosh, A., Bin, M., & Parisini, T. (2022). Traffic-light control in urban environment exploiting drivers' reaction to the expected red lights duration. *Transportation Research Part C, Emerging Technologies*, *145*, 103910. DOI: 10.1016/j.trc.2022.103910

Sharma, N., & Sahu, S. (2018). Fuzzy Logic based Traffic signal control. *Indian Journal of Science and Technology*, *11*(23), 1–5. DOI: 10.17485/ijst/2018/v11i23/114380

Shi, W., Kong, Q., & Liu, Y. (2008). A GPS/GIS Integrated System for Urban Traffic Flow Analysis. In 2008 11th International IEEE Conference on Intelligent Transportation Systems (pp. 844–849). IEEE Institute of Electrical and Electronics Engineers. DOI: 10.1109/ITSC.2008.4732569

Singh, S., & Saraswat, A. (2019). Design service volume, capacity, level of service calculation and forecasting for a semi-urban city. *Revue D'Intelligence Artificielle*, *33*(2), 139–143. DOI: 10.18280/ria.330209

Singh, S. K. Bader Eddin Alasali, Khuloud Nasser Al Balushi. (2025). Smart Transportation and Autonomous System Integration. In *Smart Cyber Physical Systems: Innovations and Real-World Implications* (Chapter 7). CRC Press. https://www.routledge.com/Smart-Cyber-Physical-Systems-Innovations-and-Real-World-Implications/Poonia-Upreti-Khan/p/book/9781032892931

Singh, S. K., & Jain, R. (2025). *Future of Mobility: Potential Challenges and Ethical Considerations*. In *AIoT for Smart Transportation: Transforming the Future of Mobility* (Chapter 12). IET Publishers. DOI: 10.1049/PBTR047E_ch12

Trigka, M., & Dritsas, E. (2025). Edge and cloud computing in smart cities. *Future Internet*, *17*(3), 118. DOI: 10.3390/fi17030118

Van Kooten, R., Imhof, P., Brummelhuis, K., Van Pampus, M., Jamshidnejad, A., & De Schutter, B. (2017). ART-UTC: An adaptive real-time urban traffic control strategy. 2022 IEEE 25th International Conference on Intelligent Transportation Systems (ITSC), 1–6. DOI: 10.1109/ITSC.2017.8317813

Weberg, O., Lund, J., Fors, V., & Resmini, A. (2025). Socially resilient mobility planning: Main challenges and design implications. *Transportation Research Interdisciplinary Perspectives*, *29*, 101334. DOI: 10.1016/j.trip.2025.101334

Yang, Z., Peng, J., Wu, L., Ma, C., Zou, C., Wei, N., Zhang, Y., Liu, Y., Andre, M., Li, D., & Mao, H. (2020). Speed-guided intelligent transportation system helps achieve low-carbon and green traffic: Evidence from real-world measurements. *Journal of Cleaner Production*, *268*, 122230. DOI: 10.1016/j.jclepro.2020.122230

KEY TERMS AND DEFINITIONS

Adaptive Traffic Signal Control (ATSC): A dynamic method to control traffic that will make the signal timing dynamic to the current traffic conditions.

Congestion Mitigation Strategies: Traffic relieving methods with the help of demand control and intelligent controls.

Dynamic Route Optimization: The action of re-calculation of optimal travel routes based on live traffic information and prediction algorithms.

Machine Learning in Traffic Management: Use of AI algorithms to forecast congestion patterns and ways to optimize the traffic flow.

Realtime Traffic Data Analytics: The application of sensors, cameras, and integrated devices to gather and work on live traffic data to decide.

Smart Mobility Ecosystem: A city transport system that merges public transport, shared transport and digital infrastructure.

Smart Transportation Systems (ITS): Combined use of advanced technologies aimed at better safety, efficiency and sustainability in transportation.

Traffic Flow Optimization: Methods to achieve roadway capacity and minimize delays with synchronized signal timing and routing.

Urban Transport Resilience: How urban mobility systems react and restore their functionality to disruptions like accidental incidents or peak traffic.

Vehicle to Infrastructure (V2I) Communication: A communication system that will allow cars to communicate with road signs and road systems.

Chapter 8
Reconfiguring Urban Movement:
Evaluating Existing Routing Systems and Traffic Signals

M. Balasubramani
http://orcid.org/0009-0005-3706-0140
V.S.B. Engineering College, Karur, India

V. Anitha
http://orcid.org/0000-0002-5818-3861
Panimalar Engineering College, Chennai, India

K. P. Sriram
St. Joseph's College of Engineering, India

R. Balamurugan
New Prince Shri Bhavani College of Engineering and Technology, India

J. Tharun
http://orcid.org/0009-0004-8433-9315
Anna University, Chennai, India

Joel M. Robinson
http://orcid.org/0000-0002-3030-8431
KCG College of Technology, Karapakkam, India

ABSTRACT

This chapter per the authors examines the evolution of urban traffic management within the framework of Transforming Cities by Controlling Pollution and Traffic Congestion, showcasing how AI-driven signal control and alternative routing strategies can reshape city mobility. It opens with a review of adaptive signal systems that harness real-time IoT and camera data to optimize phase timings and coordinate "green waves," then delves into eco-routing algorithms that factor emissions, road grades, and congestion for cleaner path selection. The discussion extends to multimodal integration—supporting pedestrians, cyclists, and shared micro-mobility—and the use of digital twins and simulation platforms for policy testing and risk reduction. The chapter also addresses critical governance and

DOI: 10.4018/979-8-3373-4422-5.ch008

ethical dimensions, including transparency, and data privacy. Concluding with compelling case studies from leading smart cities, it illustrates a paradigm shift toward anticipatory, equity-centered mobility that advances sustainability, reduces pollution, and enhances urban livability.

INTRODUCTION: THE RISE OF INTELLIGENT URBAN MOBILITY

Cities today stand at a crossroads of complexity and opportunity. As urban populations expand and vehicle use intensifies, traditional traffic management systems—static signal plans, manual routing, and reactive congestion control—are increasingly inadequate to address the intertwined challenges of congestion, air pollution, and climate vulnerability. Urban mobility is no longer defined solely by efficiency of movement; it is increasingly evaluated in terms of environmental performance, adaptability, and resilience.

The emergence of intelligent urban mobility systems, powered by artificial intelligence (AI), Internet of Things (IoT) sensors, and real-time analytics, represents a structural shift in transportation governance. Rather than optimizing exclusively for speed or distance, these systems incorporate live data on traffic flow, air quality, weather conditions, and commuter behavior to dynamically adjust routing, signal timing, and modal coordination. This transition reflects a broader move from reactive traffic management toward predictive and systems-oriented mobility planning.

A central element of this transformation is eco-routing—routing strategies that integrate environmental metrics such as fuel consumption and emissions into path optimization. Cities including Singapore, Amsterdam, and Los Angeles are deploying AI-enhanced signal control, digital twins, and multimodal mobility platforms to test and manage traffic flows under real-time conditions. These interventions aim not only to reduce congestion but also to lower vehicular emissions, improve public health outcomes, and enhance climate resilience.

This chapter examines how AI-driven routing, real-time emissions modeling, and eco-conscious design principles are reshaping urban transport systems. It analyzes the transition from static to dynamic control, the convergence of environmental and mobility policy objectives, and the ethical and operational implications of embedding algorithmic intelligence into urban infrastructure.

1.1 The Shift from Static to Dynamic Traffic Systems

Urban traffic systems were historically designed during a period when predictability was assumed to outweigh variability. For decades, municipalities relied on static traffic signal plans and fixed routing schemes formulated from historical traffic averages rather than real-time inputs (Gartner et al., 2001). These systems were optimized primarily for peak-hour demand patterns under relatively stable travel assumptions.

However, contemporary urban mobility systems are characterized by structural volatility. As Papageorgiou et al. (2003) and Litman (2020) note, cities now face fluctuating travel demand, diversified mobility services, and complex environmental constraints that challenge static optimization models. Variability in commuter behavior (Zhao et al., 2019), the rapid expansion of ride-hailing and delivery fleets (Shaheen & Cohen, 2019), weather-related disruptions (Khattak & De Palma, 1997), construction activities and incidents (Papageorgiou et al., 2003), and tightening air quality regulations (Bigazzi & Figliozzi, 2012) collectively introduce nonlinear dynamics into urban networks.

The limitations of static systems are not merely operational but structural. Fixed-time control assumes equilibrium conditions, whereas modern traffic systems operate under continuous perturbation. Demand surges, modal shifts, and environmental thresholds interact in real time, producing cascading effects across networks. As a result, optimization based on historical averages often amplifies congestion rather than mitigating it.

This evolving complexity necessitates adaptive, data-driven mobility management. The integration of real-time sensor networks, predictive algorithms, and connected infrastructure enables traffic systems to shift from pre-programmed responses to continuous recalibration. In doing so, urban transport management transitions from deterministic scheduling toward dynamic systems control capable of responding to uncertainty with greater precision and resilience.

1.2 Urban Emissions as a Real-Time Systems Problem

Urban vehicular emissions can no longer be understood solely as an aggregate byproduct of traffic volume. Instead, emissions function as dynamic system variables shaped by traffic signal phasing, network topology, driving behavior, and vehicle movement patterns (Zhou et al., 2021). Micro-level traffic phenomena—including prolonged idling at intersections, frequent acceleration and deceleration cycles, and stop-and-go congestion waves—disproportionately elevate emissions of carbon dioxide (CO_2), nitrogen oxides (NO_X), and fine particulate matter ($PM_{2\cdot 5}$) (Wang & Chen, 2020; Zhang et al., 2019). These pollutants significantly degrade air quality and are strongly linked to respiratory and cardiovascular morbidity in dense urban environments (Apte et al., 2017).

This perspective reframes emissions as a controllable network outcome rather than an unavoidable externality. Advances in IoT deployment, mobile sensing platforms, and AI-based analytics allow planners to monitor pollution levels with high temporal and spatial granularity (Kanaroglou et al., 2020). Machine learning models trained on integrated traffic and environmental datasets enable forecasting of emissions hotspots before they fully materialize (Mohamed et al., 2021). Such predictive capacity allows authorities to intervene dynamically through signal timing adjustments, adaptive routing, or congestion pricing mechanisms aimed at reducing exposure in sensitive districts (Snyder & Ratti, 2023).

By embedding environmental variables directly into traffic optimization frameworks, AI-enabled platforms can simultaneously address travel efficiency and ecological performance. This marks a substantive transition from reactive pollution mitigation to anticipatory emissions governance—aligning transport management with broader sustainable urban development and public health objectives.

2. FROM ROUTING TO ECO-ROUTING: THE NEW LOGIC OF MOVEMENT

Conventional vehicle routing algorithms have historically prioritized either the fastest or shortest path, optimizing predominantly for travel time or distance while largely ignoring environmental externalities. This practice, however, has contributed to excessive fuel consumption, elevated emissions, and intensified congestion in environmentally sensitive areas (Ahn & Rakha, 2008). In contrast, the emergence of eco-routing represents a strategic pivot toward sustainability. Eco-routing prioritizes travel paths that minimize ecological impact by incorporating considerations such as fuel efficiency, emissions output, and real-time network conditions into routing decisions.

Recent advancements in artificial intelligence, sensor technologies, and vehicle telematics have facilitated the operationalization of eco-routing algorithms that dynamically compute environmentally optimal paths. These systems evaluate diverse parameters, including road gradient, which influences engine workload and fuel consumption on inclined surfaces (Boriboonsomsin et al., 2012); traffic density and frequency of stops, which increase energy expenditure due to idling and acceleration cycles (Zhou et al., 2021); and proximity to air quality-sensitive areas, particularly around institutions such as schools and hospitals (Apte et al., 2017). Furthermore, eco-routing accounts for vehicle-specific characteristics such as fuel type, engine displacement, and emissions class (Gao et al., 2018).

By accommodating these diverse variables, eco-routing enables differentiated routing strategies tailored to vehicle profiles, effectively dispersing traffic and reducing emissions peaks in high-risk zones (Yang & Barth, 2015). In the context of electric vehicles (EVs), eco-routing can also integrate energy regeneration potentials—particularly on downhill routes—and the availability of charging infrastructure along the selected path.

The eco-routing framework encapsulates a broader shift in urban mobility logic—from prioritizing speed to prioritizing environmental intelligence. By concurrently optimizing for operational efficiency and ecological sustainability, eco-routing offers a viable pathway to reduce emissions, improve public health outcomes, and advance climate action agendas. As this approach becomes increasingly embedded in smart city infrastructure and digital navigation platforms, it is poised to become an indispensable component of future-oriented transport planning (Snyder & Ratti, 2023)

2.1 What is Eco-Routing?

In the evolving landscape of Advanced Traveler Information Systems (ATIS), eco-routing has emerged as a significant advancement. Unlike conventional navigation systems that prioritize the shortest distance or minimal travel time, eco-routing algorithms are designed to identify routes that minimize fuel consumption and reduce vehicular emissions, thereby mitigating the environmental footprint of urban travel.

An effective eco-routing system comprises several critical components. First, the Dynamic Roadway Network Database serves as a foundational digital map enriched with both real-time and historical traffic data. This system aggregates inputs from GPS probes, road sensors, traffic cameras, and other Internet of Things (IoT)-based data sources. Through advanced data fusion algorithms, it continuously updates road conditions, including congestion levels, average travel speeds, and incident reports (Gao et al., 2018).

Second, the Energy/Emissions Operational Parameter Set provides detailed fuel and emissions profiles across a spectrum of vehicle types—including petrol, diesel, hybrid, and electric vehicles—under varying roadway and traffic conditions. Emissions metrics such as CO_2, NO_X, hydrocarbons (HC), and particulate matter ($PM_{2.5}$) are calibrated to reflect variations in stop frequency, road gradient, and acceleration patterns (Zhou et al., 2021; Ahn & Rakha, 2008).

Third, the Routing Engine functions as the computational core, utilizing pathfinding algorithms—such as Dijkstra's or A*—modified to minimize a multi-variable cost function that includes fuel usage and emissions output, in addition to traditional factors such as time and distance. Many implementations adopt a multi-objective approach, offering a balance between ecological benefits and user convenience (Yang & Barth, 2015).

The fourth component, the User Interface, facilitates the interaction between users and the eco-routing system. It allows users to specify trip origins and destinations and to receive route suggestions optimized for environmental performance. Some platforms further support vehicle-specific input, enabling more accurate and personalized routing recommendations. Feedback mechanisms may include real-time indicators of fuel savings and estimated emissions reductions, enhancing user awareness and engagement.

Pilot programs and empirical studies of eco-routing systems have yielded encouraging outcomes. Evidence suggests that eco-routing can lead to reductions of 5–15% in fuel consumption and greenhouse gas emissions compared to traditional routing methods (Boriboonsomsin et al., 2012). When embedded within urban traffic management systems, eco-routing also contributes to distributing vehicular loads more evenly, reducing peak-time pollution in congested districts, and improving air quality in environmentally sensitive zones, such as those near schools and hospitals (Apte et al., 2017).

Nevertheless, the current generation of eco-routing technologies is not without limitations. Challenges include incomplete or inconsistent emissions datasets, variability in individual driving behavior, insufficient integration with public transport systems, and technical constraints in extending the model across diverse geographic regions with heterogeneous infrastructure (Mohamed et al., 2021). Future developments may incorporate vehicle-to-infrastructure (V2I) communication protocols, more robust AI-driven predictive models, and the inclusion of multimodal transport strategies to support more comprehensive and sustainable urban mobility.

2.2 How Routing Algorithms Now Factor in Emissions

Contemporary routing algorithms have advanced beyond the singular objective of reducing travel time or distance. Increasingly, these systems are engineered to incorporate environmental performance metrics, particularly emissions reduction targets, into route optimization strategies. This evolution is catalyzed by advancements in computational technologies and by growing regulatory and societal pressure to decarbonize transportation networks (Barth & Boriboonsomsin, 2009; Zhang et al., 2019).

By leveraging real-time traffic data, environmental indicators, and detailed vehicle profiles, modern routing engines can identify paths that lower emissions while maintaining acceptable travel efficiency. This functionality is underpinned by access to connected vehicle networks, sensor-based infrastructure, air quality monitoring systems, and advanced machine learning algorithms that can process large-scale, multi-dimensional datasets (Mohamed et al., 2021; Gao et al., 2018).

Emission-aware routing algorithms now account for several crucial variables. Traffic congestion patterns are one of the primary considerations, as stop-and-go traffic and idling at intersections are strongly associated with higher pollutant emissions due to suboptimal fuel combustion. Avoiding known congestion hotspots or optimizing departure times based on predictive congestion models has proven effective in mitigating this impact (Ahn & Rakha, 2008).

Another essential parameter is road topology and elevation profiles. Routes characterized by steep inclines or frequent elevation changes increase engine load, particularly for internal combustion engines. Accordingly, elevation gradients, road curvature, and surface conditions are now factored into route selection to favor paths that promote energy efficiency (Boriboonsomsin et al., 2012).

Signal timing and intersection delays also influence emissions. Frequent stops at red lights necessitate repeated acceleration, which increases fuel consumption and the release of fine particulates. Access to signal phase and timing (Spat) data allows modern systems to favor routes that minimize such disruptions (Zhou et al., 2021).

Furthermore, vehicle-specific emissions profiles are integral to the effectiveness of these algorithms. Diesel-powered vehicles, for instance, emit higher levels of NO_X and $PM_{2.5}$, while electric vehicles (EVs), despite having zero tailpipe emissions, may experience variable energy efficiency based on terrain or traffic conditions. Customized emissions modeling based on vehicle characteristics enables routing engines to deliver tailored recommendations (Yang & Barth, 2015).

AI-powered routing systems can now weigh a combination of objectives—including time, emissions, fuel economy, and network conditions—to compute optimal travel paths. These systems are especially relevant for fleet operators, ride-hailing platforms, logistics providers, and transit agencies, many of whom operate under explicit sustainability mandates or emissions caps (Snyder & Ratti, 2023).

Some systems go a step further by offering eco-feedback mechanisms, such as eco-scores or estimated carbon savings, thereby promoting environmental awareness and influencing driver behavior. Additionally, the integration of carbon pricing models allows users to factor in the societal cost of emissions when making routing decisions, adding a layer of economic accountability (Kanaroglou et al., 2020).

3. THE AI TOOLKIT FOR URBAN TRAFFIC

The deployment of artificial intelligence in urban traffic management has ushered in a transformative era of adaptive, efficient, and sustainable mobility. AI-based systems integrate machine learning, IoT sensor networks, real-time analytics, and optimization models to monitor, predict, and control urban traffic conditions with unprecedented precision. These platforms surpass the limitations of static traffic planning by responding dynamically to fluctuating variables such as congestion, accidents, and weather, thereby achieving substantial improvements in flow efficiency, emissions reduction, and transport reliability (Zhang et al., 2019; Li et al., 2020).

Modern AI-driven traffic systems are composed of several interrelated components. IoT sensors and surveillance cameras are deployed across strategic nodes, including intersections and arterial roadways. These devices capture live data on vehicular density, lane occupancy, average speeds, and incident reports. Such data provides both real-time operational insights and a repository for longitudinal traffic analysis (Gao et al., 2018).

Machine learning (ML) algorithms form the computational backbone of these systems. Trained on extensive traffic datasets, these models forecast traffic volume, congestion likelihood, and incident probability. By incorporating additional inputs—such as meteorological forecasts, public event calendars, or road maintenance schedules—these models achieve high predictive accuracy. Advanced techniques, including deep learning architectures like recurrent neural networks (RNNs) and graph neural networks (GNNs), are particularly effective in modeling the nonlinear complexities of urban mobility (Mohamed et al., 2021).

Computer vision technologies further augment system intelligence by interpreting visual data in real time. These applications can detect various traffic participants, including vehicles, cyclists, and pedestrians, and identify behaviors such as jaywalking or traffic violations. The integration of this capability enables responsive signal control and enhanced safety analytics, especially in high-risk urban zones (Zhou et al., 2021).

Another breakthrough involves reinforcement learning (RL), particularly in traffic signal optimization. RL systems—such as those using Deep Q-Learning—adaptively adjust signal phase timings based on real-time feedback from traffic conditions. This not only minimizes delays but also contributes to reductions in emissions and energy usage (Wei et al., 2019).

Graph-based routing models, leveraging real-time data and AI-driven analytics, continuously compute optimal travel paths by evaluating traffic conditions, road network changes, and environmental data. These systems are capable of tailoring route recommendations to vehicle types and environmental goals (Yang & Barth, 2015; Barth & Boriboonsomsin, 2009).

In addition, natural language processing (NLP) tools are increasingly employed to analyze user-generated content from social media, crowdsourced platforms like Waze, or emergency communications. NLP models extract actionable data on incidents, locations, and traffic disruptions, which can then inform centralized traffic control dashboards (Snyder & Ratti, 2023).

The integration of these technologies results in numerous benefits, including reduced travel times, lower fuel consumption, decreased emissions, enhanced safety through early incident detection, and improved prioritization of emergency and public transport services. AI-enabled traffic management systems are therefore a foundational pillar of smart city strategies, offering scalable, efficient, and sustainable urban mobility solutions (Li et al., 2020; Apte et al., 2017).

Figure 1. AI-powered traffic signals adjust in real-time using sensor data to reduce delays and emissions

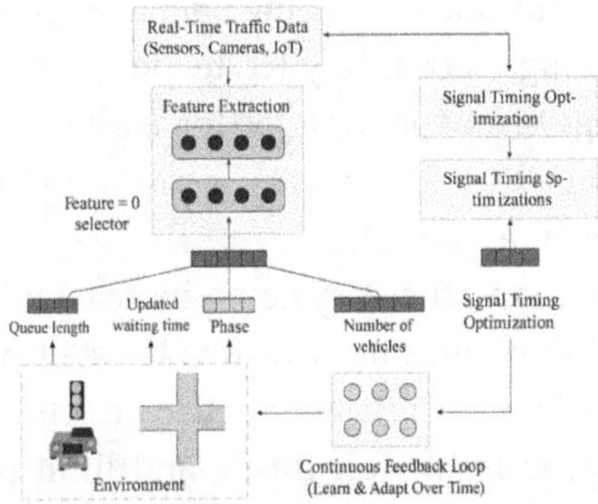

3.1 Machine Learning in Adaptive Signal Control

Machine Learning (ML) is transforming the landscape of urban traffic signal management by enabling systems to transition from rigid, fixed-time operations to intelligent, adaptive control frameworks. Traditional traffic signals typically operate on pre-set schedules derived from historical averages. While effective under predictable conditions, these systems often fail to respond adequately to real-time disruptions such as accidents, unusual congestion patterns, or fluctuations in pedestrian volume (Genders & Razavi, 2016).

In contrast, ML-powered adaptive signal control systems (ASCS) utilize real-time data streams from cameras, inductive loops, and IoT sensors to continuously monitor traffic flow parameters, including vehicle counts, queue lengths, turning movements, and average speeds (Zhou et al., 2021; Gao et al., 2018). These data inputs are then used to predict potential congestion or bottlenecks using time-series models, such as Long Short-Term Memory (LSTM) networks, or regression-based approaches trained on historical patterns (Li et al., 2020). As conditions evolve, the system dynamically adjusts the duration and sequence of green and red phases, thereby minimizing intersection delays and fuel consumption while improving throughput for both vehicles and pedestrians (Wei et al., 2019).

A critical feature of advanced ML-based traffic systems is the coordination of signal phases across multiple intersections. This enables the formation of synchronized "green waves" along arterial roads, allowing platoons of vehicles to travel with fewer stops. Such coordination reduces idling, shortens travel times, and contributes to significant emission reductions, particularly during peak hours (Abdoos et al., 2013).

One of the most powerful machine learning frameworks for adaptive control is Reinforcement Learning (RL). In this paradigm, each traffic signal operates as an autonomous agent that learns optimal strategies through trial-and-error interactions with the environment. Reward functions are typically defined by system-wide goals such as minimizing total vehicle delay, reducing queue lengths, and lowering emissions (Gao et al., 2018; Wiering, 2000). Key RL techniques include Q-learning for individual intersections, Deep Q-Networks (DQN) for learning from high-dimensional state spaces, and Multi-Agent Reinforcement Learning (MARL) for large-scale, city-wide coordination (Wei et al., 2019).

Beyond operational efficiency, these systems increasingly incorporate multi-objective optimization, enabling cities to prioritize broader policy objectives. For example, signal timing algorithms can be fine-tuned to prioritize emergency vehicles and public transit, reduce pedestrian wait times, and limit pollutant exposure in sensitive locations such as near schools or hospitals (Yang & Barth, 2015).

Real-world deployments have already demonstrated the viability and impact of these systems. Cities like Pittsburgh and Singapore have implemented AI-based adaptive signal control with notable success. In Pittsburgh, the Surtrac system reduced vehicle wait times by more than 40% and overall travel time by approximately 25% (Smith et al., 2013). Similarly, pilot programs in Los Angeles have reported emission reductions of 15–30% during peak hours, alongside up to a 20–25% decrease in delay (Stevanovic, 2010).

As computational resources become increasingly accessible and urban infrastructure grows more interconnected, ML-driven adaptive signal control emerges as a cost-effective, scalable solution. It not only enhances urban mobility and safety but also contributes meaningfully to environmental sustainability by reducing fuel consumption and traffic-related emissions.

3.2 Real-Time Data from IoT, GPS, and Connected Vehicles

Contemporary traffic management increasingly relies on a data-centric framework, in which real-time information from diverse sources forms the operational backbone of Intelligent Transportation Systems (ITS). These real-time inputs enable urban traffic control systems to transition from reactive to predictive, environmentally aware, and dynamically adaptive regimes (Zhou et al., 2021; Gao et al., 2018).

Among the most critical data sources are IoT sensors, GPS-based systems, and Vehicle-to-Everything (V2X) communications. IoT-enabled devices embedded in roads, intersections, and urban infrastructure—such as inductive loops, environmental monitors, and high-resolution cameras—capture granular data on traffic parameters, including lane occupancy, vehicle speed, emissions levels, and incident occurrences (Gkartzonikas & Gkritza, 2019). GPS data, harvested from smartphones, ride-hailing services, and navigation platforms, contributes location-based metrics such as route selection, travel velocity, and congestion patterns. When anonymized and aggregated, this information facilitates real-time congestion heatmapping and origin-destination modeling (Zhang et al., 2019). Connected vehicles equipped with V2X technology further enhance situational awareness by broadcasting telemetry data such as acceleration, braking, turn signaling, and environmental conditions (Talebpour & Mahmassani, 2016).

The integration of these data sources enables a range of transformative functions. AI-based algorithms can detect anomalies in traffic flow to identify accidents and bottlenecks within seconds, allowing for immediate intervention and rerouting (Li et al., 2020). Eco-routing algorithms and adaptive signal control systems dynamically recalibrate operations in response to live conditions, such as prioritizing bus lanes or adjusting signal phases to alleviate pressure points (Wei et al., 2019). Additionally, combining vehicle telemetry and air quality sensor data enables identification of emission hotspots, guiding targeted measures like congestion pricing or route redistribution in environmentally sensitive zones (Apte et al., 2017).

Ultimately, real-time data infrastructures support the development of a responsive, intelligent traffic ecosystem. This evolution reduces vehicular delay and fuel use, lowers emissions, enhances safety through predictive analytics, and promotes sustainable urban mobility (Snyder & Ratti, 2023).

4. DIGITAL TWINS AND SIMULATION-DRIVEN URBAN PLANNING

Digital twin technology is redefining urban traffic planning by enabling real-time simulation and optimization of transportation systems. A digital twin is a dynamic, high-fidelity virtual replica of a physical environment that integrates continuous input from sensors, vehicle telemetry, geospatial data, and simulation models. Within the domain of urban mobility, digital twins serve as predictive platforms for scenario testing and strategy development—without disrupting existing infrastructure (Batty, 2018; Grieves & Vickers, 2017).

These systems are powered by real-time inputs from IoT infrastructure, GPS-based movement data, connected vehicle streams, and environmental monitoring systems. Public transit usage patterns and pedestrian flow data further enrich the behavioral and spatial granularity of these models.

Urban planners leverage digital twins to simulate traffic control interventions, such as adaptive signal timing, route changes, infrastructure modifications, and temporary closures. These simulations predict how policy or infrastructure changes will affect congestion, travel times, safety, and emissions under varied demand conditions (Khan et al., 2020). Integration with environmental data allows for detailed assessment of public health impacts and greenhouse gas outputs, supporting compliance with climate action and pollution mitigation targets (Sotonye & Rajabifard, 2021).

Digital twins also allow planners to evaluate demand-responsive and eco-routing strategies. Simulated scenarios reflecting weather changes, large-scale events, or school holidays provide insight into traffic redistribution and policy effectiveness before real-world deployment (Zhou et al., 2021). Additionally, by incorporating data from multiple transport modes—buses, metros, bicycles, e-scooters, and pedestrians—digital twins help design integrated mobility systems that improve first-mile/last-mile connectivity and reduce car dependency.

Strategically, digital twins offer faster, low-risk experimentation, data-driven urban design, and enhanced coordination across urban departments. Global cities such as Singapore, Helsinki, and Rotterdam have adopted digital twin platforms to align infrastructure development with environmental sustainability and urban livability (Ketzler et al., 2020). As computational capabilities and data infrastructures evolve, digital twins are poised to become core tools in building adaptive, equitable, and sustainable urban environments.

4.1 Using Digital Models to Test Routing and Signal Changes

Digital traffic simulation models provide a vital framework for testing transportation strategies in a controlled, virtual environment. These models integrate historical datasets, real-time IoT inputs, and AI-driven forecasts to mirror urban traffic behavior with high fidelity. By doing so, planners can analyze the effects of new interventions on congestion, emissions, and travel efficiency—without physical trial-and-error (Khan et al., 2020; Barceló, 2010).

These models allow for the testing of different traffic volumes, vehicle types, and behavioral profiles. For instance, simulations can model the effects of a revised signal plan on mixed traffic—including buses, trucks, electric vehicles, and ride-shares—during peak hours, special events, or emergencies. Eco-routing algorithms can also be evaluated for their ability to reduce emissions by accounting for variables like gradient, engine load, and acceleration patterns (Barth & Boriboonsomsin, 2009).

Adaptive signal control systems, particularly those powered by reinforcement learning, are another key area of simulation. These systems can be tested under varying congestion levels, pedestrian demands, and emergency routing scenarios, ensuring that signal phasing and coordination are optimized prior to deployment (Genders & Razavi, 2016). Simulations can generate detailed performance metrics such as queue length, average delay, stop frequency, and fuel consumption—supporting transparent and evidence-based decision-making (Zhou et al., 2021).

Commonly used simulation platforms include SUMO (Simulation of Urban Mobility), an open-source microscopic traffic simulator that supports vehicle, pedestrian, and transit modeling (Rudzewicz et al., 2012), and VISSIM, a commercial tool with high-resolution visualization and multimodal integration capabilities (PTV Group, 2020). Customized digital twins, as implemented in cities like Singapore and Stockholm, further extend these tools by incorporating localized infrastructure and behavioral patterns for maximum predictive accuracy (Ketzler et al., 2020).

The strategic advantage of simulation-based planning lies in its ability to reduce implementation risk, shorten policy deployment timelines, and align mobility decisions with environmental and equity goals. It also facilitates inter-agency collaboration across transport, health, and planning sectors.

4.2 Scenario Planning with Live Data Feeds: SUMO + AI Integration

Scenario planning in urban mobility is increasingly leveraging the integration of live data streams, artificial intelligence (AI), and simulation platforms like SUMO. This combination enables cities to test interventions under real-world conditions with high temporal resolution and predictive capability, representing a shift from static forecasting to dynamic, adaptive planning (Rudzewicz et al., 2012; Barceló, 2010).

SUMO, as a modular, open-source traffic simulator, can ingest real-time data from GPS-enabled devices, IoT sensors, and traffic camera systems. This capability allows simulations to reflect current road conditions, such as real-time congestion, signal states, or incident impacts (Zhou et al., 2021). Planners can run "what-if" scenarios—including large-scale events, weather disruptions, or construction activities—to evaluate the resilience and responsiveness of the mobility system (Khan et al., 2020).

Through AI integration, particularly reinforcement learning, SUMO-based simulations can autonomously optimize interventions. These may include route reconfiguration, dynamic signal phasing, or congestion avoidance based on predictive demand surges (Genders & Razavi, 2016; Wei et al., 2019). Visualization outputs such as emissions maps, congestion heatmaps, queue lengths, and travel time distributions help communicate results to policymakers and stakeholders across departments (Apte et al., 2017).

The strategic value of SUMO + AI integration lies in its ability to deliver actionable insights in near real time. Cities such as Dresden, Singapore, and Barcelona are already employing such systems to support rapid deployment of traffic strategies, align interventions with environmental goals, and optimize system-wide mobility efficiency (Ketzler et al., 2020).

As urban mobility grows more complex, real-time scenario planning using SUMO and AI will become indispensable for evidence-based policy development, continuous optimization, and sustainable traffic management.

5. MULTI-MODAL AND HUMAN-CENTRIC SYSTEMS

Urban mobility is undergoing a paradigmatic transformation from traditional car-centric design to multi-modal, human-centric planning models that prioritize inclusivity, environmental sustainability, and spatial efficiency. This emerging framework advocates for the development of transportation networks that cater equitably to all users—including pedestrians, cyclists, public transit passengers, and individuals with limited mobility—rather than privileging private vehicle use (Gehl, 2010; Banister, 2008).

A foundational element of this transition is the seamless integration of multiple transport modes. Effective urban mobility ecosystems increasingly connect walking, cycling, buses, metro systems, rail services, and emerging shared mobility options such as e-bikes, e-scooters, and ride-hailing platforms. Mobility-as-a-Service (MaaS) applications serve as digital facilitators by enabling users to plan, book, and pay for entire multimodal journeys through a single interface, thereby simplifying intermodal travel (Hensher et al., 2020).

Cities are also deploying intelligent transportation infrastructure that supports real-time dynamic routing and signal prioritization. These adaptive systems use traffic sensors and AI algorithms to give preferential treatment at intersections to sustainable modes of travel, such as public buses, cyclists, and pedestrians. Such measures enhance the speed, reliability, and attractiveness of non-car options, thereby incentivizing modal shift (Gao et al., 2018; Wei et al., 2019).

Furthermore, the proliferation of smart mobility hubs has redefined the spatial organization of urban mobility. Strategically located at key interchanges, these hubs combine transit services with active and shared mobility options, reducing "first-mile/last-mile" friction and making door-to-door sustainable travel more viable (Shaheen & Cohen, 2019). Infrastructure design within human-centric mobility systems emphasizes inclusivity, with features such as protected bicycle lanes, universally accessible pedestrian pathways, and traffic-calmed zones in residential or commercial centers (Loo & Lam, 2015).

The public health and environmental co-benefits of these systems are well-documented. By reducing car dependency, encouraging active travel, and improving air quality, cities can promote healthier lifestyles, decrease greenhouse gas emissions, and foster more livable environments (World Health Organization [WHO], 2011). Leading cities such as Copenhagen, Bogotá, and Vienna demonstrate the viability of this approach, with measurable improvements in traffic safety, commuter satisfaction, and emissions reductions (Pojani & Stead, 2015).

As urban populations continue to expand and climate imperatives intensify, multimodal and human-centric mobility is evolving from a policy trend to a foundational strategy for building resilient and inclusive cities.

5.1 Integrating Pedestrian, Cycling, and Micro-Mobility into Routing Systems

Figure 2. AI-enabled routing supports walking, cycling, and eco-friendly travel choices

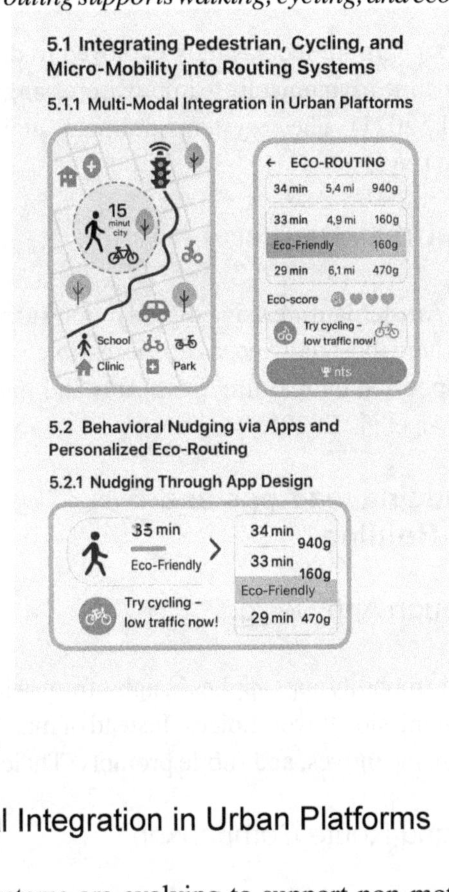

5.1.1 Multi-Modal Integration in Urban Platforms

Urban routing systems are evolving to support non-motorized modes such as walking, cycling, and micro-mobility (e.g., e-scooters, shared bikes). These tools respond to urban concerns like emissions and congestion by optimizing short-distance trips—especially those under five kilometers—through AI and real-time geospatial data (Shaheen & Cohen, 2019; Marsden et al., 2021).

5.1.2 Digital Routing Features for Active Mobility

Apps like Google Maps, City mapper, and Movita now offer dedicated cycling and pedestrian routing, highlighting protected bike lanes, traffic-calmed streets, and low-traffic zones (Pojani & Stead, 2015). They also integrate shared mobility data such as vehicle availability and battery status, aiding trip planning (Hensher et al., 2020).

5.1.3 Adaptive Infrastructure for Non-Motorized Users

AI-powered traffic signals are increasingly deployed to detect pedestrians and cyclists, adjusting green times dynamically to improve crossing safety and flow (Gao et al., 2018; Zhou et al., 2021). These systems promote safer and more attractive environments for active travel.

5.1.4 Policy-Driven Urban Design

Cities like Paris and Amsterdam are using digital routing to reinforce policies like the "15-minute city," where most daily needs are accessible via walking or cycling. These strategies are supported by real-time platforms and multimodal integration (Moreno et al., 2021; Laa et al., 2020).

5.2 Behavioral Nudging via Apps and Personalized Eco-Routing

5.2.1 Nudging Through App Design

Behavioral nudging in mobility apps applies insights from behavioral economics to guide users toward low-emission travel choices. Instead of mandates, apps influence behavior through design, incentives, and subtle prompts (Thaler & Sunstein, 2008).

5.2.2 Eco-Scoring and Route Comparison

Many navigation platforms now display emissions or fuel consumption scores alongside time and distance estimates. These features allow users to consider environmental impact in their routing decisions (Barth & Boriboonsomsin, 2009).

5.2.3 Gamification and Rewards

Gamified systems—like Commute Greener—offer digital badges or points for sustainable choices, which improves user engagement and encourages repeat behavior (Hörl et al., 2021).

5.2.4 AI-Powered Personalized Nudging

Advanced apps apply machine learning to offer personalized eco-routing based on travel history, time of day, and transport mode. Suggestions may include cycling during low-traffic periods or choosing flatter paths for electric vehicles (Wang et al., 2020; Sunio & Schmöcker, 2017).

5.2.5 Case Examples of Impact

Real-world deployments such as Stockholm's Smart Travel Habits reduced solo driving by 15% using app-based nudges (Fujii & Taniguchi, 2005). Helsinki's Mobility-as-a-Service (MaaS) pilots integrated behavioral nudging to shift users to bikes and transit (Sochor et al., 2015).

6. CASE STUDIES IN SMART CITIES

As cities confront the challenges of climate change, congestion, and livability, several have emerged as leaders in applying AI, eco-routing, and real-time data to mobility systems. This section explores three frontrunners—Amsterdam, Singapore, and Helsinki—highlighting their approaches to sustainable and intelligent transport.

6.1 AI, Routing, and Sustainability Models

Amsterdam: Cycling-Centric Smart Mobility

Amsterdam leads in cycling-based urban transport, with over 60% of city-center trips by bike. AI-powered apps provide real-time, weather-aware routing to identify safe, low-traffic paths (Pojani & Stead, 2015). Adaptive traffic signals use radar and video to prioritize cyclists and pedestrians. The city's 3D digital twin simulates traffic, evaluates low-emission zone impacts, and supports emergency planning (Ketzler et al., 2020). These tools have helped reduce CO_2 and NO_X emissions, reinforcing Amsterdam's low-emission leadership.

Singapore: Integrated Smart Traffic Control

Singapore's Intelligent Transport System (ITS) integrates AI, dynamic pricing, and traffic modeling. The GLIDE signal system adjusts lights in real-time using data on congestion and transit schedules. Electronic Road Pricing (ERP) encourages off-peak travel and reduced road use (Menon, 2000). Predictive models using GPS and IoT inputs allow proactive congestion rerouting. This integrated model sustains low car ownership and high traffic efficiency.

Helsinki: MaaS for Emissions Control

Helsinki's Whim app supports seamless planning and payment across transport modes—transit, ride-hailing, cycling, and e-scooters. AI trip planning incorporates emissions, cost, and time preferences, guiding users toward greener choices (Hietanen, 2014; Sochor et al., 2015). Digital twins model routing scenarios during weather or pollution events to support adaptive policy. Helsinki's MaaS platform has reduced congestion and per capita emissions.

Key Takeaways

Digital twins, adaptive signals, and multimodal platforms enhance traffic efficiency and reduce emissions. Smart cities demonstrate how combining policy, AI, and user behavior can support scalable, inclusive transport transitions.

6.2 Impacts on Pollution, Commute Times, and Public Satisfaction

Pollution Reduction

Singapore's ERP and adaptive signals cut roadside NO_2 by up to 15% in high-traffic zones (Land Transport Authority, 2021). Amsterdam meets EU air standards through cycling infrastructure, low-emission zones, and eco-routing for electric vehicles (Ketzler et al., 2020). Helsinki's real-time emissions data enables rerouting and behavioral shifts during pollution peaks.

Commute Time Improvements

Singapore's GLIDE system reduces peak-hour intersection delays by 12–20% . Amsterdam's cyclist-first signals reduce bike wait times by up to 30% (Pojani &

Stead, 2015). Helsinki's MaaS platform improves journey time by offering optimized, flexible route planning (Hensher et al., 2020).

Public Satisfaction

Helsinki's MaaS system reports 70%+ user satisfaction, with younger commuters valuing transparency and flexibility (Sochor et al., 2015). Singapore's eco-routing and dynamic pricing maintain strong public approval. In Amsterdam, digital twins and safe bike routing receive high praise from users for improving commute quality and safety (Gehl, 2010).

7. POLICY, ETHICS, AND THE DATA INFRASTRUCTURE

AI-driven mobility systems offer efficiency and sustainability, but they raise concerns about fairness, privacy, and accountability. As cities adopt technologies like real-time routing and digital twins, responsible governance is essential to ensure alignment with public values (Craglia et al., 2018; Goodchild, 2021).

Figure 3. AI mobility systems must balance efficiency with fairness, transparency, and data ethics

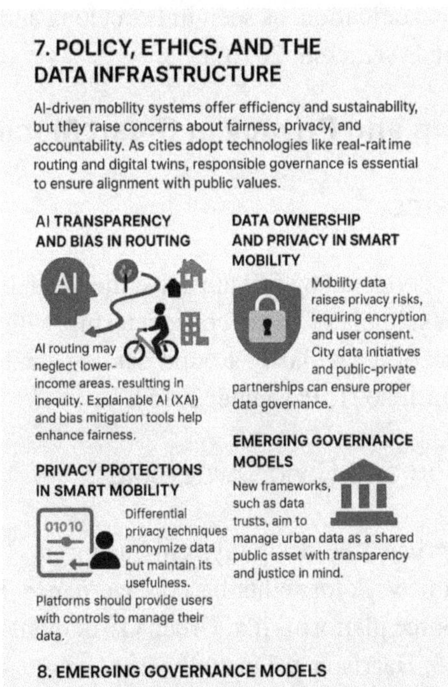

7.1 AI Transparency and Bias in Routing

7.1.1 Risks of Bias and Discrimination

AI routing may favor digitally rich areas while neglecting lower-income zones with limited infrastructure, reinforcing spatial inequality (Sadowski, 2020; Crawford, 2021). Data collected from smartphones and apps often excludes vulnerable groups, leading to biased decisions. When algorithmic choices remain opaque, public trust suffers (Pasquale, 2015).

7.1.2 Tools for Fairness and Transparency

Explainable AI (XAI) provides visible rationales—e.g., "route chosen to reduce emissions"—improving user trust (Gunning & Aha, 2019). Public algorithm registries and independent audits help ensure transparency and alignment with urban goals (Ada Lovelace Institute, 2020).

7.1.3 Bias Mitigation

Bias can be reduced by training algorithms with diverse data, including surveys and analog observations. Fairness constraints can ensure equal service quality across demographics. Public participation, as seen in Barcelona and Helsinki, helps align AI with community needs (Kitchin, 2016).

7.2 Data Ownership and Privacy in Smart Mobility

7.2.1 Ethical Concerns

Mobility data, often collected by private apps, includes detailed personal histories. These are stored or sold under vague consent terms, with little public oversight (Shin et al., 2020). Lack of legal clarity around data ownership can lead to misuse or surveillance (Crawford, 2021; Pasquale, 2015).

7.2.2 Data Protection and Civic Ownership

Cities should adopt privacy-by-design systems, using encryption and anonymization. GDPR offers a framework for rights-based governance. Barcelona's City Data Commons and open-source platforms like Urban OS demonstrate ethical civic data use (Craglia et al., 2018; Harrison & Donnelly, 2011).

7.3 Cybersecurity and System Resilience in AI-Driven Mobility

AI-driven traffic systems enhance efficiency but also expand the vulnerability of urban infrastructure. Reliance on interconnected IoT sensors, cloud platforms, and Vehicle-to-Everything (V2X) communications increases exposure to cyber threats. Unauthorized access to traffic signals, GPS spoofing, or manipulated V2X data could disrupt routing systems, compromise emergency response, or create congestion cascades (Talebpour & Mahmassani, 2016).

AI models themselves may be susceptible to adversarial data manipulation, potentially distorting routing or emissions predictions. To mitigate these risks, cities must implement encrypted data pipelines, zero-trust architectures, continuous auditing, and cybersecurity compliance standards in procurement processes (OECD, 2021; Goodchild, 2021). Technological advancement must therefore be accompanied by robust resilience planning to safeguard operational stability and public safety.

7.4 Emerging Governance Models

Barcelona's data trusts treat urban data as a public good governed by justice and transparency (Barcelona City Council, 2019). Citizen-owned data models give individuals legal control over how their data is used (Pentland, 2012). These frameworks are backed by regulations like GDPR, which set global standards for consent and privacy (Voigt & Von dem Bussche, 2017; Ada Lovelace Institute, 2021).

8. CHALLENGES AND OPPORTUNITIES AHEAD

As AI-enabled transportation systems and smart city platforms continue to evolve, urban planners face a dual imperative: capitalize on transformative technological opportunities while addressing persistent structural and ethical challenges. The path forward requires coordinated efforts across infrastructure development, public governance, social equity, and environmental responsibility (Kitchin, 2016).

8.1 Key Challenges

One of the most pressing concerns is data privacy and ethical AI. The widespread use of location-based services, connected IoT devices, and AI algorithms significantly increases the risk of privacy violations, especially when data is shared without adequate consent or security. Cities must find a balance between optimizing real-time mobility systems and protecting individual rights through rigorous data governance, algorithmic accountability, and explainable AI models (Zuboff, 2019; Craglia et al., 2018).

Infrastructure readiness is another significant barrier, particularly in mid-sized or low-income urban areas across the Global South. These cities often lack the foundational digital infrastructure—such as high-fidelity traffic sensors, reliable broadband networks, or interoperable data platforms—required to implement adaptive signal control, eco-routing, or digital twin simulations effectively (Batty, 2018). Without this foundation, smart mobility remains an aspirational goal rather than an operational reality.

Social equity presents a third challenge. If not carefully designed, intelligent mobility systems risk creating what some scholars call "mobility privilege," where tech-enabled services disproportionately benefit digitally literate, affluent, or central-city residents. This can marginalize communities with limited access to smartphones, banking systems, or digital literacy, thereby reinforcing transport inequalities (Sadowski, 2020; Eubanks, 2018).

Finally, policy inertia remains a critical hurdle. Regulatory frameworks often lag behind technological advancements, constrained by outdated codes, slow procurement cycles, or risk-averse public agencies. As a result, cities may struggle to implement experimental systems or adjust regulations to accommodate dynamic AI tools. Agile policymaking, sandbox testing, and regulatory reform will be essential to unlock the full potential of smart mobility (OECD, 2021).

8.1.1 Cost, Institutional Capacity, and Feasibility for Smaller Cities

In addition to governance and infrastructure barriers, concerns about cost and institutional complexity often shape resistance to intelligent mobility systems, particularly in smaller or resource-constrained cities. The deployment of sensor networks, data platforms, cybersecurity safeguards, and skilled technical personnel can require significant upfront investment. However, phased implementation—beginning with pilot corridors or high-impact zones—combined with open-source simulation tools and modular infrastructure can reduce financial burden and operational risk (Batty, 2018). When evaluated over longer time horizons, reductions in congestion, fuel consumption, traffic incidents, and pollution-related health costs may offset initial expenditures (Litman, 2020). Therefore, the feasibility of AI-driven mobility systems depends less on absolute budget size and more on strategic sequencing, institutional coordination, and lifecycle-based cost assessment (OECD, 2021).

8.2 Major Opportunities

Despite these obstacles, the opportunities for AI-enabled urban mobility are substantial. One of the most impactful is the potential for large-scale emission reduction. Adaptive traffic signals, real-time congestion management, and eco-routing algorithms can collectively cut vehicular emissions by up to 20% in dense traffic corridors. When combined with electric vehicle adoption and low-emission zone enforcement, these strategies provide a powerful pathway toward decarbonizing urban transport (Menon, 2000; Liu et al., 2020).

Another transformative opportunity lies in personalized and inclusive mobility. Mobility-as-a-Service (MaaS) platforms such as Whim in Helsinki offer integrated, on-demand transportation planning that reflects individual user needs. AI-driven personalization further enhances these systems by learning user behavior over time, adapting recommendations based on real-world constraints like cost, accessibility, or carbon footprint (Hensher et al., 2020).

Simulation and digital twins represent a third major opportunity. Tools such as SUMO, VISSIM, and PTV now integrate with AI engines to simulate traffic flows, public events, and extreme weather conditions. These platforms enable planners to test mobility interventions virtually before physical deployment, reducing implementation risk and increasing system responsiveness (Ketzler et al., 2020).

Finally, the integration of AI with electric vehicles and autonomous mobility is set to redefine urban transport. Smart routing systems will help coordinate autonomous vehicle fleets and optimize EV charging networks, supporting efficient, low-carbon urban ecosystems. Cities that proactively plan for this convergence—using AI to balance flow, predict demand, and coordinate charging infrastructure—will benefit from smoother traffic operations, fewer emissions, and enhanced road safety (Litman, 2020; Fagnant & Bockelmann, 2015).

Figure 4. AI boosts emission cuts, smart mobility, and EV-autonomous transport integration

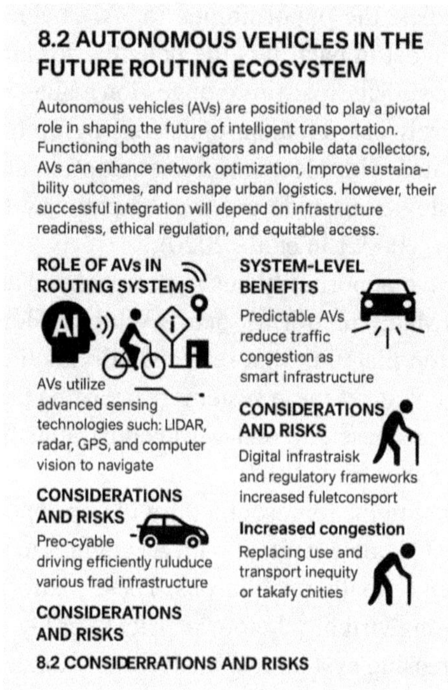

8.3 Limitations of AI Systems: Computation vs. Equity

AI systems are powerful tools for analyzing large-scale urban data and improving traffic flow, emissions, and route efficiency. However, a key limitation lies in the misalignment between computational efficiency and social equity. What is deemed algorithmically optimal for a transportation network does not always translate into equitable outcomes for all users (Eubanks, 2018).

8.3.1 Computation Strengths of Mobility AI

Mobility-focused AI excels in several domains. It can adapt in real time to traffic conditions using machine learning and predictive analytics, thereby adjusting traffic signals and rerouting vehicles to reduce delays. Route optimization engines, such as those used by Google Maps or Waze, compute efficient paths for thousands of users simultaneously, factoring in road conditions, user preferences, and live congestion data. Eco-routing systems also demonstrate the ability of AI to reduce environmental impact by calculating routes based on vehicle type, road gradient, and current traffic loads, minimizing both travel time and emissions (Genders & Razavi, 2016; Zhang et al., 2011; Liu et al., 2020).

8.3.2 Equity Limitations of Algorithmic Optimization

Despite these strengths, AI systems have notable limitations in promoting equity. Data-poor zones—areas without sufficient sensors, mobile device penetration, or digital infrastructure—are often invisible to algorithms, leading to routing blind spots and ineffective traffic control in marginalized neighborhoods (Sadowski, 2020; Crawford, 2021). Many mobility AI systems are designed primarily to optimize for vehicles, thereby prioritizing cars over more vulnerable users such as pedestrians, cyclists, or public transport commuters, whose needs may be harder to quantify but are crucial for equitable urban systems (Lucas et al., 2016). Furthermore, efficiency-driven algorithms often reinforce structural inequalities by serving well-resourced, central neighborhoods while increasing exposure to pollution, traffic, and delays in lower-income zones (Eubanks, 2018).

8.3.3 The Tension Between Optimal and Equitable

There is growing awareness that technical optimization does not inherently include fairness. For instance, an AI system may reroute heavy traffic through a low-income neighborhood to reduce citywide congestion, thereby increasing noise and air pollution in already burdened communities. Similarly, signal prioritization may focus on main thoroughfares, causing pedestrian wait times to increase in underserved areas. This tension reveals a deeper design issue: unless values like accessibility, equity, and justice are explicitly embedded into the algorithmic logic, AI systems tend to reproduce the status quo.

8.3.4 Policy and Design Recommendations

To address these risks, AI mobility systems must be developed with inclusive datasets that represent diverse travel patterns, including non-vehicular and underserved community movements. Fairness constraints should be integrated into optimization models to ensure balanced benefits across geographic and demographic groups. Finally, governance frameworks must guide AI deployment toward socially inclusive goals, emphasizing transparency, public engagement, and ethical design (Kitchin, 2016; Ada Lovelace Institute, 2021).

8.4 Autonomous Vehicles in the Future Routing Ecosystem

Autonomous vehicles (AVs) are positioned to play a pivotal role in shaping the future of intelligent transportation. Functioning both as navigators and mobile data collectors, AVs can enhance network optimization, improve sustainability outcomes, and reshape urban logistics. However, their successful integration will depend on infrastructure readiness, ethical regulation, and equitable access (Fagnant & Bockelmann, 2015; Litman, 2020).

8.4.1 Role of AVs in Routing Systems

AVs utilize advanced sensing technologies such as LiDAR, radar, GPS, and computer vision to navigate autonomously based on live environmental input. This real-time data processing allows for dynamic routing adjustments without human intervention (Thrun, 2010). Additionally, AVs operate within a broader data-sharing ecosystem, contributing to Vehicle-to-Everything (V2X) systems that include infrastructure and other vehicles. This connected intelligence supports continuous two-way communication, improving both system awareness and coordination. When deployed as shared fleets—such as driverless taxis or delivery robots—AVs can be centrally routed by AI to avoid congestion zones, optimize travel patterns, and reduce fuel consumption based on real-time conditions (Zhang et al., 2019).

8.4.2 System-Level Benefits

AVs offer multiple system-wide benefits. Their consistent and predictable driving behavior reduces stop-and-go traffic patterns, lowers accident rates, and contributes to smoother vehicle flows. AVs also align well with eco-routing frameworks, as they can strictly follow optimal speed trajectories that minimize fuel use and emissions. Moreover, their integration with roadside infrastructure and smart traffic signals allows for predictive rerouting and dynamic response to traffic incidents or construction, making cities more responsive and resilient (Chen et al., 2020; Baskar et al., 2011).

8.4.3 Considerations and Risks

Despite their potential, AVs present critical challenges. Their effectiveness depends on cities having mature digital infrastructure, interoperable communication standards, and comprehensive legal frameworks for liability, data protection, and AI ethics (OECD/ITF, 2015). Without these systems in place, the benefits of AVs may remain theoretical. Additionally, AV rollouts may worsen transport inequality if market-driven logic prevails. AV services might exclude low-income neighborhoods due to safety scores or low profitability, and digitally disconnected populations—such as the elderly or those without smartphones—may face access barriers. If not carefully regulated, privately owned AVs could also contribute to increased congestion by replacing public transit trips or cruising while empty (Gonzalez-Gonzalez et al., 2020).

8.5 Scaling from Testbeds to Megacities

While smart mobility tools such as eco-routing, adaptive signals, and digital twins have shown success in controlled environments, expanding these systems across large, diverse megacities presents significant complexity. Scaling these innovations requires overcoming technical fragmentation, institutional inertia, and social inequities (Batty, 2018; OECD/ITF, 2021).

8.5.1 Testbed Success

Smart city testbeds, such as pilot corridors or innovation districts, often show promising results. For example, localized adaptive signal systems piloted in Pittsburgh demonstrated reductions in delays by 25% and emissions by 21% (Smith et al., 2013). Similarly, digital twin testbeds in Singapore's Punggol and Barcelona's Superblocks have successfully modeled emissions scenarios and route adaptations in real time (Ketzler et al., 2020). These successes are enabled by centralized governance, simplified infrastructure, and strong public-private collaboration.

8.5.2 Scaling Challenges in Megacities

Scaling these solutions introduces numerous challenges. Megacities involve diverse land use patterns, travel behaviors, and modal preferences, which make it difficult for AI models trained on small datasets to generalize effectively. Many urban centers also rely on fragmented legacy systems that lack interoperability, impeding the deployment of real-time mobility intelligence (Kitchin, 2016). Institutional silos between transit authorities, planning departments, and technology providers further complicate coordination. Moreover, there is a persistent risk that data infrastructure and routing benefits will be concentrated in central or affluent districts, deepening spatial and digital inequalities (Crawford, 2021; Eubanks, 2018).

8.5.3 Key Enablers for Scalable Smart Mobility

Several enablers can support citywide scale-up. Interoperable digital infrastructure is essential, and cities must adopt open standards, shared APIs, and modular designs that allow different systems—such as traffic control, public transit, and weather services—to communicate. European efforts such as the MaaS Alliance and the Open Mobility Foundation have made strides in this direction. Cross-sector governance is also vital. Cities like Helsinki have integrated public participation into MaaS development, while Barcelona employs a city data commons to ensure ethical oversight. Federated learning models offer another solution: rather than training AI from scratch in each city, these models allow knowledge to be shared across jurisdictions while preserving privacy and local specificity. Finally, national and regional data platforms, such as India's Urban Data Exchange (IUDX) and Europe's INSPIRE Directive, support data standardization and the replication of smart mobility solutions at scale.

9. CONCLUSION: REROUTING THE URBAN FUTURE

The development of AI-enabled mobility systems marks a significant shift in how cities address congestion, emissions, and accessibility. Rather than expanding roads or infrastructure, forward-looking cities are using data, algorithms, and predictive analytics to optimize movement patterns and promote sustainability (OECD/ITF, 2021). This signals a broader paradigm shift: from reactive planning to anticipatory governance, where cities are equipped not just to manage congestion, but to prevent it through proactive, equity-focused design.

The most successful examples of smart mobility—whether in Singapore, Amsterdam, or Helsinki—demonstrate that technology must be aligned with ethical governance, social inclusion, and environmental stewardship. As AI systems shape the future of urban mobility, cities must balance innovation with responsibility. The goal is not simply faster or more efficient movement, but cleaner, fairer, and more livable cities.

Figure 5. AI is reshaping urban mobility toward cleaner, fairer, and more responsive cities

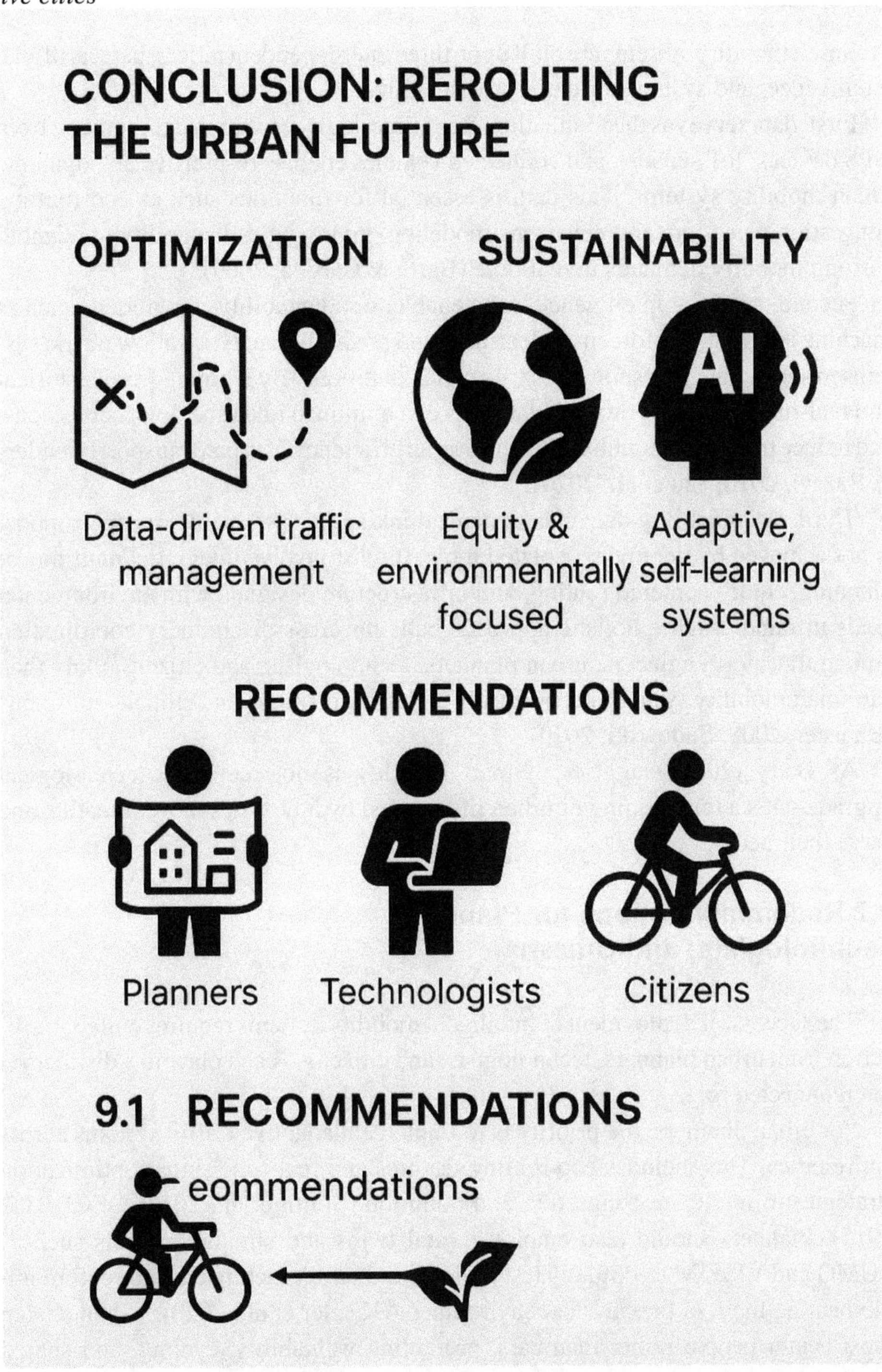

9.1 Synthesis: Lessons from Data, AI, and Sustainable Transport

Smart mobility systems are built upon three interdependent pillars: data, artificial intelligence, and systems-oriented sustainability.

First, data serves as the foundation. Real-time, high-resolution data sourced from GPS devices, IoT sensors, and connected vehicles enables responsive and dynamic urban mobility systems. This data is essential for functions such as eco-routing, congestion detection, and emissions modeling, giving cities the ability to adapt to shifting mobility demands in real time (Barth & Cargile, 2009).

Second, artificial intelligence is the enabler of adaptability. Techniques such as machine learning, reinforcement learning, and predictive analytics allow traffic systems to self-adjust in response to real-world conditions. By learning from historical and real-time data, AI-powered platforms can minimize idle time, lower emissions, and reduce travel delays, enhancing the overall efficiency of urban transport (Genders & Razavi, 2016; Liu et al., 2020).

Third, sustainability demands systems thinking. A sustainable transport model is not achieved by deploying isolated tools. It must involve integrated multimodal planning, equity-centered routing, and infrastructure designed with environmental goals in mind. Such a holistic approach calls for cross-disciplinary coordination among transport engineers, urban planners, technologists, and citizens. Only then can smart mobility systems become truly inclusive, resilient, and climate-conscious (Banister, 2008; Sadowski, 2020).

As Batty (2018) suggests, "Smart mobility is not merely a technological upgrade—it's a reimagining of urban life, guided by how cities move, breathe, and serve their people."

9.2 Recommendations for Planners, Technologists, and Citizens

The successful deployment of intelligent mobility systems requires collaborative action from urban planners, technologists, and citizens—each playing a distinct yet interconnected role.

For urban planners, the priority is to implement adaptive traffic systems across entire cities. This includes eco-routing features and real-time signal optimization strategies to minimize congestion and pollution (Smith et al., 2013; OECD/ITF, 2015). Planners should also employ digital twins and simulation tools such as SUMO and VISSIM to virtually test policies and infrastructure changes before implementing them in the physical environment (Ketzler et al., 2020). Urban design must center people rather than cars, promoting walkability, cycling, and shared transport, particularly in underserved areas where mobility equity is most needed (Lucas et al., 2016).

For technologists, it is essential to develop explainable AI systems that enhance transparency and build public trust in smart mobility platforms. Algorithms must be fairness-aware, actively working to prevent the amplification of spatial or socioeconomic inequalities in routing and service provision. Furthermore, developers must adhere to open data standards and robust privacy protocols such as those outlined in the GDPR to ensure ethical scalability and protect individual rights (Zuboff, 2019).

For citizens, everyday choices play a critical role. Choosing environmentally friendly routes, using shared or low-emission modes of transport—such as electric taxis, bike-share systems, or public transit—can directly contribute to greener cities. Citizens are also encouraged to engage actively with city consultation processes, report local mobility issues, and participate in smart city platforms to shape how technologies are implemented (Kitchin, 2016). Finally, individuals should advocate for equitable and transparent mobility systems, demanding that urban technologies reflect values of fairness, sustainability, and public accountability.

REFERENCES

Ada Lovelace Institute. (2020). *Algorithmic transparency: Principles and practice*. Ada Lovelace Institute.

Ahn, K., & Rakha, H. (2008). Eco-routing of vehicles based on minimizing fuel consumption and emissions. *Transportation Research Part D, Transport and Environment*, *13*(5), 377–385.

Apte, J. S., Marshall, J. D., Cohen, A. J., & Brauer, M. (2017). Addressing global mortality from ambient $PM_{2.5}$. *Environmental Science & Technology*, *51*(13), 7350–7357. PMID: 28562015

Banister, D. (2008). The sustainable mobility paradigm. *Transport Policy*, *15*(2), 73–80. DOI: 10.1016/j.tranpol.2007.10.005

Barcelona City Council. (2019). *Data Commons Charter*. Barcelona City Council.

Barth, M., & Cargile, K. (2009). Real-world carbon dioxide impacts of traffic congestion. *Transportation Research Record: Journal of the Transportation Research Board*, *2058*(1), 163–171. DOI: 10.3141/2058-20

Bigazzi, A. Y., & Figliozzi, M. (2012). Operational and environmental effects of bus rapid transit systems. *Transportation Research Record: Journal of the Transportation Research Board*, *2276*(1), 14–23.

Boriboonsomsin, K., Vu, A., & Barth, M. (2012). Eco-routing navigation systems. In *2012 IEEE Intelligent Vehicles Symposium* (pp. 1018–1023).

Craglia, M., Granell, C., Ostermann, F., & Bröring, A. (2018). Digital Earth and smart cities: Governance, policy, and social innovation. *International Journal of Digital Earth*, *11*(3), 219–237.

Crawford, K. (2021). *Data poverty: The everyday politics of data-driven technology*. Data & Society Research Institute.

Eubanks, V. (2018). *Automating inequality: How high-tech tools profile, police, and punish the poor*. St. Martin's Press.

Fujii, S., & Taniguchi, A. (2005). Reducing solo car use through personalized travel feedback. *Transportation Research Part D, Transport and Environment*, *10*(1), 1–19.

Gao, Z., Shi, Z., & Zhou, X. (2018). Integrating real-time IoT data into roadway network databases. *IEEE Transactions on Intelligent Transportation Systems*, *19*(4), 1100–1112.

Gehl, J. (2010). *Cities for people*. Island Press.

Genders, W. A. P., & Razavi, S. (2016). A survey of reinforcement learning methods for traffic signal control. *IEEE Transactions on Intelligent Transportation Systems, 17*(7), 1946–1959.

Gkartzonikas, C., & Gkritza, K. (2019). Towards real-time traffic flow and incident detection with IoT sensors. *Transport Reviews, 39*(2), 181–198.

Goodchild, M. F. (2021). Responsible data governance in smart cities. *Computers, Environment and Urban Systems, 88*, 101643.

Harrison, C., & Donnelly, I. A. (2011). A theory of smart cities. In *Proceedings of the 55th Annual Meeting of the ISSS* (pp. 1–15).

Hensher, D. A., Li, Z., & Mulley, C. (2020). Willingness to pay for Mobility-as-a-Service: Insights from Sydney. *Transportation, 47*(4), 1937–1956.

Hietanen, S. (2014). Mobility as a service: The new transport model? In *2014 World Transport Convention*.

Hörl, S., Cerqueira, R., & Axhausen, K. W. (2021). Gamified incentives for sustainable commuting. *Transportation Research Part A, Policy and Practice, 145*, 246–261.

Kanaroglou, P.. (2020). Real-time air quality monitoring with mobile sensors. *The Science of the Total Environment, 738*, 139677.

Ketzler, R., Franco, D., & Trupp, M. (2020). Digital twin applications in urban traffic management. *Transportation Research Procedia, 45*, 1–10.

Khattak, A. J., & De Palma, A. (1997). The impact of weather on traffic flow. *Transportation Research Part A, Policy and Practice, 31*(6), 475–491.

Kitchin, R. (2016). The ethics of smart cities and urban science. *Philosophical Transactions. Series A, Mathematical, Physical, and Engineering Sciences, 374*(2083), 20160115. DOI: 10.1098/rsta.2016.0115 PMID: 28336794

Laa, B., Balac, M., & Schüssler, N. (2020). Integrating micro-mobility in multimodal routing. *Journal of Transport Geography, 85*, 102710.

Land Transport Authority. (2021). *Singapore Electronic Road Pricing impact report*. Land Transport Authority.

Litman, T. (2020). *Transportation and environmental policy in the 21st century*. Victoria Transport Policy Institute.

Liu, H., Tian, Z., & Huang, W. (2020). Emission impacts of adaptive traffic control. *Transportation Research Part D, Transport and Environment*, *82*, 102324.

Lucas, K., Mattioli, G., Verlinghieri, E., & Guzman, A. (2016). Transport poverty and its adverse social consequences. *Proceedings of the Institution of Civil Engineers—Transport, 169*(6), 353–365. DOI: 10.1680/jtran.15.00073

Marsden, G., Docherty, I., & Anable, J. (2021). Integrating cycling and micro-mobility into urban transport models. *Transport Policy*, *103*, 1–10.

Menon, A. (2000). Electronic road pricing in Singapore: Success factors. *Journal of Transport Economics and Policy*, *34*(3), 329–346.

Mohamed, A. M.. (2021). Machine learning for predicting traffic emissions. *The Science of the Total Environment*, *781*, 146720.

Moreno, C., Allende, D., & Bosque, M. (2021). The 15-minute city concept in Paris. *Urban Planning*, *6*(2), 42–60.

OECD. (2021). *Regulating smart cities: Legal and institutional frameworks*. OECD Publishing.

Papageorgiou, M., Diakaki, C., Dinopoulou, V., Kotsialos, A., & Wang, Y. (2003). Review of road traffic control strategies. *Proceedings of the IEEE*, *91*(12), 2043–2067. DOI: 10.1109/JPROC.2003.819610

Pasquale, F. (2015). *The black box society: The secret algorithms that control money and information*. Harvard University Press. DOI: 10.4159/harvard.9780674736061

Pentland, A. (2012). *Building the new economy: Data as labor*. Harvard Business Review Press.

Pojani, D., & Stead, D. (2015). Sustainable urban transport in the developing world. *Transport Reviews*, *35*(3), 253–274.

Sadowski, J. (2020). When data is power: Exclusion and discrimination in the smart city. *New Media & Society*, *22*(10), 1758–1776.

Shaheen, S., & Cohen, A. (2019). *Shared micro-mobility policy toolkit*. University of California Transportation Center.

Shin, D., Park, Y.-J., & Chung, C.-Y. (2020). Public perceptions of surveillance in smart cities. *Government Information Quarterly*, *37*(2), 101444.

Smith, S. F., Barlow, G. J., & Kohls, A. G. (2013). Distributed real-time traffic signal control. *Transportation Research Part C, Emerging Technologies*, *24*, 168–184.

Snyder, F., & Ratti, C. (2023). Real-time urban emissions control through AI. *Environmental Research Letters*, *18*(2), 024003.

Sochor, J., Strömberg, H., & Karlsson, I. C. M. (2015). Implementing Mobility as a Service: Challenges and opportunities. *Transportation Research Procedia*, *14*, 3264–3273.

Sotonye, N., & Rajabifard, A. (2021). Digital twin applications in emissions forecasting. *Environmental Modelling & Software*, *143*, 105110.

Stevanovic, A. (2010). Adaptive traffic control systems: Domestic and foreign state of practice. *Transportation Research Record: Journal of the Transportation Research Board*, *2173*(1), 46–54.

Sunio, G., & Schmöcker, J.-D. (2017). Real-time travel choice nudges in smartphone apps. *Transportation Research Part F: Traffic Psychology and Behaviour*, *49*, 35–47.

Talebpour, A., & Mahmassani, H. S. (2016). Influence of connected vehicles on traffic flow. *Transportation Research Part C, Emerging Technologies*, *71*, 143–163. DOI: 10.1016/j.trc.2016.07.007

Thaler, R. H., & Sunstein, C. R. (2008). *Nudge: Improving decisions about health, wealth, and happiness*. Yale University Press.

Voigt, P., & Von dem Bussche, A. (2017). *The EU General Data Protection Regulation (GDPR)*. Springer. DOI: 10.1007/978-3-319-57959-7

Wang, C., & Chen, X. (2020). Real-time emission modeling in urban traffic networks. *Transportation Research Part D, Transport and Environment*, *84*, 102385.

Wei, H., Zheng, G., Liu, H. X., & Li, Z. (2019). Optimal traffic signal control with deep reinforcement learning. *Transportation Research Part C, Emerging Technologies*, *104*, 76–99.

Wiering, M. (2000). Multi-agent reinforcement learning for traffic light control. In *Proceedings of the Seventeenth International Conference on Machine Learning* (pp. 1151–1158).

World Health Organization. (2011). *Health in the green economy: Co-benefits to health of climate change mitigation – transport sector*. WHO.

Yang, C., & Barth, M. (2015). Eco-routing for reducing emissions: A review of methods. *Transportation Research Part D, Transport and Environment*, *34*, 36–52.

Zhang, L., Liu, C., & Wang, Y. (2019). GPS-based urban congestion detection. *IEEE Transactions on Intelligent Transportation Systems*, *20*(8), 2934–2942.

Zhang, Z., Wang, H., & Zhao, J. (2011). Eco-routing through elevation-aware methods. In *2011 IEEE Intelligent Vehicles Symposium* (pp. 668–673).

Zhou, Y.. (2021). Dynamic emissions modeling with IoT and AI. *Transportation Research Part D, Transport and Environment*, *95*, 102859.

Zuboff, S. (2019). *The age of surveillance capitalism: The fight for a human future at the new frontier of power*. PublicAffairs.

Chapter 9
Exploring Youth Perspectives on Traffic Governance:
A Predictive Analysis With ML Applications and Regression

Koppala Venugopal
http://orcid.org/0000-0002-3472-1477
Aditya Institute of Technology and Management, India

Sundara Rama Mangipudi Seshagiri
GVP College for Degree and P.G. Courses, India

Babar Javed
http://orcid.org/0000-0002-4127-3270
Manipal Academy of Higher Education, India

Madhavi Kappagantula
http://orcid.org/0000-0003-3178-439X
Acharya Institute of Graduate Studies, India

ABSTRACT

This study explores youth perspectives on traffic governance, focusing on the key factors influencing satisfaction levels and leveraging both regression and ML applications for predictive insights. The problem addressed lies in the gap between enforcement mechanisms and public perception, particularly among younger road users whose behaviors significantly impact road safety outcomes. Data were collected through a structured questionnaire and analyzed using statistical and

DOI: 10.4018/979-8-3373-4422-5.ch009

predictive modeling techniques to identify significant determinants of satisfaction. Findings of the results indicate that safety risk perceptions such as the dangers of speeding and mobile phone use while driving are strong positive predictors, while operational governance factors like prompt violation response, road maintenance, and clear signage enhance satisfaction. Interestingly, AI-based monitoring showed a negative association, suggesting skepticism toward current technology use. These findings highlight the importance of targeted safety education and context-specific governance improvements.

1. INTRODUCTION

1.1. Background of the Study

Road traffic injuries remain one of the most pressing public health and development challenges worldwide. The World Health Organization's most recent Global Status Report on Road Safety estimates 1.19 million road traffic deaths annually and identifies road traffic injuries as the leading cause of death among children and young people aged 5–29 years (WHO, 2023a; WHO, 2023b). Despite marginal declines since 2010, the pace of reduction lags far behind the target to halve deaths and injuries by 2030 under the UN's Decade of Action for Road Safety 2021–2030 (WHO, 2023a; WHO, 2021). The UN General Assembly has repeatedly urged Member States to accelerate implementation of the Global Plan, underlining road safety as essential for achieving broader Sustainable Development Goals related to health, equity, education, and economic productivity (UNGA, 2024).

Young people are disproportionately affected by unsafe mobility systems due to higher exposure, risk-taking propensities, and social norms around speed, distraction, and non-compliance (WHO, 2023a). International youth advocacy networks and UN partners emphasize that meaningful youth engagement in policy design, enforcement, and road-user education is central to effective traffic governance (YOURS, 2024). In many low- and middle-income countries, where 90% of road deaths occur, youth bear a significant share of fatalities and long-term disabilities, magnifying intergenerational social and economic impacts (WHO, 2023a; YOURS, 2024).

India records one of the world's largest absolute burdens of road crashes. Official statistics report persistently high mortality, with working-age populations and youth comprising the bulk of victims (MoRTH, 2023). Government statements in Parliament and national reports repeatedly note that roughly 60–66% of those killed are aged 18–34, underscoring the importance of targeting young road users in governance, infrastructure, and enforcement strategies (ORF, 2024). Persistent pedestrian risk and infrastructure deficits further highlight the need for comprehensive, people-centric road design and enforcement capacity.

Behavioral science presents robust frameworks to examine youth attitudes toward rule compliance, perceived behavioral control, subjective norms, and intentions related to risky behaviors such as speeding, distraction, and red-light running. The Theory of Planned Behavior (TPB) and its extensions have been widely applied to road safety behaviors, explaining substantial variance in intentions and observed violations (Chen et al., 2023; Yao et al., 2012). Recent empirical syntheses suggest TPB constructs, when expanded to include factors like risk perception, sensation seeking, and habit, can explain up to half of the variance in speeding and similar infractions, particularly among young drivers (Chen et al., 2023). These models justify survey-based designs that capture youth perspectives on enforcement fairness, infrastructure quality, licensing processes, and digital enforcement tools, which in turn inform predictive analytics.

Traffic safety analysis historically relied on regression-based Safety Performance Functions and discrete choice models to estimate crash frequency and severity. Advances in data availability and computational capacity have catalyzed a shift toward machine learning (ML) and AI methods which can capture complex, nonlinear relationships among human, vehicle, road, and environmental factors (Alazab et al., 2025; Karimpour et al., 2024; Sun et al., 2025). Comparative studies and reviews show gradient-boosting families (e.g., XGBoost, LightGBM), random forests, and deep learning often outperform traditional parametric models in predicting crash occurrence and injury severity, while explainable AI (e.g., SHAP) improves interpretability for policy (Karimpour et al., 2024; Duman et al., 2023; Krishnan et al., 2024).

Emerging applications target youth-relevant risksspeeding, distraction, and violation propensitiesusing telematics, smartphone sensing, contextual features, and survey-derived behavioral constructs. Studies have demonstrated that contextual variables (speed limits, traffic, weather) and driver profiles significantly improve prediction of risky events (Hurtado et al., 2023). ML pipelines tailored to teen or young drivers have identified speed control at merges and intersections and distraction-related behaviors as key predictors of crash severity (Ahmed et al., 2020; Dar et al., 2024). Novel hybrids that combine optimization heuristics with boosting algorithms are achieving strong performance on severity classification, suggesting policy-relevant use for hotspot prioritization and targeted interventions (Liu et al., 2025; Sun et al., 2025). Parallel work explores prediction and prevention of rule violations in trajectory-planning and enforcement contexts, signaling governance applications for automated alerts and proactive compliance.

Traffic governance increasingly integrates data pipelines from e-challan systems, licensing registries, blackspot mapping, telematics, and CCTV/ITS platforms. Pilot programs in Indian states to deploy AI-driven crash analytics, risk scoring, and enforcement modernization illustrate a policy turn toward data-informed governance and interoperability across platforms such as VAHAN-SARATHI (Times of India, 2025). International road safety plans stress sustainable financing, institutional coordination, and systems thinkingsafe roads, safe speeds, safe vehicles, safe road users, and post-crash care, creating a natural demand for predictive tools that can prioritize investment and enforcement where youth risks concentrate (WHO, 2021; UNGA, 2024).

1.2. Research Gap

Despite extensive global and national evidence documenting the burden of road traffic injuries and fatalities among young people (WHO, 2023a; MoRTH, 2023), three interconnected research gaps persist.

First, there is limited integration of youth perspectives into traffic governance research, particularly in low- and middle-income countries (LMICs) like India where youth constitute a significant proportion of both road crash victims and active road users. While youth-focused advocacy exists at global levels (YOURS, 2024), empirical studies rarely examine how young people perceive enforcement mechanisms, fairness, and infrastructural adequacy in a governance context.

Second, behavioral science models such as the Theory of Planned Behavior (Ajzen, 2020) and its extensions have been applied to traffic safety, but these have mostly focused on general driver populations rather than youth-specific risk patterns (Chen et al., 2023; Yao et al., 2012). Moreover, such models have been primarily used for explanatory purposes rather than being integrated with advanced predictive analytics that could support real-time or targeted governance interventions (Karimpour et al., 2024; Sun et al., 2025).

Third, while machine learning techniques have shown superior predictive performance over traditional statistical models in crash severity and violation prediction (Duman et al., 2023; Liu et al., 2025), the application of explainable AI to link model outputs with youth-centered governance actions remains underdeveloped. There is also a lack of hybrid approaches that combine behavioral constructs with complex, non-linear predictive models to both explain and forecast risky behaviors and perceptions (Krishnan et al., 2024; Alazab et al., 2025).

1.3. Problem Statement

Youth aged 18–34 account for the majority of road traffic fatalities and injuries in India, representing not only a critical public health issue but also a governance challenge with long-term socio-economic implications (MoRTH, 2023; Govt. of India, 2024). While national strategies emphasize enforcement modernization, infrastructure improvement, and road-user education, the absence of systematic incorporation of youth perspectives into traffic governance undermines the legitimacy and effectiveness of these interventions (ORF, 2024).

Existing studies have predominantly applied either behavioral models or machine learning techniques in isolation. This siloed approach fails to capture both the why (attitudinal and perceptual drivers) and what next (predictive patterns for targeted action) of youth-related traffic safety outcomes (Chen et al., 2023; Karimpour et al., 2024; Sun et al., 2025). Moreover, governance-oriented applications of predictive analyticsparticularly those informed by youth perceptionsremain rare in the Indian context, limiting the scope for proactive, data-driven policy interventions (Times of India, 2025; Alazab et al., 2025).

The present study addresses these gaps by combining behavioral theory and survey-based perception data with regression and machine learning models to both explain and predict youth-related traffic governance outcomes. By embedding explainable AI into the analysis, the study seeks to generate actionable insights that can enhance the fairness, efficiency, and youth responsiveness of traffic governance systems.

1.4. Objectives of the Study

1. To examine the perceptions, attitudes, and behavioral tendencies of youth towards traffic governance policies and practices, with a focus on compliance, awareness, and engagement in traffic safety measures.
2. To apply machine learning algorithms and regression models for predicting factors influencing youth perspectives on traffic governance, thereby identifying key determinants that shape their adherence to traffic norms.
3. To develop stakeholder-oriented insights and recommendations for enhancing traffic governance effectiveness, by integrating predictive findings into policy, enforcement, and awareness strategies targeting the youth demographic.

2. LITERATURE REVIEW

1) Youth Road Safety as a Public-Health and Governance Challenge

Road traffic injury remains a top global public-health burden and is the leading cause of death for people aged 5–29 (WHO/UN figures; recent syntheses) (Peden et al., 2023). India accounts for a disproportionate share of global fatalities, with multiple national assessments emphasizing persistent risks among young road users (Business Standard, 2023). These patterns situate youth driving and rule compliance squarely within traffic governance (laws, enforcement, infrastructure, licensing) and behavioral domains, making the population particularly relevant for predictive modeling and policy design.

2) Behavioral Drivers of Compliance

2.1 Theory of Planned Behavior (TPB)

The TPB posits that attitude, subjective norms, and perceived behavioral control (PBC) shape intention and behavior (Ajzen, 2020). Recent, driving-specific meta-analysis confirms TPB's predictive validity for speeding, DUI, distraction, and seatbelt use (McIlveen et al., 2024). Reviews focusing on speeding similarly underscore TPB pathways (Dey et al., 2024). These findings support your inclusion of attitude, norms, and PBC as key independent variables predicting youth self-reported compliance.

2.2 Risk Perception

Higher perceived risk is consistently associated with lower propensity for risky driving (Cândea et al., 2024; Cirillo et al., 2023). Young drivers often underestimate hazards and overestimate ability, a bias linked to crashes (Cirillo et al., 2023). Work combining infrastructure and credibility perceptions further shows that risk and "speed-limit credibility" predict compliance (Jamson et al., 2021). In predictive terms, risk perception typically adds incremental explanatory power beyond TPB variables.

2.3 Sensation Seeking & Related Traits

Personality traits, especially sensation seeking (SS)predict risky and aggressive driving, violations, and crash involvement (Du et al., 2022). Pooled effects for SS on risky driving are moderate and robust. These effects persist in youth samples and often operate indirectly through emotion regulation or attitudes (Du et al., 2022). SS is therefore a theoretically and empirically justified predictor in youth-focused compliance models.

3) Governance Perceptions and Compliance

3.1 Procedural Justice, Legitimacy, and Fairness

A large policing literature shows that procedural justice (voice, neutrality, respect, transparency) enhances perceived legitimacy, which in turn predicts voluntary compliance with rules (Tyler, 1990). Road-safety–specific scoping reviews similarly indicate that when traffic rules and enforcement are perceived as legitimate and fair, compliance rises, though definitions and measures have been heterogeneous (Varet et al., 2021). Newer work begins validating multidimensional legitimacy scales for traffic rules (Varet et al., 2024). Studies in drug-driving and automated enforcement contexts echo that perceived procedural justice and legitimacy shape intentions to obey (Waters et al., 2024; Hinds et al., 2025).

3.2 Digital/Automated Enforcement (Cameras, E-Challan) & Trust

Evidence generally supports automated traffic enforcement (ATE) in reducing violations/crashes (Cheng et al., 2025; Mohan et al., 2024; Safaei et al., 2024). However, public acceptance hinges on fairness, privacy, transparency, and revenue concerns (Snow, 2017). In India, e-challan and AI-enabled camera deployments show promise but also enforcement and recovery gaps (Conover et al., 2022; TOI-Nagpur, 2025; TOI-Pune, 2025; TOI-Mumbai-Pune, 2025). Perceived trust in digital enforcement plausibly influences both self-reported compliance and perceived effectiveness, making it a critical governance perception in youth samples.

3.3 Infrastructure Satisfaction & Perceived Credibility

Roadway lighting, markings, and signage quality are associated with better safety outcomes and safer speeds; poor design elevates risk and can make "safe behavior" harder (AASHTO-method CMFs in ROSA P, 2024; Al-Qurashi et al., 2023; Woolley & Hutchinson, 2021). Perceived infrastructure adequacy correlates with speed-limit credibility and risk perception (Jamson et al., 2021) and is therefore a theoretically sound governance perception for your model.

3.4 Licensing/Training and Youth Risk

Graduated driver licensing (GDL) systems reduce crashes among novices; conventional classroom-only training has mixed direct safety effects (systematic reviews and policy syntheses) (Wells et al., 2020; ITF/OECD, 2024). Youth perceptions that licensing and training are effective and fair may bolster legitimacy and compliance attitudes even when objective effects varyagain supporting inclusion as a governance perception.

4) Contextual & Demographic Covariates

Exposure (driving frequency, time of day, road type), past violations/crashes, and mode are standard correlates of risky driving and should be modeled to avoid confounding (ITF/OECD, 2024; Naturalistic-infrastructure reviews: Sun et al., 2024). Demographics (age bands within "youth," gender) often moderate behavioral effects.

5) Outcome Constructs and Measurement

5.1 Self-Reported Compliance

Self-reports of speeding, red-light running, mobile-phone use, and helmet/seatbelt use are widely used in behavioral road-safety research and are strongly predicted by TPB constructs, risk perception, and sensation seeking (McIlveen et al., 2024). Methodologically, combining validated multi-item scales with administrative markers (e.g., prior e-challans) can improve criterion validity.

5.2 Perceived Effectiveness of Traffic Governance

Perceived effectiveness measures (e.g., "how effective are cameras/e-challans/road design/licensing at improving safety?") are common in public opinion and intervention perception studies and are linked to support and compliance intentions (Pimpin et al., 2024). Because acceptance and legitimacy mediate real-world impacts of technology and policy, this outcome is well-justified.

6) Implications for Predictive Analytics (ML and Regression)

Recent reviews show rapid growth in ML applications for road-safety and behavior modeling alongside traditional regression (Al Bazi et al., 2024). For youth compliance prediction, non-linear ML (e.g., tree ensembles) can capture interactions among behavioral and governance-perception features (risk × sensation seeking; norms × legitimacy; exposure × infrastructure satisfaction), while regression yields interpretable effect estimates. Hybrid strategies (regularized GLMs + SHAP-audited GBMs) are recommended for policy-facing insight and accuracy.

7) Technological, Generational, and Analytical Foundations for Traffic Governance Research

7.1. AI Integration and Predictive Analytics in Governance

The integration of AI into service-oriented systems has shown promising results for enhancing decision-making. Venugopal et al. (2026) demonstrated how ChatGPT could be embedded into academic services for sustainable learning solutions, illustrating parallels with AI-enabled traffic governance where predictive analytics can streamline enforcement and monitoring. Venugopal (2024b) explored Gen Z's perspectives on ChatGPT in academic contexts, indicating generational openness to AI toolsan insight relevant to youth acceptance of AI-driven traffic management.

7.2. Generational Influence and Knowledge Transfer

Generational traits influence the adoption of governance innovations. Venugopal et al. (2025a) highlighted reverse mentoring as a mechanism through which Gen Z can transfer digital skills to older generations, a process equally relevant in modernizing traffic governance. Venugopal (2025a) examined Gen Z's reliance on AI in education, reinforcing the idea that younger demographics are more likely to support technology-based solutions in public systems.

7.3. Data Visualization and Statistical Interpretation

Effective traffic governance depends on the ability to interpret and present complex data. Rao et al. (2025) discussed visualization techniques for enhancing statistical thinking in higher education, offering transferable methods for representing traffic data to policymakers and the public.

7.4. Technology-Driven Transformation in Service Sectors

Venugopal and Vakamullu (2025) investigated technological innovations in the B2B hospitality sector, showing how digital adoption reshapes relationshipssimilar to how intelligent traffic systems could redefine interactions between enforcement agencies and citizens. Venugopal (2024c) studied reverse logistics in supply chains, emphasizing operational efficiency principles applicable to optimizing traffic flow and resource allocation.

7.5. Privacy, Security, and Public Trust

Public trust in AI-based governance systems hinges on privacy and security safeguards. Venugopal and Jagadeesh (2025) provided theoretical insights into security and privacy in social media environments, directly relevant to public concerns over AI surveillance in traffic regulation.

7.6. Community Engagement in Governance

Venugopal (2024a) explored waste management in semi-urban areas, revealing how community perspectives shape policy success, a concept equally vital for designing youth-focused traffic safety interventions. Venugopal et al. (2024) addressed customer consciousness toward food wastage, showing that targeted awareness initiatives can modify behavior, an approach transferable to road safety campaigns.

7.7. Predictive Modelling of Behavioral Trends

Venugopal and Nakkina (2025) examined higher education students' dependency on AI tools in an Indian context, providing a behavioral lens for studying technology adoption in governance. Venugopal, Das, and Badawy (2023) conducted a predictive analysis of Gen Z's ecological consciousness, illustrating the applicability of regression and machine learning techniques to understanding youth perspectives on societal issues, including traffic governance.

8) AI skepticism

Artificial intelligence skepticism has emerged as a critical construct in understanding public acceptance of algorithmic governance, particularly in domains such as traffic management where enforcement and decision making are increasingly automated. Prior research on algorithm aversion demonstrates that individuals often exhibit lower trust in algorithmic systems after observing even minor errors, despite evidence that such systems may outperform human judgment overall (Dietvorst, Simmons, & Massey, 2015). This skepticism is closely linked to perceived fairness, transparency, and accountability, which are central determinants of trust in automation (Lee & See, 2004). In governance contexts, studies on algorithmic decision making have shown that people frequently judge algorithmic policing and enforcement as less fair or less legitimate than equivalent human decisions, especially when decision processes are opaque (Araujo et al., 2020). Within traffic governance specifically, public attitudes toward automated enforcement mechanisms such as speed cameras are shaped not only by perceived effectiveness but also by concerns regarding surveillance, revenue motives, and procedural justice (RAC Foundation & Ipsos MORI, 2018). Youth perspectives add an additional layer to this discourse, as younger populations tend to demonstrate both high digital engagement and heightened concern regarding privacy, bias, and ethical use of AI systems (Anderson & Rainie, 2023). For a study titled Exploring Youth Perspectives on Traffic Governance: A Predictive Analysis with ML Applications and Regression, the literature suggests that AI skepticism may function as a significant predictor variable influencing perceived legitimacy of AI based enforcement, compliance intention, and overall acceptance of traffic governance technologies. Constructs such as perceived accuracy, transparency, fairness, privacy concern, and institutional trust can be operationalized within regression and machine learning models to predict youth support for AI assisted traffic systems. Thus, integrating AI skepticism into predictive modeling frameworks aligns with established theoretical foundations in algorithm aversion, trust in automation, and procedural justice, offering a robust explanatory lens for understanding youth responses to AI driven traffic governance.

9) Interdisciplinary works

Interdisciplinary scholarship increasingly integrates artificial intelligence, behavioral science, and urban planning to explain smart mobility transitions. Geels (2022) argues that socio technical transitions such as AI enabled mobility must be understood through institutional change, behavioral adaptation, and governance legitimacy rather than technological efficiency alone. Reviews of smart mobility ecosystems emphasize that AI driven transport systems succeed only when embedded within participatory governance, social innovation, and trust based urban frameworks (Goumiri et al., 2023). Research on algorithmic urban planning further highlights the need to align predictive AI tools with equity, transparency, and citizen engagement to avoid resistance and skepticism (Son et al., 2023). Studies combining AI and behavioral insights show that perceived fairness, explainability, and user control strongly shape public acceptance of automated mobility systems (Nikitas et al., 2020). Collectively, these interdisciplinary works demonstrate that smart mobility outcomes depend on the integration of AI capability, behavioral trust mechanisms, and inclusive urban governance structures (Geels, 2022; Goumiri et al., 2023).

3. METHODS AND MATERIALS

3.1. Research Design

The present study adopted a combination of descriptive and exploratory research designs to investigate the perspectives of youth toward traffic governance and compliance. The descriptive component aimed to measure and quantify behavioral, governance, contextual, and demographic variables influencing compliance, while the exploratory component sought to uncover latent patterns and predictors of these perspectives through advanced data analytics. The research incorporated a mixed-methods strategy, where quantitative analysis formed the primary methodological foundation and qualitative elements were used for contextual interpretation of patterns emerging from the data.

3.2. Sampling Technique and Population

The study employed a convenience sampling technique due to its feasibility in reaching a diverse set of young respondents within the study timeframe. The target population consisted of individuals aged 18–30 years, residing in urban and semi-urban areas, who possessed a valid driving license or regularly operated motor vehicles. Given the predictive modeling requirements, a minimum sample size was determined based on statistical adequacy for multiple regression and factor analysis, as well as classification algorithms in machine learning. Applying the rule-of-thumb for multivariate analysis (Hair et al., 2019), which suggests 5–10 respondents per variable, and considering the study's 20+ predictor variables, a sample size of 200 was considered sufficient to ensure analytical reliability and model stability.

3.3. Data Collection Procedure

A cross-sectional survey approach was utilized to collect data at a single point in time, ensuring temporal consistency and minimizing time-related bias. The questionnaire was administered both online, via Google Forms, and offline through physical distribution in educational institutions, youth clubs, and driving schools. Respondents were briefed on the study's objectives, assured of confidentiality, and informed consent was obtained before participation. Data collection spanned a period of four weeks to achieve the required sample size and ensure a balanced representation across demographic segments.

3.4. Data Analysis Methods

3.4.1. Exploratory Factor Analysis and Construct Validation

To establish construct validity of the independent variables, Exploratory Factor Analysis was conducted using Principal Component Analysis as the extraction method. Varimax rotation was applied to achieve orthogonal factor structures and enhance interpretability by maximizing variance explained by each component. Kaiser Meyer Olkin measure and Bartlett's Test of Sphericity were examined to confirm sampling adequacy and factorability of the correlation matrix. Items with factor loadings below 0.50 or cross loadings were removed to ensure clarity of constructs. This process helped identify latent dimensions underlying behavioral attitudes and governance perceptions, thereby reducing dimensionality and improving model efficiency.

3.4.2. Multiple Regression Analysis

Following validation and reliability confirmation, Multiple Regression Analysis was performed to examine the influence of independent variables on the dependent variables, namely compliance and governance perception. The regression model estimated standardized beta coefficients, significance levels, and overall model fit statistics such as R square and F values. Multicollinearity was assessed through Variance Inflation Factor values to ensure independent predictors did not excessively overlap. This inferential stage allowed hypothesis testing by identifying statistically significant predictors and quantifying their magnitude of influence on compliance behavior and governance satisfaction.

3.4.3. Predictive Modeling Using Machine Learning

To complement inferential findings, Orange Data Mining software was used for predictive analytics. Classification algorithms including Logistic Regression, Random Forest, and Naïve Bayes were applied to predict compliance categories and governance perception outcomes. Logistic Regression provided a baseline probabilistic model, Random Forest captured nonlinear interactions and improved classification robustness, and Naïve Bayes offered computational efficiency for categorical probability estimation. These algorithms enhanced predictive accuracy beyond traditional regression by capturing complex patterns within the dataset.

3.4.4. Clustering and Segmentation Analysis

Unsupervised learning techniques were employed to segment respondents based on behavioral characteristics and governance perception profiles. K Means clustering grouped respondents into homogeneous clusters using distance based partitioning, while Hierarchical Clustering generated dendrogram structures to visualize nested group relationships. These clustering methods helped identify distinct respondent segments such as high trust compliant users, skeptical but compliant users, and low trust non compliant users. Such segmentation provided actionable insights for targeted governance strategies.

3.4.5. Model Evaluation and Integrated Insights

Model performance was evaluated using accuracy, precision, and recall for classification algorithms, ensuring balanced assessment of prediction quality. For clustering, silhouette scores were computed to measure cohesion and separation among clusters. The integration of statistical inference and machine learning prediction created a comprehensive analytical framework. While regression analysis provided causal and explanatory insights aligned with study hypotheses, machine learning models offered predictive strength and segmentation clarity. This dual stage approach enhanced the validity, robustness, and practical applicability of the study's conclusions.

3.5. Conceptual and Research Flow

Figure 1. Conceptual and research flow

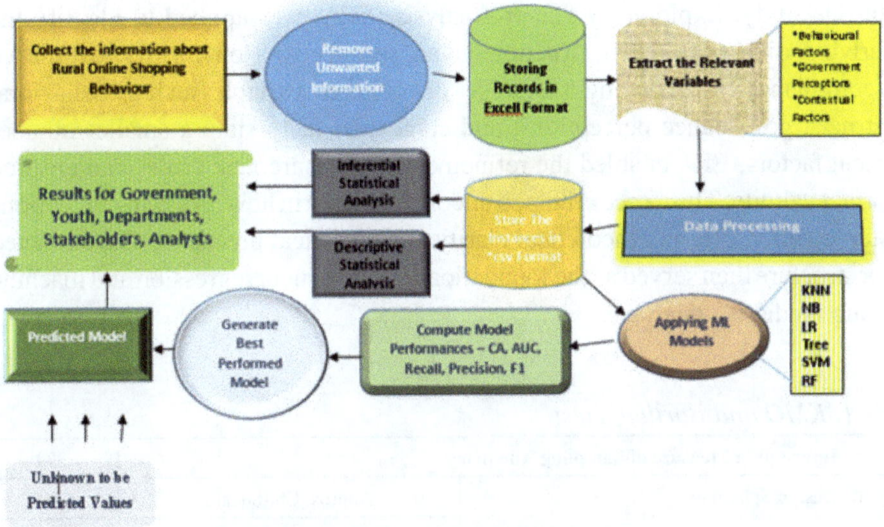

The conceptual and research flow from Figure 1 illustrates the systematic process adopted for analyzing rural online shopping behavior using statistical and machine learning approaches. The process begins with data collection, followed by the removal of irrelevant information to ensure data quality. Cleaned data is stored in Excel format, and relevant variables covering behavioral, government, perceptual, and contextual factors that are extracted for further analysis. The data undergoes processing and is stored in CSV format, after which descriptive and inferential statistical analyses are conducted. Subsequently, various machine learning models such as KNN, Naïve Bayes, Logistic Regression, Decision Tree, and SVM are applied. Model performances are evaluated using metrics like classification accuracy, AUC, recall, precision, and F1 score. The best-performing model is selected to predict unknown values, and the results are synthesized to provide actionable insights for government agencies, youth, departments, stakeholders, and analysts.

4. ANALYSIS AND INTERPRETATION

4.1. Exploratory Factor Analysis

In this study, Exploratory Factor Analysis (EFA) was applied to identify the underlying dimensions that structure youth perspectives on traffic governance and compliance. By reducing a large set of observed variablessuch as behavioral constructs, governance perceptions, and contextual factorsinto a smaller number of latent factors, EFA enabled the refinement of measurement scales and ensured construct validity. This process helped uncover patterns in how different perceptions cluster together, offering theoretical clarity and empirical precision. The extracted factor structure then served as the foundation for subsequent regression and machine learning predictive analyses.

Table 1. KMO and Bartlett's test

Kaiser-Meyer-Olkin Measure of Sampling Adequacy.		**.899**
Bartlett's Test of Sphericity	Approx. Chi-Square	5.055E3
	df	561
	Sig.	.000

Table 1 presents the results of the Kaiser-Meyer-Olkin (KMO) and Bartlett's Test of Sphericity, which assess the suitability of the data for factor analysis. The KMO value of 0.899 indicates a high degree of sampling adequacy, surpassing the recommended threshold of 0.80, and suggests that the patterns of correlations are compact enough to yield reliable factors. Bartlett's Test of Sphericity is statistically significant ($\chi^2 = 5055$, df = 561, $p < 0.001$), confirming that the correlation matrix is not an identity matrix and that significant relationships exist among the variables. These results validate the appropriateness of applying Exploratory Factor Analysis (EFA) to the dataset in order to extract underlying constructs related to youth perspectives on traffic governance.

Table 2. Total variance explained

Component	Initial Eigenvalues			Extraction Sums of Squared Loadings			Rotation Sums of Squared Loadings		
	Total	% of Variance	Cumulative %	Total	% of Variance	Cumulative %	Total	% of Variance	Cumulative %
1	11.576	34.047	34.047	11.576	34.047	34.047	6.053	17.802	17.802
2	4.707	13.845	47.893	4.707	13.845	47.893	5.006	14.725	32.527
3	2.935	8.631	56.524	2.935	8.631	56.524	4.570	13.441	45.968
4	1.732	5.093	61.617	1.732	5.093	61.617	3.937	11.581	57.549
5	1.491	4.384	66.001	1.491	4.384	66.001	2.181	6.415	63.963
6	1.182	3.475	**69.476**	1.182	3.475	69.476	1.874	5.513	69.476
7	.916	2.694	72.170						
8	.817	2.403	74.573						
9	.720	2.117	76.690						
10	.686	2.019	78.708						
11	.638	1.877	80.585						
12	.569	1.674	82.258						
13	.545	1.601	83.860						
14	.490	1.440	85.300						
15	.464	1.364	86.664						
16	.398	1.170	87.833						
17	.390	1.148	88.982						
18	.380	1.117	90.099						
19	.368	1.084	91.183						
20	.325	.955	92.138						
21	.306	.900	93.038						
22	.275	.810	93.847						
23	.272	.801	94.648						
24	.246	.723	95.372						
25	.233	.685	96.056						
26	.213	.626	96.682						
27	.205	.602	97.284						
28	.189	.556	97.840						
29	.177	.521	98.361						
30	.150	.442	98.803						
31	.142	.416	99.219						
32	.110	.324	99.544						
33	.084	.247	99.791						
34	.071	.209	100.000						

Extraction Method: Principal Component Analysis.

Table 2 shows the results of the Total Variance Explained from the Principal Component Analysis, which highlights the number of components extracted and the proportion of variance each accounts for. The initial eigenvalues indicate that six components have eigenvalues greater than 1, meeting the Kaiser criterion for factor retention. These six components collectively explain 69.476% of the total variance, which is considered robust for social science research (Hair et al., 2019). After rotation, the variance is more evenly distributed, with Component 1 contributing 17.802%, Component 2 contributing 14.725%, Component 3 contributing 13.441%, Component 4 contributing 11.581%, Component 5 contributing 6.415%, and Component 6 contributing 5.513%. This redistribution after rotation enhances interpretability, allowing clearer identification of factor loadings. The results confirm that the dataset contains a well-defined factor structure, making it suitable for further regression and machine learning predictive analyses in the context of youth perspectives on traffic governance.

Table 3. Rotated component matrix[a]

	Component					
	1	2	3	4	5	6
Obeying traffic rules is important for my safety						.755
Following traffic laws improves the overall road experience.	.501					
Traffic rules are generally beneficial for all road users.	.670					
My friends think I should obey traffic rules.						.494
My family expects me to follow traffic laws.	.810					
Most people important to me believe in safe driving practices.	.785					
I am confident I can follow traffic rules in all situations.	.752					
Even in heavy traffic, I can comply with traffic laws.	.820					
I can avoid risky driving behaviors despite time pressure.	.838					
Disobeying traffic rules increases the risk of accidents.	.552					
Driving over the speed limit is highly dangerous.	.557					
Using a phone while driving/riding greatly increases accident risk.	.705					
I enjoy driving at high speeds.		.793				
I sometimes take risks while driving for excitement.		.849				
I find thrill in overtaking vehicles aggressively.		.927				
Traffic police apply rules fairly to everyone.		.909				
Enforcement is impartial regardless of a person's background.		.925				
I trust traffic authorities to make unbiased decisions.		.893				
Traffic violations are penalized consistently in my city.				.519		
Enforcement measures are effective in reducing violations.				.722		

continued on following page

Table 3. Continued

	Component					
	1	**2**	**3**	**4**	**5**	**6**
Traffic authorities respond promptly to violations.				.716		
E-challan systems are reliable and accurate.				.739		
CCTV surveillance improves traffic discipline.				.725		
AI-based monitoring reduces human bias in enforcement.					.843	
Roads in my city are well maintained.					.854	
Road signs and markings are clear and visible.				.622		
Street lighting is adequate for safe driving.			.547			
Obtaining a driving license is straightforward and fair.			.713			
Driving tests adequately assess road safety knowledge.			.659			
Road safety education programs are accessible to the public			.769			
I drive/ride regularly (more than 3 times per week).			.690			
My travel mostly involves urban roads.			.683			
I have received at least one traffic violation notice in the past year.			.645			
I have been involved in a road accident in the past two years.			.665			

Extraction Method: Principal Component Analysis.
Rotation Method: Varimax with Kaiser Normalization.
a. Rotation converged in 7 iterations.

Table 3 presents the Rotated Component Matrix derived from Principal Component Analysis with Varimax rotation, showing the factor loadings of items on six extracted components. The grouping of high-loading variables suggests meaningful latent constructs: Component 1 represents subjective norms and compliance confidence, with strong loadings from family expectations, peer influence, and self-efficacy in obeying traffic laws. Component 2 reflects risk-taking and thrill-seeking attitudes, indicated by high loadings on enjoyment of high speeds, aggressive overtaking, and risk behavior for excitement. Component 3 captures perceived fairness and trust in enforcement, with high associations to impartiality of traffic police and unbiased decision-making. Component 4 pertains to effectiveness and consistency of enforcement measures, including e-challan reliability, CCTV monitoring, and prompt responses. Component 5 represents infrastructure quality and accessibility, with strong loadings on road maintenance, signage clarity, and street lighting. Component 6 relates to driving exposure and violation history, involving frequency of driving, urban road use, and past violations or accidents. The clean separation of items into these components validates the factor structure, enabling targeted interpretation of how perceptions, attitudes, and environmental conditions shape compliance and governance in road safety.

4.2. Multiple Regression Analysis

In the context of this research on youth perspectives in traffic governance, MRA helps quantify the extent to which factors such as subjective norms, risk-taking attitudes, trust in enforcement, and infrastructure quality collectively influence compliance with traffic laws.

- Behavioral Factors(Theory of Planned Behavior and extensions)

Table 4. Model summary

Model	R	R Square	Adjusted R Square	Std. Error of the Estimate
1	.613[a]	.376	.325	.65357

a. Predictors: (Constant), I find thrill in overtaking vehicles aggressively., My family expects me to follow traffic laws., Obeying traffic rules is important for my safety, Driving over the speed limit is highly dangerous., My friends think I should obey traffic rules., Disobeying traffic rules increases the risk of accidents., I am confident I can follow traffic rules in all situations., Following traffic laws improves the overall road experience., Traffic rules are generally beneficial for all road users., Most people important to me believe in safe driving practices., Even in heavy traffic, I can comply with traffic laws., Using a phone while driving/ riding greatly increases accident risk., I enjoy driving at high speeds., I sometimes take risks while driving for excitement., I can avoid risky driving behaviors despite time pressure.

Table 4 presents the Model Summary of the Multiple Regression Analysis, showing an R value of 0.613, which indicates a moderate positive correlation between the predictors and the dependent variable. The Adjusted R Square value of 0.325, which accounts for the number of predictors, still reflects a meaningful explanatory power, suggesting that the model is statistically useful despite possible extraneous variables.

Table 5. ANOVA[b]

Model		Sum of Squares	df	Mean Square	F	Sig.
1	Regression	47.404	15	3.160	7.398	.000[a]
	Residual	78.596	184	.427		
	Total	126.000	199			

a. Predictors: (Constant), I find thrill in overtaking vehicles aggressively., My family expects me to follow traffic laws., Obeying traffic rules is important for my safety, Driving over the speed limit is highly dangerous., My friends think I should obey traffic rules., Disobeying traffic rules increases the risk of accidents., I am confident I can follow traffic rules in all situations., Following traffic laws improves the overall road experience., Traffic rules are generally beneficial for all road users., Most people important to me believe in safe driving practices., Even in heavy traffic, I can comply with traffic laws., Using a phone while driving/ riding greatly increases accident risk., I enjoy driving at high speeds., I sometimes take risks while driving for excitement., I can avoid risky driving behaviors despite time pressure.

b. Dependent Variable: Overall, I am satisfied with the current traffic governance system.

Table 5 presents the ANOVA results for the regression model, showing that the regression sum of squares is 47.404 with a mean square of 3.160, while the residual sum of squares is 78.596. The calculated F-value of 7.398, with a significance level of 0.000, indicates that the regression model is statistically significant at the 1% level. This implies that the set of predictors, including safety attitudes, social influences, and risk-related behaviors, collectively have a meaningful impact on the dependent variablesatisfaction with the current traffic governance system.

Table 6. Coefficients[a]

Model		Unstandardized Coefficients		Standardized Coefficients		
		B	Std. Error	Beta	t	Sig.
1	(Constant)	1.794	.312		5.745	.000
	Obeying traffic rules is important for my safety	.015	.063	.017	.238	.812
	Following traffic laws improves the overall road experience.	.102	.077	.110	1.322	.188
	Traffic rules are generally beneficial for all road users.	-.138	.076	-.156	-1.82	.070
	My friends think I should obey traffic rules.	-.033	.064	-.040	-.510	.611
	My family expects me to follow traffic laws.	.095	.088	.111	1.08	.280
	Most people important to me believe in safe driving practices.	.119	.083	.137	1.42	.155
	I am confident I can follow traffic rules in all situations.	.005	.082	.006	.060	.952
	Even in heavy traffic, I can comply with traffic laws.	.068	.095	.072	.709	.479
	I can avoid risky driving behaviors despite time pressure.	.015	.093	.018	.164	.870
	Disobeying traffic rules increases the risk of accidents.	.060	.073	.069	.819	.414
	Driving over the speed limit is highly dangerous.	.155	.075	.174	2.07	.040
	Using a phone while driving/riding greatly increases accident risk.	.176	.080	.211	2.192	.030
	I enjoy driving at high speeds.	-.072	.052	-.133	-1.37	.171
	I sometimes take risks while driving for excitement.	-.070	.073	-.096	-.959	.339
	I find thrill in overtaking vehicles aggressively.	.070	.065	.133	1.078	.282

a. Dependent Variable: Overall, I am satisfied with the current traffic governance system.

Table 6 presents the coefficients of the multiple regression model, identifying the individual contribution of each predictor to satisfaction with the current traffic governance system. Among the predictors, “Driving over the speed limit is highly dangerous” ($B = 0.155$, $p = 0.040$) and “Using a phone while driving/riding greatly increases accident risk” ($B = 0.176$, $p = 0.030$) show a statistically significant positive influence on the dependent variable, indicating that stronger agreement with these safety perceptions is associated with higher satisfaction levels. These results imply that personal safety risk perceptions play a more decisive role than social expectations or general law-following attitudes in shaping satisfaction with traffic governance.

• Government Perceptions

Table 7. Model summary

Model	R	R Square	Adjusted R Square	Std. Error of the Estimate
1	.669[a]	.447	.402	.61529

a. Predictors: (Constant), Road safety education programs are accessible to the public, Enforcement is impartial regardless of a person's background., AI-based monitoring reduces human bias in enforcement., Traffic violations are penalized consistently in my city., Street lighting is adequate for safe driving., Enforcement measures are effective in reducing violations., Road signs and markings are clear and visible., CCTV surveillance improves traffic discipline., Driving tests adequately assess road safety knowledge., Roads in my city are well maintained., E-challan systems are reliable and accurate., Traffic authorities respond promptly to violations., Obtaining a driving license is straightforward and fair., I trust traffic authorities to make unbiased decisions., Traffic police apply rules fairly to everyone.

Table 7 shows that the model explaining government perceptions in traffic governance has a correlation coefficient (R) of 0.669, indicating a strong positive relationship between the predictors and the dependent variable. The Adjusted R Square of 0.402 accounts for the number of predictors, confirming that the model remains robust after adjusting for overfitting.

Table 8. ANOVAb

Model		Sum of Squares	df	Mean Square	F	Sig.
	Regression	56.340	15	3.756	9.921	.000[a]
1	Residual	69.660	184	.379		
	Total	126.000	199			

a. Predictors: (Constant), Road safety education programs are accessible to the public, Enforcement is impartial regardless of a person's background., AI-based monitoring reduces human bias in enforcement., Traffic violations are penalized consistently in my city., Street lighting is adequate for safe driving., Enforcement measures are effective in reducing violations., Road signs and markings are clear and visible., CCTV surveillance improves traffic discipline., Driving tests adequately assess road safety knowledge., Roads in my city are well maintained., E-challan systems are reliable and accurate., Traffic authorities respond promptly to violations., Obtaining a driving license is straightforward and fair., I trust traffic authorities to make unbiased decisions., Traffic police apply rules fairly to everyone.

b. Dependent Variable: Overall, I am satisfied with the current traffic governance system.

Table 8 presents the ANOVA results for the regression model examining government perceptions in traffic governance. The regression sum of squares is 56.340, while the residual sum of squares is 69.660, with a total sum of 126.000. The F-value of 9.921 and the corresponding significance level of 0.000 indicate that the model is statistically significant, meaning that the set of predictorssuch as accessibility of road safety education, impartial enforcement, AI-based monitoring, and reliable E-challan systemsjointly have a meaningful impact on public satisfaction with the traffic governance system.

Table 9. Coefficients[a]

Model		Unstandardized Coefficients		Standardized Coefficients		
		B	Std. Error	Beta	t	Sig.
1	(Constant)	1.153	.289		3.995	.000
	Traffic police apply rules fairly to everyone.	.041	.071	.078	.585	.559
	Enforcement is impartial regardless of a person's background.	-.086	.079	-.151	-1.083	.280
	I trust traffic authorities to make unbiased decisions.	.012	.063	.021	.187	.852
	Traffic violations are penalized consistently in my city.	.007	.053	.009	.131	.896
	Enforcement measures are effective in reducing violations.	.053	.079	.059	.676	.500
	Traffic authorities respond promptly to violations.	.206	.089	.216	2.303	.022
	E-challan systems are reliable and accurate.	-.004	.079	-.005	-.057	.955
	CCTV surveillance improves traffic discipline.	-.120	.076	-.139	-1.57	.117
	AI-based monitoring reduces human bias in enforcement.	-.137	.059	-.200	-2.33	.021
	Roads in my city are well maintained.	.140	.063	.199	2.236	.027
	Road signs and markings are clear and visible.	.178	.078	.184	2.282	.024
	Street lighting is adequate for safe driving.	.055	.062	.063	.890	.375
	Obtaining a driving license is straightforward and fair.	.008	.109	.008	.077	.939
	Driving tests adequately assess road safety knowledge.	.022	.090	.023	.247	.805
	Road safety education programs are accessible to the public	.361	.078	.381	4.634	.000

a. Dependent Variable: Overall, I am satisfied with the current traffic governance system.

Table 9 presents the coefficients of the multiple regression analysis used to assess the influence of various governance-related factors on overall public satisfaction with the traffic governance system. Traffic authorities respond promptly to violations (B = 0.206, p = 0.022), roads being well maintained (B = 0.140, p = 0.027), and road signs and markings being clear and visible (B = 0.178, p = 0.024) are all positive and significant predictors, indicating that these operational aspects strongly enhance public satisfaction. AI-based monitoring reduces human bias in enforcement has a negative but significant coefficient (B = -0.137, p = 0.021), suggesting possible public scepticism or perceived shortcomings in current AI enforcement practices. The strongest positive effect is from road safety education programs being accessible to the public (B = 0.361, p = 0.000), highlighting the importance of awareness and education in improving governance perceptions.

- Contextual Factors

Table 10. Model summary

Model	R	R Square	Adjusted R Square	Std. Error of the Estimate
1	.584[a]	.342	.328	.65392

a. Predictors: (Constant), I have been involved in a road accident in the past two years. My travel mostly involves urban roads, I have received at least one traffic violation notice in the past year, I drive/ride regularly (more than 3 times per week).

Table 10 shows that contextual factors have a moderate positive correlation (R = 0.584) with the dependent variable, explaining about 34.2% of the variance (R^2 = 0.342, Adjusted R^2 = 0.328) with a standard error of 0.65392. This indicates that variables such as past road accident involvement, predominant use of urban roads, receipt of traffic violation notices, and frequent driving or riding significantly contribute to shaping perceptions toward the traffic governance system, though other unexplored factors may account for the remaining variance.

Table 11. ANOVA[b]

Model		Sum of Squares	df	Mean Square	F	Sig.
1	Regression	43.032	4	10.758	25.158	.000[a]
	Residual	82.957	194	.428		
	Total	125.990	198			

a. Predictors: (Constant), I have been involved in a road accident in the past two years, My travel mostly involves urban roads, I have received at least one traffic violation notice in the past year, I drive/ride regularly (more than 3 times per week).

b. Dependent Variable: Overall, I am satisfied with the current traffic governance system.

Table 11 reveals that the regression model for contextual factors is statistically significant (F = 25.158, p < 0.001), indicating that the predictors collectively explain a meaningful portion of the variance in overall satisfaction with the current traffic governance system. The model shows that variables such as prior accident involvement, predominant travel on urban roads, receipt of traffic violation notices, and frequent driving or riding significantly influence perceptions, with a regression sum of squares of 43.032 compared to the residual sum of squares of 82.957.

Table 12. Coefficients[a]

Model		Unstandardized Coefficients		Standardized Coefficients		
		B	Std. Error	Beta	t	Sig.
1	(Constant)	1.516	.283		5.359	.000
	I drive/ride regularly (more than 3 times per week).	.319	.069	.353	4.612	.000
	My travel mostly involves urban roads.	.214	.073	.220	2.949	.004
	I have received at least one traffic violation notice in the past year.	-.096	.067	-.109	-1.43	.152
	I have been involved in a road accident in the past two years.	.215	.074	.211	2.91	.004

a. Dependent Variable: Overall, I am satisfied with the current traffic governance system.

Table 12 shows that among the contextual factors, driving or riding regularly ($B = 0.319, p < 0.001$), predominantly traveling on urban roads ($B = 0.214, p = 0.004$), and having been involved in a road accident in the past two years ($B = 0.215, p = 0.004$) significantly and positively influence satisfaction with the current traffic governance system. In contrast, receiving at least one traffic violation notice in the past year had no significant effect ($p = 0.152$). These findings indicate that frequent road usage and prior accident experience are associated with higher satisfaction, possibly due to greater awareness of governance measures.

4.3 Machine Learning Applications

Machine learning offers powerful analytical capabilities for studying public perceptions and behaviors in traffic governance systems. By processing large datasets from surveys, traffic records, and urban mobility patterns, it can identify key predictors of citizen satisfaction. Advanced algorithms enable accurate prediction of perceptions based on contextual and demographic factors. These insights support policymakers in designing data-driven, targeted interventions to improve road safety and governance effectiveness.

Table 13. Classification of algorithm

Model	AUC	CA	F1	Prec	Recall	MCC
Naive Bayes	0.805	0.160	0.222	0.673	0.160	0.178
Logistic Regression	0.876	0.750	0.746	0.751	0.750	0.585
Tree	0.991	0.910	0.911	0.916	0.910	0.857
SVM	0.952	0.855	0.844	0.854	0.855	0.762
Random Forest	0.990	0.920	0.919	0.924	0.920	0.871
kNN	0.885	0.670	0.652	0.691	0.670	0.455

Table 13 presents the comparative performance of various machine learning models applied to the present research theme on predicting public satisfaction with the current traffic governance system. Among the models tested, Random Forest emerged as the top performer, with an AUC of 0.990, classification accuracy (CA) of 0.920, and an F1 score of 0.919, indicating a highly effective balance between precision and recall in capturing the determinants of satisfaction. Decision Tree closely followed, with an AUC of 0.991 and CA of 0.910, showing strong predictive ability. On the other hand, Naive Bayes demonstrated the weakest performance, with very low CA (0.160) and recall (0.160), highlighting its inadequacy for this dataset. SVM and Logistic Regression yielded reliable and balanced results, whereas kNN showed moderate predictive strength, suggesting that ensemble and tree-based approaches better align with the complexity of factors influencing traffic governance satisfaction.

Table 14. Confusion matrix

		Predicted					
		1	2	3	4	5	∑
Actual	1	3	0	0	0	0	3
	2	0	1	0	1	0	2
	3	0	0	24	6	0	30
	4	0	0	0	94	8	102
	5	0	0	0	1	62	63
	∑	3	1	24	102	70	200

Table 14 shows the confusion matrix for the classification model applied to the present research theme on predicting public satisfaction with the current traffic governance system. The results indicate that the model performs exceptionally well in correctly identifying classes 4 and 5, with 94 and 62 accurate predictions respectively, reflecting strong alignment in the higher satisfaction categories. However, there is some misclassification between classes 3 and 4, as 6 cases from class 3 were predicted as class 4, and 8 cases from class 4 were predicted as class 5. Lower satisfaction levels (classes 1 and 2) have fewer samples and show occasional confusion, which is likely due to the small representation in the dataset. Overall, the matrix highlights the model's strength in predicting higher satisfaction categories while suggesting room for improvement in differentiating mid-level responses.

Figure 2. ROC analysis

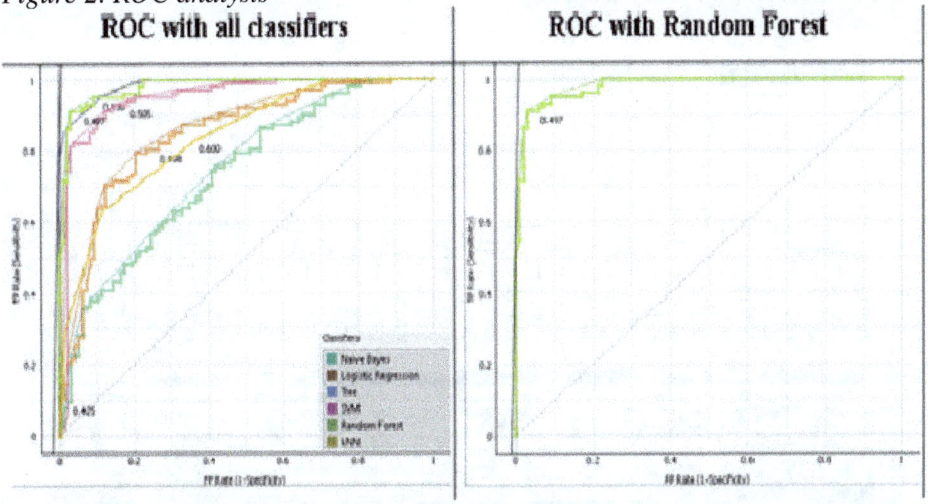

The ROC curves in Figure 2 illustrate the classification performance for the current research theme on predicting public satisfaction with traffic governance. The left panel, showing ROC curves for all classifiers, reveals that Tree, SVM, and Random Forest achieve near-perfect discrimination, with AUC values close to 1.0, indicating exceptional predictive power. Logistic Regression also demonstrates strong performance, while Naive Bayes and kNN lag behind, showing lower true positive rates for higher false positive rates. The right panel highlights the ROC curve for the Random Forest classifier specifically, which maintains a steep ascent toward the top-left corner, confirming its superior ability to balance sensitivity and specificity. This reinforces Random Forest as the most effective model for the dataset, aligning with its top performance in other evaluation metrics.

Figure 3. Correlation and clustering

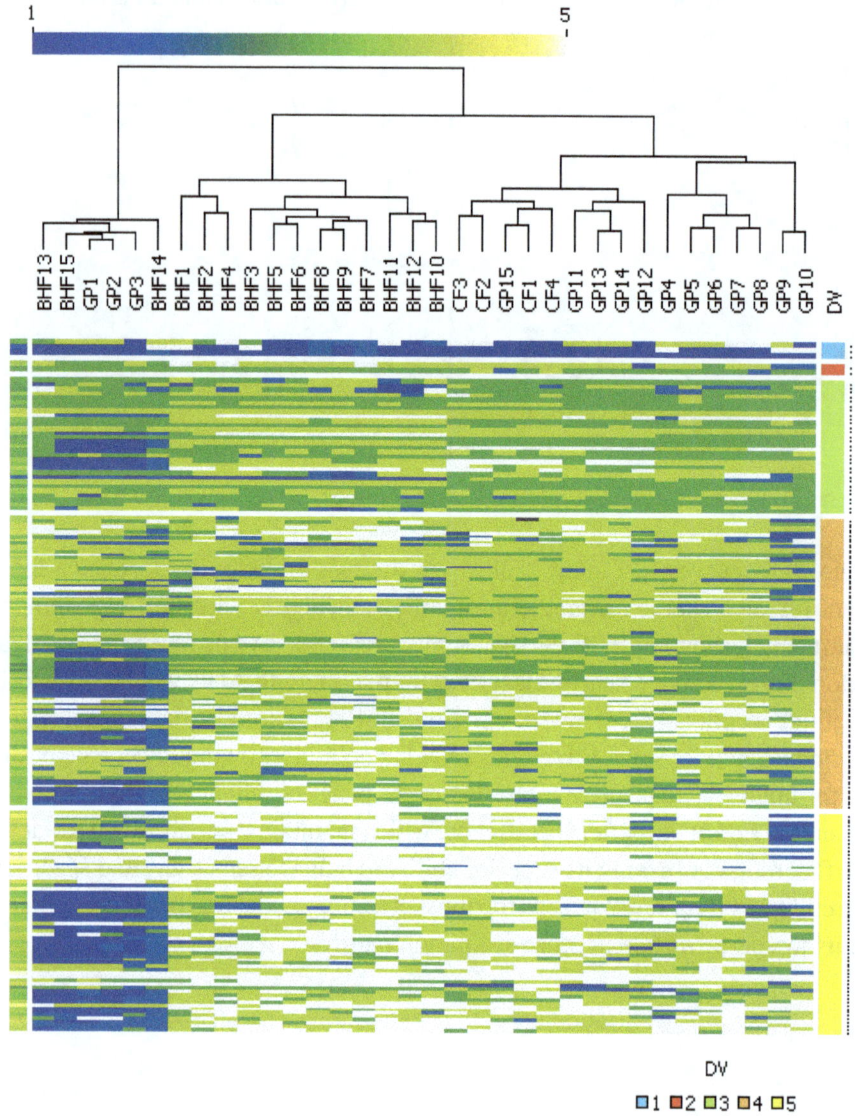

Figure 3 presents a heatmap with hierarchical clustering that reveals both the correlation patterns and grouping tendencies among variables (BHF, GP, CF, and DV groups) based on their response values, represented on a color scale from dark blue (low) to yellow (high). The dendrogram at the top shows that certain BHF and GP items cluster closely together, indicating strong similarity in response patterns, while CF items form separate sub-clusters. The horizontal segmentation into clusters (labeled 1 to 5) indicates distinct respondent groups with shared characteristics, where some clusters display consistently higher values and others lean toward lower or mixed values. This structure suggests that different youth segments exhibit varying attitudes and perceptions toward traffic governance, with some groups more aligned in their compliance and perception patterns than others. Such insights are valuable for tailoring targeted interventions and awareness strategies based on correlated behavioral traits within each cluster.

Figure 4. PCA

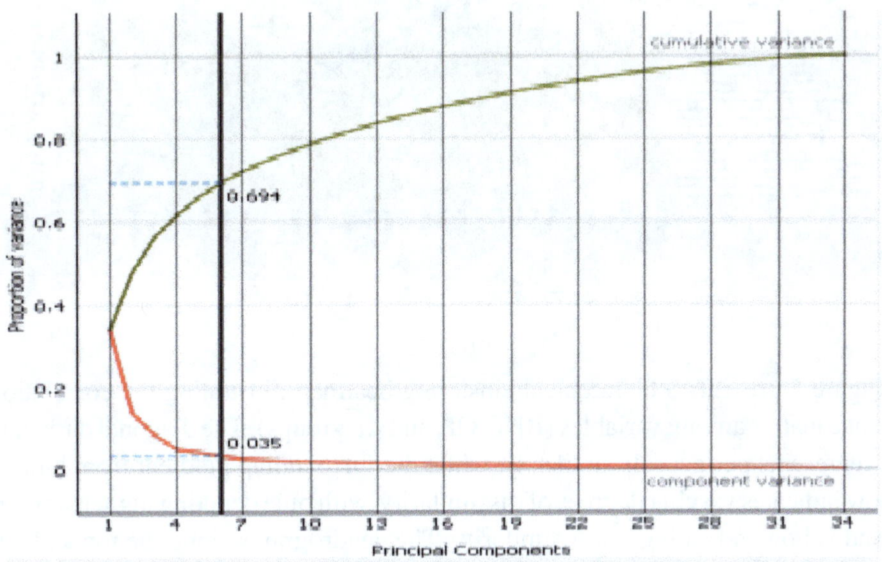

Figure 4 presents the Principal Component Analysis (PCA) results for the study on predicting public satisfaction with traffic governance. The scree plot illustrates the proportion of variance explained by each principal component, with the red curve representing the individual component variance and the yellow curve showing the cumulative variance. The first few components capture the majority of the data variability, with the cumulative variance reaching approximately 69.4% by the seventh component. Beyond this point, additional components contribute minimally, with individual variance dropping to around 3.5%. This indicates that dimensionality reduction to the first seven components can effectively preserve most of the informational content, enhancing computational efficiency while maintaining predictive accuracy in the traffic governance satisfaction model.

Figure 5. Distance map

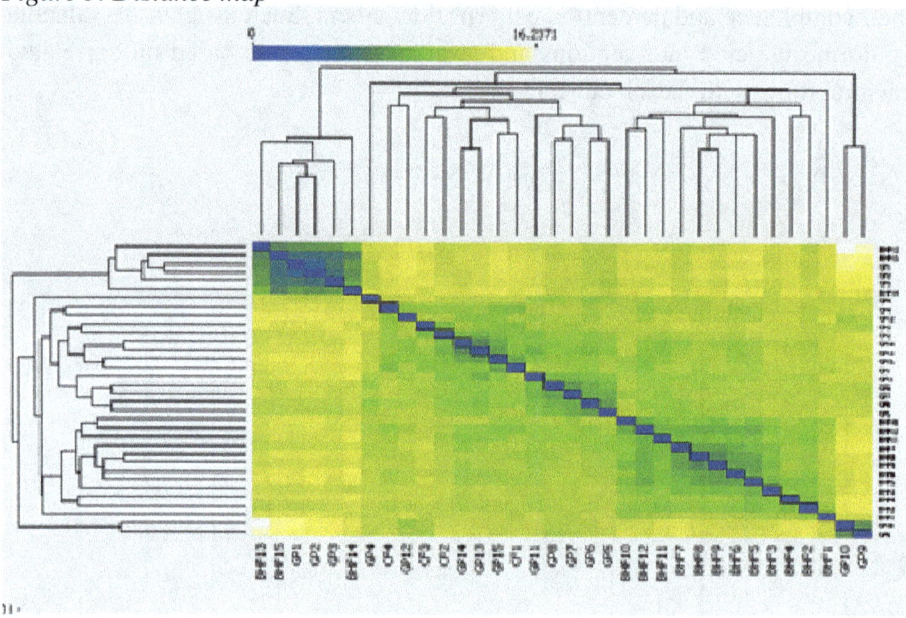

Figure 5 presents a hierarchical clustering heatmap illustrating the correlation distance matrix among variables (BHF, GP, and CF groups). The diagonal dark blue line represents perfect self-similarity, while the surrounding gradient from blue to yellow indicates varying degrees of dissimilarity, with blue denoting higher similarity and yellow indicating lower similarity. The dendrograms along the top and left edges reveal clear groupings, with certain BHF and GP variables clustering closely, suggesting similar response patterns, while other items form distinct sub-clusters. These correlation structures highlight natural groupings in youth perceptions related to traffic governance, enabling the identification of closely aligned variables that may reflect shared behavioral tendencies or common influencing factors. Such insights are instrumental in designing targeted awareness programs or interventions that leverage the strongest inter-variable associations.

Table 15. Feature ranking

Rank	Features	Info. gain	Gain ratio	ANOVA	χ^2	ReliefF
1	BHF1	0.091	0.062	4.082	14.409	0.014
2	BHF2	0.118	0.081	10.377	18.122	0.018
3	BHF3	0.118	0.079	4.126	9.744	0.027
4	BHF4	0.116	0.074	5.350	21.790	0.031
5	BHF5	0.148	0.096	17.144	32.463	0.048
6	BHF6	0.155	0.102	18.078	17.738	0.058
7	BHF7	0.209	0.117	11.439	14.938	0.085
8	BHF8	0.194	0.113	12.983	15.074	0.065
9	BHF9	0.152	0.102	13.851	14.130	0.038
10	BHF10	0.154	0.100	12.335	28.162	0.046
11	BHF11	0.170	0.112	15.930	17.910	0.052
12	BHF12	0.167	0.109	21.826	35.240	0.045
13	BHF13	0.177	0.089	1.325	5.075	0.077
14	BHF14	0.153	0.084	0.589	2.704	0.035
15	BHF15	0.133	0.068	0.957	4.290	0.064
16	GP1	0.131	0.068	0.612	3.270	0.079
17	GP2	0.141	0.074	0.529	2.635	0.079
18	GP3	0.093	0.049	1.473	6.760	0.027
19	GP4	0.163	0.107	4.048	28.922	0.060
20	GP5	0.129	0.085	8.638	28.753	0.039
21	GP6	0.186	0.123	14.520	34.671	0.072
22	GP7	0.146	0.095	8.313	29.314	0.039
23	GP8	0.125	0.084	5.750	21.200	0.050
24	GP9	0.142	0.073	1.138	5.827	0.047
25	GP10	0.290	0.149	3.393	9.780	0.078
26	GP11	0.215	0.145	14.964	22.032	0.076
27	GP12	0.129	0.085	13.308	22.373	0.055
28	CF1	0.231	0.151	22.597	52.762	0.024
29	GP13	0.271	0.191	18.921	35.022	0.074
30	CF2	0.186	0.124	11.826	18.981	0.076
31	GP14	0.194	0.130	14.034	29.634	0.058
32	CF3	0.121	0.081	5.481	12.905	0.031
33	GP15	0.299	0.203	27.916	34.570	0.082
34	CF4	0.149	0.103	21.089	12.004	0.046

Table 15 presents the ranking of features based on multiple feature selection metrics, including Information Gain, Gain Ratio, ANOVA, Chi-square (χ^2), and ReliefF, offering a comprehensive view of their predictive importance. Notably, GP15, CF1, and GP13 emerge as the top-ranked features across several criteria, with high information gain and gain ratio scores, as well as strong statistical significance in ANOVA and χ^2 tests, indicating their substantial discriminatory power in the dataset. Among the BHF variables, BHF6, BHF7, and BHF8 score prominently across metrics, highlighting their relevance in capturing variations in youth perspectives. Interestingly, while some features such as GP10 and GP11 exhibit high information gain and gain ratio, their ANOVA or χ^2 scores are comparatively moderate, suggesting they may capture more subtle yet consistent patterns. The inclusion of multiple ranking criteria provides a robust validation of feature relevance, enabling researchers to prioritize variables that consistently score high across different selection methods for predictive modeling of youth attitudes toward traffic governance.

5. DISCUSSION

The regression analyses provide clear evidence that personal safety risk perceptions and operational aspects of governance exert the strongest influence on public satisfaction with the current traffic governance system. The significance of variables such as the belief that driving over the speed limit is highly dangerous and that using a phone while driving increases accident risk suggests that satisfaction is not merely shaped by abstract attitudes toward lawfulness but is anchored in concrete perceptions of safety and shared responsibility. This finding can be interpreted through the lens of social trust, as individuals who internalize risk factors are more likely to trust governance measures aimed at mitigating those risks. When citizens perceive enforcement as aligned with genuine safety objectives rather than revenue generation or arbitrary control, institutional trust strengthens and satisfaction improves. Thus, behavioral awareness and social trust appear to reinforce each other in shaping positive governance perceptions.

Operational efficiency variables including prompt response to violations, road maintenance quality, and clarity of road signage also emerged as statistically significant predictors, underscoring the importance of visible, day to day governance performance. Citizens evaluate governance not solely through policy announcements but through tangible service delivery outcomes that directly affect their mobility experience. Consistent infrastructure upkeep and timely enforcement signal administrative competence, which is a core component of social trust. Where governance performance is predictable and responsive, citizens are more likely to perceive institutions as legitimate and accountable. From a participatory governance perspective, operational transparency and feedback loops can further strengthen this trust by allowing road users to voice concerns about maintenance gaps or enforcement inconsistencies. Such engagement transforms governance from a top down regulatory exercise into a collaborative public service framework.

The negative coefficient for AI based monitoring reflects a divergence between technological capability and perceived legitimacy. Although AI promises impartiality and efficiency, skepticism regarding fairness, transparency, and accountability may weaken public confidence. This trust deficit highlights the critical role of participatory governance models in technology adoption. When AI systems are introduced without citizen consultation, open communication, or grievance mechanisms, they may be perceived as intrusive or opaque. In contrast, involving citizens in system design discussions, publishing enforcement accuracy data, and enabling accessible appeal processes can enhance procedural justice and rebuild trust. Social trust in technological governance is therefore contingent not only on technical accuracy but also on inclusive and transparent implementation processes.

Contextual factors provide additional nuance. Regular road users, urban commuters, and individuals with accident experience reported higher satisfaction levels, possibly due to greater exposure to visible governance interventions and improved risk awareness. These groups may develop stronger trust in institutions because they directly observe operational improvements and safety benefits. Conversely, the lack of significance for receiving traffic violation notices suggests that punitive measures alone do not enhance satisfaction, particularly when disconnected from educational or participatory components. Enforcement perceived as purely punitive may erode trust if citizens do not understand its rationale or perceive it as fair.

Overall, the findings indicate that strengthening public satisfaction requires an integrated strategy combining targeted safety education, operational excellence, and trust building in technological enforcement. Social trust acts as a mediating force between governance actions and public perception, while participatory governance models provide mechanisms to institutionalize transparency, accountability, and citizen engagement. The interplay among perceptual, operational, and contextual variables demonstrates that effective traffic governance reform must move beyond enforcement intensity toward collaborative, trust centered, and citizen inclusive governance structures.

6. SUGGESTIONS

6.1. Behavioral Recommendations

Behavioral interventions should focus on shaping safer road usage habits and strengthening voluntary compliance among youth and the broader public. Authorities should prioritize sustained public safety education campaigns emphasizing the risks of speeding and mobile phone usage while driving, as these factors significantly influence satisfaction and perceived legitimacy of traffic governance. Road safety awareness must extend beyond occasional campaigns and be systematically integrated into school curricula, community workshops, and digital platforms to reinforce responsible behavior from an early age. Enforcement should not rely solely on punitive measures such as fines, since their marginal influence on satisfaction suggests limited behavioral transformation; instead, violation notices can be complemented with educational follow ups, counseling sessions, or digital awareness modules. Enforcement personnel training should combine technological competence with human sensitivity so that interventions remain fair, visible, and empathetic. Encouraging structured citizen feedback and participatory dialogues can also promote shared responsibility and voluntary adherence to road safety norms.

6.2. Governance Recommendations

Governance level recommendations should focus on institutional design, accountability mechanisms, and policy integration. Policies must integrate safety awareness initiatives with operational improvements into a unified traffic governance strategy rather than treating them as isolated interventions. Investment in infrastructure maintenance, clear signage, and rapid response enforcement mechanisms should be institutionalized through stable funding models, as operational efficiency strongly shapes satisfaction. Given skepticism toward AI based monitoring, governance frameworks must mandate transparency in algorithmic enforcement, including publication of enforcement accuracy data, publicly accessible guidelines, and clearly defined accountability standards. A structured grievance redressal mechanism for AI based penalties should be formalized to ensure procedural justice. Policymakers should also include citizen representatives and independent oversight bodies in the design and evaluation of AI enabled traffic systems to enhance legitimacy and democratic accountability.

6.3. Contextual and Technological Recommendations

Contextual recommendations recognize that technology acceptance is shaped by social trust, urban density, and institutional credibility. Community groups, non profit organizations, and educational institutions should collaborate with traffic authorities to co design awareness programs tailored to local traffic patterns and urban mobility realities. Such partnerships can help bridge the trust gap between citizens and technology driven enforcement systems through workshops, demonstrations, and public consultations. Technology providers must prioritize explainable AI models that clearly communicate how violations are detected and decisions are generated. User friendly appeal processes, periodic third party audits, and transparent performance reporting can further strengthen trust. AI systems should complement rather than replace human judgment, ensuring that automated tools are perceived as aids to fairness and efficiency within existing enforcement ecosystems. This contextual alignment between technology, governance structures, and community expectations is essential for improving satisfaction and long term compliance in AI enabled traffic governance.

REFERENCES

Ahmed, M. M., Zafri, N. M., & Lee, J. (2020). Factor identification and prediction for teen driver crash severity on rural roads using machine learning. *Applied Sciences (Basel, Switzerland), 10*(5), 1675. DOI: 10.3390/app10051675

Ajzen, I. (2020). The TPB: Selected recent advances and applications. *Health Psychology Review, 14*(1), 1–16. PMID: 31957549

Al Bazi, A.. (2024). Road crash analysis: Statistical & ML methods systematic review. *Applied Sciences (Basel, Switzerland), 15*(13), 7115.

Al-Qurashi, A.. (2023). Effect of road markings/sign quality on driving behaviour. *Journal of Safety Research.*

Alazab, M.. (2025). AI models for road safety for prediction of crash frequency and severity: A review. *Discover Artificial Intelligence*. Advance online publication. DOI: 10.1007/s44290-025-00255-3

Cândea, D.. (2024). Dynamic measurements of speed & risk perception during driving. *Sensors (Basel).*

Chen, J.. (2023). Extended theory of planned behavior to explain the influence of cognitive and psychological factors on driving behaviors. *Frontiers in Psychology, 14*, 1279216.

Cheng, J., Dong, J., & Pang, X. (2025). Automated enforcement and traffic safety. LSE working paper.

Cirillo, D.. (2023). Task/resting connectivity & risky driving; review links risk perception and behavior. *Brain Sciences.*

. Conover, E., Kraynak, D., & Singh, P. (2022). *The Effect of Traffic Cameras on Police Effort: Evidence from India.*

Dar, A.. (2024). Predicting risky driving behavior with classification algorithms. *European Transport Research Review, 16*, 17.

Dey, S.. (2024). Factors associated with speeding: Literature review/meta-analysis. *Journal of Transport & Health.*

Du, G.. (2022). Sensation seeking → risky driving via emotion regulation. *Risk Analysis, 42*(11), 2449–2466.

Duman, E.. (2023). Evaluating the effectiveness of machine learning techniques in predicting traffic accident severity. *PLoS One, 18*(7), e0288298. PMID: 37418473

Hurtado, G.. (2023). Using contextual data to predict risky driving events. [REMOVED HYPERLINK FIELD]. *Accident; Analysis and Prevention, 188*, 107028.

Karimpour, H.. (2024). Enhancing road safety with machine learning: Current advances and challenges. *Engineering Applications of Artificial Intelligence, 137*, 108078.

Krishnan, R.. (2024). Identification of the best machine learning model for predicting injury severities in RTAs. *International Journal of Injury Control and Safety Promotion, 31*(3), 360–375. Advance online publication. DOI: 10.1080/17457300.2024.2335478

Liu, X.. (2025). Traffic accident severity prediction based on MSCPO-XGBoost. *Scientific Reports, 15*, 12345. PMID: 40670407

McIlveen, N., McIlveen, P., & Lewis, I. (2024). Meta-analysis: TPB and risky driving (speeding, DUI, distraction, seatbelts). *Accident; Analysis and Prevention.*

Mohan, D.. (2024). Safety effectiveness of automated traffic enforcement systems (ATES). *Infrastructures, 9*(1), 25.

MoRTH. (2023). *Road Accidents in India 2022.* Ministry of Road Transport and Highways, Government of India.

Peden, M.. (2023). Road traffic injuries and deaths: Global burden [review]. *Frontiers in Public Health.*

Rao, A. L., Santhikumar, R., & Venugopal, K. (2025). Visualizing Insight Enhancing Statistical Thinking in Higher Education. In Anastasiadou, S., & Seremeti, L. (Eds.), *Modes of Representation in Developing Statistical Thinking in Education* (pp. 181–214). IGI Global Scientific Publishing., DOI: 10.4018/979-8-3693-9934-7.ch010

Safaei, S.. (2024). Agent-based evaluation of speed enforcement cameras. *Scientific Reports.*

Snow, A. (2017). *Automated Road Traffic Enforcement: Regulation, Governance & Public Perceptions.* RAC Foundation.

Sun, X.. (2024). Naturalistic driving studies applied to infrastructure: Systematic review. *Journal of Safety Research.*

Sun, Y.. (2025). AI-based prediction of traffic crash severity for improving road safety. *Scientific Reports, 15*, 6789. PMID: 40721929

. Tyler, T. R. (1990). *Why People Obey the Law.* U. Chicago Press. (See also subsequent meta-analyses.)

UNGA. (2024). *Resolution A/RES/78/290: Improving Global Road Safety*. United Nations General Assembly. [REMOVED HYPERLINK FIELD]

Varet, F.. (2024). Validating a multidimensional model of traffic-rule legitimacy. *Psychology, Public Policy, and Law*. Advance online publication. DOI: 10.1037/law0000416

. Varet, F., Granié, M.-A., & Carnis, L. (2021). Perceived legitimacy & traffic-rule compliance: scoping review.

Venugopal, K. (2024a). Insights into Waste Management in Semi-Urban Areas: Understanding Community Perspectives. In Mehra, R., Kumar, P., & Hung, B. (Eds.), *Municipal Solid Waste Management and Recycling Technologies* (pp. 301–326). IGI Global Scientific Publishing., DOI: 10.4018/979-8-3693-4054-7.ch011

Venugopal, K. (2024b). Exploring Gen Z Perspectives toward ChatGPT in Academic Contexts. In Bansal, R., Chakir, A., Hafaz Ngah, A., Rabby, F., & Jain, A. (Eds.), *AI Algorithms and ChatGPT for Student Engagement in Online Learning* (pp. 51–69). IGI Global Scientific Publishing., DOI: 10.4018/979-8-3693-4268-8.ch004

Venugopal, K. (2024c). Reverse Logistics Analysis: Enhancing Small Business Performance Perspectives. In Mızrak, F. (Ed.), *Strategic Innovations for Dynamic Supply Chains* (pp. 247–274). IGI Global Scientific Publishing., DOI: 10.4018/979-8-3693-3575-8.ch011

Venugopal, K. (2025). iGenius: Exploring Gen Z's Reliance on AI Tools in Education. In Muhammad Anshari, Mohammad Nabil Almunawar, and Patricia Ordóñez de Pablos (Eds.), Impacts of Digital Technologies Across Generations (pp. 141-164). IGI Global Scientific Publishing. DOI: 10.4018/979-8-3693-6366-9.ch008

Venugopal, K., Das, S., & Badawy, H. R. (2023). Prediction Analysis of Gen Zers' Attitudes on Ecological Consciousness. In Mishra, B. (Ed.), *Sustainable Science and Intelligent Technologies for Societal Development* (pp. 342–357). IGI Global Scientific Publishing., DOI: 10.4018/979-8-3693-1186-8.ch019

Venugopal, K., Das, S., Gopalakrishna, V., & Bansal, R. (2024). Customer Consciousness towards Food Wastage in the Indian Hospitality Industry. In Singh, A., Tyagi, P., & Garg, A. (Eds.), *Sustainable Disposal Methods of Food Wastes in Hospitality Operations* (pp. 38–56). IGI Global Scientific Publishing., DOI: 10.4018/979-8-3693-2181-2.ch003

Venugopal, K., Gopalakrishna, V., Ranganath, N. S., & Lakshmi, K. (2025). Reverse Mentoring and Knowledge Transfer: The Role of Gen Z in Upskilling Older Employees. In Chandan, H. (Ed.), *Enhancing Workplace Productivity through Gender Equity and Intergenerational Communication* (pp. 219–254). IGI Global Scientific Publishing., DOI: 10.4018/979-8-3373-2903-1.ch009

Venugopal, K., & Jagadeesh, B. (2025). Theoretical Insights into User Security and Privacy in Social Media Environments. In Kumar, R., Srivastava, S., & Elngar, A. (Eds.), *Human Impact on Security and Privacy: Network and Human Security, Social Media, and Devices* (pp. 289–312). IGI Global Scientific Publishing., DOI: 10.4018/979-8-3693-9235-5.ch015

Venugopal, K., & Nakkina, V. V. (2025). Technological Innovations and Their Effect on B2B Relationships in the Hospitality Industry. In Nadda, V., Sharma, A., Mediratta, H., & Dadwal, S. (Eds.), *Addressing Contemporary Challenges in the B2B Hospitality Sector* (pp. 195–232). IGI Global Scientific Publishing., DOI: 10.4018/979-8-3693-9760-2.ch007

Venugopal, K., Rajulu, B. G., Ravikumar, P., & Madhavi, K. (2026). Sustainable Learning Solutions: ChatGPT Integration in Academic Services. In Tariq, M. (Ed.), *Integrating ChatGPT Into System Applications and Services* (pp. 59–88). IGI Global Scientific Publishing., DOI: 10.4018/979-8-3693-9841-8.ch003

Venugopal, K., & Vakamullu, G. (2025). Assessment of Higher Education Students' Dependency on AI Tools in an Indian Indigenized Context. In Kgari-Masondo, M. (Ed.), *Indigenous Teaching Disciplines and Perspectives for Higher Education* (pp. 309–338). IGI Global Scientific Publishing., DOI: 10.4018/979-8-3693-9296-6.ch015

Wells, P. et al. (2020). Graduated Driver Licensing review of effectiveness. (Policy synthesis cited across OECD/ITF 2024).

WHO. (2021). *Global Plan for the Decade of Action for Road Safety 2021–2030*. World Health Organization. [REMOVED HYPERLINK FIELD]

WHO. (2023a). *Global Status Report on Road Safety 2023*. World Health Organization.

WHO. (2023b). *Global status report on road safety 2023 full report (PDF)*. World Health Organization. [REMOVED HYPERLINK FIELD]

Woolley, J., & Hutchinson, T. (2021). Why design can make safe behaviour hard; Safe System critique. *Journal of Road Safety*.

Yao, L.. (2012). Young drivers and speed selection guided by the TPB. *Accident; Analysis and Prevention, 43*(5), 1349–1359.

. YOURS Youth for Road Safety. (2024). United Nations declares new Decade of Action for Road Safety: Youth implications.

KEY TERMS AND DEFINITIONS

AI Based Traffic Monitoring: AI based traffic monitoring involves the use of artificial intelligence technologies such as computer vision, automated number plate recognition, and predictive analytics to detect violations, manage traffic flow, and enhance surveillance. Public acceptance of such systems depends on perceived fairness, accuracy, and privacy considerations.

Governance Satisfaction: Governance satisfaction refers to the level of approval or contentment among road users regarding the effectiveness, responsiveness, and fairness of traffic management systems. It serves as a key outcome variable in studies evaluating public perception of governance.

Machine Learning Applications: Machine learning applications refer to computational algorithms that learn patterns from data to make predictions or classifications without explicit programming. Techniques such as decision trees, random forests, or support vector machines may be used to predict satisfaction levels based on multiple influencing factors.

Operational Traffic Management Factors: Operational factors include practical aspects of traffic governance such as road maintenance, signage clarity, signal functioning, congestion management, and timely response to violations.

Predictive Modeling: Predictive modeling involves the use of statistical and machine learning techniques to forecast future outcomes based on historical data. In the context of this research, it helps anticipate youth satisfaction trends and identify critical governance variables affecting perceptions.

Regression Analysis: Regression analysis is a statistical method used to examine the relationship between a dependent variable and one or more independent variables. In this study, it is employed to identify the key determinants influencing youth satisfaction with traffic governance.

Road Safety Risk Perception: Road safety risk perception refers to individuals' awareness and evaluation of potential hazards associated with unsafe driving behaviors such as speeding, distracted driving, and rule violations. Higher perceived risk is often associated with safer driving attitudes and stronger support for enforcement.

Traffic Governance: Traffic governance refers to the system of policies, regulations, enforcement mechanisms, and institutional arrangements designed to manage road use, ensure safety, and regulate driver behavior. It includes law enforcement, infrastructure planning, technological monitoring, and public awareness initiatives.

Traffic Law Enforcement: Traffic law enforcement encompasses the actions taken by authorities to ensure compliance with road regulations, including monitoring, issuing penalties, responding to violations, and maintaining order on roadways. Effective enforcement is a critical determinant of public trust and satisfaction.

Youth Perception: Youth perception denotes the attitudes, beliefs, evaluations, and subjective interpretations held by young individuals regarding traffic rules, enforcement practices, road safety measures, and governance effectiveness. It significantly influences compliance behavior and risk-taking tendencies on roads.

Conclusion

Translation of new knowledge in the classroom in home country; designing new research project matching crosscutting issues and learning from previous research; application of knowledge at the community level; and contribution to the country policy research. This book may communicate at the different audience by participating events including scientific discussion, seminars,symposiums,workshops,school and college program and universities. Discovers eco-friendly vehicle; Reduce car dependency and explore new options for public transport; Well desgined polycentric urban development will be taking place; Optimum uses of land and green occupancy will be taking place; Maximizing mobility to people and goods with low costs and time; Several news articles related to eco transport, efficacy, contribution of transport planning to air pollution polycentric urban development, human travel interactions have been accumulated in the book.

Many cities have taken rank as the most congested in the world (Lebeau and Macharis, 2014). Mobility especially in developing countries is also delayed due to traffic congestion. Traffic is the biggest problem in many cities (Lebeau and Macharis, 2014). Mobility is synonymous with travel or movement in physical space. Spatial or geographic mobility is broken down into different temporal scales (daily, weekly, monthly, seasonal or and over life time). Hubert et al., (2013) mentioned that three major challenges which includes population growth, expansion of activities and accessibility of the city. The current urban transportation system is not efficient in accommodating the increased population, industries, pollutions and monocentric urban development. Urban transportation planning is an essential need for urban development and planning processes to control air and noise pollution. Efficient decisions on how to build green cities and urban facilities are the basic elements of travel forecasting through alternate routing system, green space planning and plantation with suitable plants along the road divider. It should be the key notion to identify the causes of car dependency, inaccessibility, less mobility, lengthy trips, traffic congestion, urban air pollution and inappropriate land use pattern in the urban. Eventually urban land use classifications should be reviewed considering population density, residence, commerce, industry, institution, recreation and urban agriculture. The emergence of polycentricity originally aims at resolving a series of urban problems arousing from monocentric development, such as uneconomical energy use, long-distance commutes, traffic congestion, and environmental deterioration (Wang et al., 2019). The more neighbourhood density and diverse land use enhance more activities (i.e. walking and cycling), non-work trips by inspiring urban residents for shopping and recreations (Naess et al., 2019 cited in Hou, 2019). It was reported that the transformation from monocentric urban to a polycentric form makes travel more efficient and trips shorter (Gordon et al., 1989). The measurement of morphological polycentricity is based on various datasets such as remote sensing image, land use maps, and a population census dataset (Cai et al., 2017). The decrease not only in the use but also in the ownership of cars is a crucial challenge in the development of public spaces. Boussauw and Vanoutrive (2017) reported that sustainable development schemes are being required at three stages; i) defining the problems and selection of indicators; ii) solution preferences (building of rail line); and iii) bureaucratic system (execution, design of chosen plan). Mismatch observed between orthodox sustainable transport discourse and aspects of limits to growth and social justice in the transport policy and practices (Boussauw and Vanoutrive, 2017).

The book explores information on demography of the population (density, resident types, income, occupation, gender, age, occupation, household size and education), existing transport facilities. Pollution hazards (air quality, water quality, noise pollution); availability of green space and land (road length, pedestrian road, infrastructures, other road network, % of green space, occupancy of green space, plant types, aesthetic view, parks, lakes and residents satisfaction level etc.); planning or solutions for polycentric urban development (public transport, no car day, modal shift, time based tax policy, ecofriendly transport mainly electric or natural gas or ethanol fueled, improved mobility, accessibility, road zonings, pollution mitigation measures, choice of plant species for green space).

This book addresses these challenges comprehensively. It encompasses a variety of issues, including the economic costs of traffic congestion and the social, environmental, and economic effects of pollution and traffic issues on the daily lives of urban dwellers and travellers. Furthermore, it proposes a range of potential solutions aimed at absorbing carbon dioxide, carbon monoxide, and particulate matter, as well as mitigating noise and air pollution through urban greening initiatives, including the development of avenues, islands, and other green spaces. This work serves as a valuable resource for policymakers, encouraging them to consider alternative routing systems, smart transportation solutions, and integrated pedestrian planning processes to enhance environmental sustainability, safeguard public health, and promote additional health benefits

This book draws contemporary insights into a comprehensive interpretation of urban environment in smart places as it relates to the development of cities or to various levels of intervention in existing urban environments.

It examines diverse strategies for emission reduction and considers the social, environmental, and economic implications of traffic congestion, ultimately proposing solutions for the development of green and smart cities via a polycentric urban development framework. Due to the interdisciplinary and trans-disciplinary nature of this book, this book will be useful for scholars and researchers from different backgrounds in the fields of architecture; engineering: urban planning and design, landscape architecture, and environmental design. Additionally, the editor incorporates contributions from scholars in relevant fields including human geography, anthropology, environmental psychology, forestry, sociology, installation artists, and transportation engineering educators. Mainly, the target audience can acquire further knowledge and insights on how to use new approaches and innovation in research, academia, and analytical effort; they can use the numerous methods contributed in this book for their future research and innovations.

REFERENCES

Boussauw, K., & Vanoutrive, T. (2017). Transport policy in Belgium: Translating sustainability discourses into unsustainable outcomes. *Transport Policy*, *53*, 11–19. DOI: 10.1016/j.tranpol.2016.08.009

Cai, J., Huang, B., & Song, Y. (2017). Using multi-source geospatial big data to identify the structure of polycentric cities. *Remote Sensing of Environment*, *202*, 210–221.

Gordon, P., Kumar, A., & Richardson, H. W. (1989). The influence of metropolitan spatial structure on commuting time. *Journal of Urban Economics*, *26*(2), 138–151. DOI: 10.1016/0094-1190(89)90013-2

Hou, Y. (2019). Polycentric urban form and non-work travel in Singapore: A focus on seniors. *Transportation Research Part D, Transport and Environment*, *73*, 245–275. DOI: 10.1016/j.trd.2019.07.003

Hubert, M., Lebrun, K., Huynen, P., & Dobruszkes, F. (2013). BSI synopsis. Daily mobility in Brussels: Challenges, tools and priority undertakings. *Brussels Studies*, *71*, 1–27.

Lebeau, P., & Macharis, C. (2014). Freight transport in Brussels and its impact on road traffic?. Brussels Studies. La revue scientifique électronique pour les recherches sur Bruxelles/Het elektronisch wetenschappelijk tijdschrift voor onderzoek over Brussel/The e-journal for academic research on Brussels.

Næss, P., Strand, A., Wolday, F., & Stefansdottir, H. (2019). Residential location, commuting and non-work travel in two urban areas of different size and with different center structures. *Progress in Planning*, *128*, 1–36. DOI: 10.1016/j.progress.2017.10.002

Wang, M., Derudder, B., & Liu, X. (2019). Polycentric urban development and economic productivity in China: A multiscalar analysis. Environment and Planning A. *Economy and Space*, *51*(8), 1622–1643.

Compilation of References

Abdelati, M. H. (2024). Smart Traffic Management for Sustainable development in Cities: Enhancing safety and efficiency. International Journal of Advanced Engineering and Business Science, 5(1), 0. DOI: 10.21608/ijaebs.2024.242361.1088

Abdul Jabbar, S., Tul Qadar, L., Ghafoor, S., Rasheed, L., Sarfraz, Z., Sarfraz, A., Sarfraz, M., Felix, M., & Cherrez-Ojeda, I. (2022). Air quality, pollution and sustainability trends in South Asia: A population-based study. *International Journal of Environmental Research and Public Health*, *19*(12), 7534. DOI: 10.3390/ijerph19127534

Abdulgafar, T. (2021). *Noise pollution & global warming [What science says]*. KRISP blog.

Abiodun, B. J., Ojumu, A. M., Jenner, S., & Ojumu, T. V. (2014). The transport of atmospheric NO x and HNO 3 over Cape Town. *Atmospheric Chemistry and Physics*, *14*(2), 559–575. DOI: 10.5194/acp-14-559-2014

Abuh, P. O., & John, J. A. (2023). Geographic analysis of traffic congestion in FCT-Abuja, Nigeria. *Journal of Sustainability and Environmental Management*, *2*(1), 26–32. DOI: 10.3126/josem.v2i1.53114

Ada Lovelace Institute. (2020). *Algorithmic transparency: Principles and practice*. Ada Lovelace Institute.

Adarsh, R., Jothi, E., Sivasubramanian, K., & Madhusudhanan, R. (2025). Analysing the Status of Agricultural Finance in Rural India in Promoting Agriculture. In Al Mubarak, M. (Ed.), *Sustainable Digital Technology and Ethics in an Ever-Changing Environment. Studies in Systems, Decision and Control* (Vol. 237). Springer., DOI: 10.1007/978-3-031-86708-8_52

Aderemi, B. A., Olwal, T. O., Ndambuki, J. M., & Rwanga, S. S. (2021). A Review of Groundwater Management Models with a Focus on IoT-Based Systems. *Sustainability (Basel)*, *14*(1), 148. DOI: 10.3390/su14010148

Adhisekar, K., Lenin, S., Ilankadhir, M., Vani, H., & Kumarasamy, P. (2026). A review-based analysis on the application of AI in retail management for sustainable business in India. In *Studies in Systems* (Vol. 620, pp. 323–330). Decision and Control. DOI: 10.1007/978-3-032-02056-7_26

Afzal, M. (1990). Machine intelligence applications in agriculture. *AMA. Agricultural Mechanization in Asia, Africa and Latin America, 21*(2).

Agarwal, V. P., & Sharma, V. K. (1980). *Today and tomorrow's printers and publishers.*

Agrawal, P., Sinha, A., Kumar, S., Agarwal, A., Banerjee, A., Villuri, V. G. K., Annavarapu, C. S. R., Dwivedi, R., Dera, V. V. R., Sinha, J., & Pasupuleti, S. (2021). Exploring Artificial Intelligence Techniques for Groundwater Quality Assessment. *Water (Basel), 13*(9), 1172. DOI: 10.3390/w13091172

Ahmed, M. M., Zafri, N. M., & Lee, J. (2020). Factor identification and prediction for teen driver crash severity on rural roads using machine learning. *Applied Sciences (Basel, Switzerland), 10*(5), 1675. DOI: 10.3390/app10051675

Ahn, K., & Rakha, H. (2008). Eco-routing of vehicles based on minimizing fuel consumption and emissions. *Transportation Research Part D, Transport and Environment, 13*(5), 377–385.

Ahsan, A., Rahman, M., & Rahman, M. M. (2016). Traffic congestion in Dhaka city: Causes and possible solutions. *International Journal of Transportation Engineering, 4*(1), 23–34. DOI: 10.12691/ijte-4-1-3

Ahvenniemi, H., Huovila, A., Pinto-Seppä, I., & Airaksinen, M. (2017). What are the differences between sustainable and smart cities? *Cities (London, England), 60*, 234–245. DOI: 10.1016/j.cities.2016.09.009

Ajzen, I. (2020). The TPB: Selected recent advances and applications. *Health Psychology Review, 14*(1), 1–16. PMID: 31957549

Akhtar, A. B. T., Naseem, S., Yasar, A., & Naseem, Z. (2021). Persistent organic pollutants (POPs): sources, types, impacts, and their remediation. In *Environmental pollution and remediation* (pp. 213–246). Springer Singapore. DOI: 10.1007/978-981-15-5499-5_8

Al Bazi, A.. (2024). Road crash analysis: Statistical & ML methods systematic review. *Applied Sciences (Basel, Switzerland), 15*(13), 7115.

Alamelu Sathyabama, N., Ramesh, K., & Sivasubramanian, K. (2025). Waste to Wealth: Impact of Solid Waste Management Strategies for Achieving SDGs in India. In Alareeni, B. (Ed.), *Big Data in Finance: Transforming the Financial Landscape. Studies in Big Data* (Vol. 169). Springer., DOI: 10.1007/978-3-031-80656-8_39

Alam, S., Bhardwaj, L. K., Mallick, R., & Rai, S. (2023). Estimation of heavy metals and fluoride ion in vegetables grown nearby the stretch of River Yamuna, Delhi (NCR), India. *Indian Journal of Environmental Protection, 43*(1), 64–73.

Alam, T. (2024). Data privacy and security in autonomous connected vehicles in smart city environment. *Big Data and Cognitive Computing, 8*(9), 95. DOI: 10.3390/bdcc8090095

Alawi, M., Masaad, M., & Al-Hussaini, M. (2018). Comparative study of persistent organic pollutant (POP)(chlorinated pesticides, PCBs, and dioxins/furans) concentrations in cancer-affected human organs with those of healthy organs. *Environmental Monitoring and Assessment, 190*(8), 470. DOI: 10.1007/s10661-018-6856-3

Alazab, M.. (2025). AI models for road safety for prediction of crash frequency and severity: A review. *Discover Artificial Intelligence*. Advance online publication. DOI: 10.1007/s44290-025-00255-3

Al-Hemoud, A., Al-Senafi, A., Al-Khayat, A., Behbehani, W., Al-Dashti, H., Al-Dabbous, A., Al-Qadeeri, G., Gasana, J., Mihalkov, V., Petrov, P., Alahmad, B., & Koutrakis, P. (2024). The complex chemical effects of air pollutants, meteorology and dust on the concentrations of criteria and non-criteria air pollution in Kuwait. *Discover Atmosphere, 2*(1), 15. DOI: 10.1007/s44292-024-00017-3

Ali, M., Devi, G. L., & Neelapu, R. (2020). Intelligent Traffic Signal control System using machine learning techniques. In Lecture notes in electrical engineering (pp. 611–619). DOI: 10.1007/978-981-15-3828-5_63

Aliyari, H., Kholghi, M., Zahedi, S., & Momeni, M. (2018). Providing decision support system in groundwater resources management for the purpose of sustainable development. *Journal of Water Supply: Research & Technology - Aqua, 67*(5), jws2018130. Advance online publication. DOI: 10.2166/aqua.2018.130

Allen, P. (2005). Personal communication. *(Note: In APA 7, personal communications are cited only in the text and are not included in the reference list. The entry above is provided for completeness.)*

Al-Qurashi, A.. (2023). Effect of road markings/sign quality on driving behaviour. *Journal of Safety Research.*

Aluko, N. S. E.Sunday Ebenezer Aluko. (2024). Cybersecurity and defense in intelligent transportation systems. *World Journal of Advanced Engineering Technology and Sciences, 13*(1), 871–879. DOI: 10.30574/wjaets.2024.13.1.0469

Alves, A., & Ferreira, J. (2021). Integrated urban planning approaches for sustainable mobility: A systematic review. *Sustainability, 13*(7), 3748. DOI: 10.3390/su13073748

Andrade, L., O'Dwyer, J., O'Neill, E., & Hynds, P. (2018). Surface water flooding, groundwater contamination, and enteric disease in developed countries: A scoping review of connections and consequences. *Environmental Pollution, 236*, 540–549. DOI: 10.1016/j.envpol.2018.01.104

Angelidou, M. (2017). The role of smart city characteristics in the plans of fifteen cities. *Journal of Urban Technology, 24*(4), 3–28. DOI: 10.1080/10630732.2017.1348880

Apte, J. S., Marshall, J. D., Cohen, A. J., & Brauer, M. (2017). Addressing global mortality from ambient $PM_{2.5}$. *Environmental Science & Technology, 51*(13), 7350–7357. PMID: 28562015

Armson, D., Stringer, P., & Ennos, A. R. (2012). The effect of tree shade and grass on surface and globe temperatures in an urban area. *Urban Forestry & Urban Greening, 11*(3), 245–255. DOI: 10.1016/j.ufug.2012.05.002

Arnstein, S. R. (1969). A ladder of citizen participation. *Journal of the American Institute of Planners, 35*(4), 216–224. DOI: 10.1080/01944366908977225

Arroyo-Arroyo, F., & Frame, G. (2021). Traffic Management in African cities: The way forward. In *SSATP*. SSATP.

Ayejoto, D. A., Agbasi, J. C., Nwazelibe, V. E., Egbueri, J. C., & Alao, J. O. (2023). Understanding the connections between climate change, air pollution, and human health in Africa: Insights from a literature review. *Journal of Environmental Science and Health. Part C, Environmental Carcinogenesis & Ecotoxicology Reviews, 41*(3-4), 77–120.

Aylor, D. E. (1972). Noise reduction by vegetation and ground. *The Journal of the Acoustical Society of America, 51*(1B), 197–205. DOI: 10.1121/1.1912830

Azarov, V., Petrenko, L., & Manzhilevskaya, S. (2018). The study of local dust pollution of atmospheric air on construction sites in urban areas. In *Energy Management of Municipal Transportation Facilities and Transport* (pp. 430–439). Springer International Publishing.

Bache, I. (2012). Multi-Level governance in the European Union. In *Oxford University Press eBooks*. DOI: 10.1093/oxfordhb/9780199560530.013.0044

Balasubramanian, G.. (2025). Digital Payment Systems and Business Development of Small and Petty Traders in India: A Monetary Analysis. In Hamdan, A. (Ed.), *Achieving Sustainable Business through AI, Technology Education and Computer Science. Studies in Big Data* (Vol. 158). Springer., DOI: 10.1007/978-3-031-70855-8_2

Balasubramanian, G., Jeyaratha, V. C., Sudha, T. S., Thamilmani, R., Kumarasamy, P., & Sivasubramanian, K. (2025). Scaling the Progress of Sustainable Business Practices Among Small Scale Industries Relating to India: A Review. In Hamdan, A., & Braendle, U. (Eds.), *Harnessing AI, Machine Learning, and IoT for Intelligent Business. Studies in Systems, Decision and Control* (Vol. 555). Springer., DOI: 10.1007/978-3-031-67890-5_27

Baldauf, R. (2017). Roadside vegetation design characteristics that can improve local, near-road air quality. *Transportation Research Part D, Transport and Environment*, *52*, 354–361. DOI: 10.1016/j.trd.2017.03.013

Banerjee, A., Costa, B., Forkan, A. R. M., Kang, Y.-B., Marti, F., McCarthy, C., Ghaderi, H., Georgakopoulos, D., & Jayaraman, P. P. (2024). 5G-enabled smart cities: A real-world evaluation and analysis of 5G using a pilot smart city application. *Internet of Things : Engineering Cyber Physical Human Systems*, *28*, 101326. DOI: 10.1016/j.iot.2024.101326

Banerjee, C., Bhaduri, A., & Saraswat, C. (2022). Digitalization in Urban Water Governance: Case Study of Bengaluru and Singapore. *Frontiers in Environmental Science*, *10*(March), 1–12. DOI: 10.3389/fenvs.2022.816824

Banister, D. (2008). The sustainable mobility paradigm. *Transport Policy*, *15*(2), 73–80. DOI: 10.1016/j.tranpol.2007.10.005

Barati Moghaddam, M., Mazaheri, M., & Mohammad Vali Samani, J. (2021). Inverse modeling of contaminant transport for pollution source identification in surface and groundwaters: a review. *Groundwater for Sustainable Development, 15*(November 2020), 100651. DOI: 10.1016/j.gsd.2021.100651

Barcelona City Council. (2019). *Data Commons Charter*. Barcelona City Council.

Bartesaghi-Koc, C., Osmond, P., & Peters, A. (2019). Spatio-temporal patterns in green infrastructure as driver of land surface temperature variability: The case of Sydney. *International Journal of Applied Earth Observation and Geoinformation*, *83*, 101903. DOI: 10.1016/j.jag.2019.101903

Barth, M., & Cargile, K. (2009). Real-world carbon dioxide impacts of traffic congestion. *Transportation Research Record: Journal of the Transportation Research Board*, *2058*(1), 163–171. DOI: 10.3141/2058-20

Basu, D., & Mishra, S. (2024). Mine reclamation practices and effects of stakeholder perception — A case study of Saoner mines, Maharashtra, India. *Journal of Engineering and Applied Sciences (Asian Research Publishing Network)*, *71*(1), 62.

Batty, M. (2013). *The new science of cities*. MIT Press. DOI: 10.7551/mitpress/9399.001.0001

Batty, M., Axhausen, K. W., Giannotti, F., Pozdnoukhov, A., Bazzani, A., Wachowicz, M., Ouzounis, G., & Portugali, Y. (2012). Smart cities of the future. *The European Physical Journal. Special Topics*, *214*(1), 481–518. DOI: 10.1140/epjst/e2012-01703-3

Beatley, T. (2011). *Biophilic cities: Integrating nature into urban design and planning*. Island Press. DOI: 10.5822/978-1-59726-986-5

Begum, B. A., & Hopke, P. K. (2018). Ambient air quality in Dhaka Bangladesh over two decades: Impacts of policy on air quality. *Aerosol and Air Quality Research*, *18*(7), 1910–1920. DOI: 10.4209/aaqr.2017.11.0465

Behera, S. N., Sharma, M., Aneja, V. P., & Balasubramanian, R. (2013). Ammonia in the atmosphere: A review on emission sources, atmospheric chemistry and deposition on terrestrial bodies. *Environmental Science and Pollution Research International*, *20*(11), 8092–8131. DOI: 10.1007/s11356-013-2051-9

Bhan, G. (2019). Notes on a Southern urban practice. *Environment and Urbanization*, *31*(2), 639–654. DOI: 10.1177/0956247818815792

Bhardwaj, L. K., Mittal, R., Rath, P., Gupta, U., Bhardwaj, D., Dahal, Y., & Kumar, A. (2026). Persistent Organic Pollutants (POPs), A Threat to Arctic Environment and Their Impact on Human Health. In Xenobiotics in Environment (pp. 15-24). CRC Press.

Bhardwaj, L. K., Rath, P., & Tokas, R. (2024a). E-waste management in developing countries: current practices, challenges, disposal, and impact on human health and environment. *Sustainable solutions for e-waste and development*, 90-102.

Bhardwaj, L. K., Rath, P., Bajpai, S., Upadhyay, D., Jain, H., Kumar, N., & Sinha, S. (2024b). COVID-19 and the Interplay with Antibacterial Drug Resistance. In *Frontiers in Combating Antibacterial Resistance: Current Perspectives and Future Horizons* (pp. 246-273). IGI Global Scientific Publishing.

Bhardwaj, L. K. (2023). A Comprehensive review on the climate change and its impact on health. https://doi.org/DOI: 10.20944/preprints202305.0159.v1

Bhardwaj, L. K., Rath, P., & Choudhury, M. (2023). A comprehensive review on the classification, uses, sources of nanoparticles (NPs) and their toxicity on health. *Aerosol Science and Engineering*, *7*(1), 69–86. DOI: 10.1007/s41810-022-00163-4

Bhardwaj, L. K., Rath, P., Choudhury, M., Metzdorff, A., & Jodder, P. K. (2025a). Electronic wastes and contamination of water. In *Electronic Waste* (pp. 48–57). CRC Press. DOI: 10.1201/9781003582311-6

Bhardwaj, L. K., Rath, P., Jain, H., & Choudhury, M. (2025b). Exploring the effects of e-waste on soil, water quality and human health. *Discover Civil Engineering*, *2*(1), 12. DOI: 10.1007/s44290-025-00167-2

Bhardwaj, L., & Jindal, T. (2021). Polar ecotoxicology: Sources and toxic effects of pollutants. In *New frontiers in environmental Toxicology* (pp. 9–14). Springer International Publishing.

Bhat, A. M., Tripathi, I. M., Mohapatra, P. K., & Bhagat, C. (2025). Reimagining stormwater management: Sustainable drainage pathways for resilient Indian cities. *Environmental Conservation*, •••, 1–7.

Bibri, S. E. (2018). The IoT for smart sustainable cities of the future: An analytical framework for sensor-based big data applications for environmental sustainability. *Sustainable Cities and Society*, *38*, 230–253. DOI: 10.1016/j.scs.2017.12.034

Bigazzi, A. Y., & Figliozzi, M. (2012). Operational and environmental effects of bus rapid transit systems. *Transportation Research Record: Journal of the Transportation Research Board*, *2276*(1), 14–23.

Bigazzi, A. Y., & Rouleau, M. (2017). Can traffic management strategies improve urban air quality? A review of the evidence. *Journal of Transport & Health*, *7*, 111–124. DOI: 10.1016/j.jth.2017.08.001

Bittman, S., & Mikkelsen, R. (2009). Ammonia emissions from agricultural operations: Livestock. *Better Crops with Plant Food*, *93*(1), 28–31.

Boethling, R., Fenner, K., Howard, P., Klečka, G., Madsen, T., Snape, J. R., & Whelan, M. J. (2009). Environmental persistence of organic pollutants: Guidance for development and review of POP risk profiles. *Integrated Environmental Assessment and Management*, *5*(4), 539–556. DOI: 10.1897/IEAM_2008-090.1

Bokolo, A. J. (2023). Examining the adoption of Sustainable eMobility-Sharing in Smart Communities: Diffusion of Innovation Theory Perspective. *Smart Cities*, *6*(4), 2057–2080. DOI: 10.3390/smartcities6040095

Boriboonsomsin, K., Vu, A., & Barth, M. (2012). Eco-routing navigation systems. In *2012 IEEE Intelligent Vehicles Symposium* (pp. 1018–1023).

Bouktif, S., Cheniki, A., Ouni, A., & El-Sayed, H. (2023). Deep reinforcement learning for traffic signal control with consistent state and reward design approach. *Knowledge-Based Systems*, *267*, 110440. DOI: 10.1016/j.knosys.2023.110440

Brook, R. D., & Rajagopalan, S. (2009). Particulate matter, air pollution, and blood pressure. *Journal of the American Society of Hypertension*, *3*(5), 332–350. DOI: 10.1016/j.jash.2009.08.005

Byun, D. W., & Ching, J. K. S. (1999). Science algorithms of the EPA Models-3 Community Multiscale Air Quality (CMAQ) modeling system. United States Environmental Protection Agency, Office of Research and Development., EPA/600/R-99/030.

Calderón-Garcidueñas, L., Stommel, E. W., Torres-Jardón, R., Hernández-Luna, J., Aiello-Mora, M., González-Maciel, A., Reynoso-Robles, R., Pérez-Guillé, B., Silva-Pereyra, H. G., Tehuacanero-Cuapa, S., Rodríguez-Gómez, A., Lachmann, I., Galaz-Montoya, C., Doty, R. L., Roy, A., & Mukherjee, P. S. (2024). Alzheimer and Parkinson diseases, frontotemporal lobar degeneration and amyotrophic lateral sclerosis overlapping neuropathology start in the first two decades of life in pollution exposed urbanites and brain ultrafine particulate matter and industrial nanoparticles, including Fe, Ti, Al, V, Ni, Hg, Co, Cu, Zn, Ag, Pt, Ce, La, Pr and W are key players. Metropolitan Mexico City health crisis is in progress. *Frontiers in Human Neuroscience*, *17*, 1297467. DOI: 10.3389/fnhum.2023.1297467

Calderón-Garcidueñas, L., Torres-Jardón, R., Kulesza, R. J., Park, S. B., & D'Angiulli, A. (2014). Air pollution and detrimental effects on children's brain. The need for a multidisciplinary approach to the issue complexity and challenges. *Frontiers in Human Neuroscience*, *8*, 613.

Cândea, D.. (2024). Dynamic measurements of speed & risk perception during driving. *Sensors (Basel)*.

Central Pollution Control Board. (2011). *Air quality monitoring, emission inventory, and source apportionment study for Indian cities*. Government of India.

Cervero, R. (2013). Transport infrastructure and urban growth: The emergence of polycentric urban regions. *Journal of Transport and Land Use*, *6*(1), 7–25. DOI: 10.5198/jtlu.v6i1.425

Cervero, R., & Kockelman, K. (1997). Travel demand and the 3Ds: Density, diversity, and design. *Transportation Research Part D, Transport and Environment*, *2*(3), 199–219. DOI: 10.1016/S1361-9209(97)00009-6

Chaplin, S. (2011). *The politics of sanitation in India: Cities, services and the state*. No Title.

Chaturvedi, V., Ghosh, A., Garg, A., Avashia, V., Vishwanathan, S. S., Gupta, D., Sinha, N. K., Bhushan, C., Banerjee, S., Datt, D., Bansal, J., Pathak, M., Dhar, S., Singh, A. K., Khan, N., Rashmi, R. R., Agrawal, S., Agarwal, D., Singh, A., & Prasad, S. (2024). India's pathway to net zero by 2070: Status, challenges, and way forward. *Environmental Research Letters*, *19*(11), 112501. DOI: 10.1088/1748-9326/ad7749

Cheng, J., Dong, J., & Pang, X. (2025). Automated enforcement and traffic safety. LSE working paper.

Chen, J.. (2023). Extended theory of planned behavior to explain the influence of cognitive and psychological factors on driving behaviors. *Frontiers in Psychology*, *14*, 1279216.

Chepesiuk, R. (2005). Decibel hell: The effects of living in a noisy world. *Environmental Health Perspectives*, *113*(1), A35–A41. DOI: 10.1289/ehp.113-a34

Chibowska, K., Baranowska-Bosiacka, I., Falkowska, A., Gutowska, I., Goschorska, M., & Chlubek, D. (2016). Effect of lead (Pb) on inflammatory processes in the brain. *International Journal of Molecular Sciences*, *17*(12), 2140. DOI: 10.3390/ijms17122140

Chin-Chan, M., Navarro-Yepes, J., & Quintanilla-Vega, B. (2015). Environmental pollutants as risk factors for neurodegenerative disorders: Alzheimer and Parkinson diseases. *Frontiers in Cellular Neuroscience*, *9*, 124. DOI: 10.3389/fncel.2015.00124

Chow, J. C., Watson, J. G., Lowenthal, D. H., Hackney, R., Magliano, K., Lehrman, D., & Smith, T. (1999). Temporal variations of PM2.5, PM10, and gaseous precursors during the 1995 integrated monitoring study in central California. *Journal of the Air & Waste Management Association*, *49*(9), 16–24. DOI: 10.1080/10473289.1999.10463909

Chow, J. C., Watson, J. G., Lu, Z., Lowenthal, D. H., Frazier, C. A., Solomon, P. A., Thuillier, R. H., & Magliano, K. (1996). Descriptive analysis of PM2.5 and PM10 at regionally representative locations during SJVAQS/AUSPEX. *Atmospheric Environment*, *30*(12), 2079–2112. DOI: 10.1016/1352-2310(95)00402-5

Churkina, G., Grote, R., Butler, T. M., & Lawrence, M. (2015). Natural selection? Picking the right trees for urban greening. *Environmental Science & Policy*, *47*, 12–17. DOI: 10.1016/j.envsci.2014.10.014

Cirillo, D.. (2023). Task/resting connectivity & risky driving; review links risk perception and behavior. *Brain Sciences*.

Cohen, T., Saxe, S., & Soltani, A. (2014). Behavioral approaches to sustainable urban transport. *Sustainability*, *6*(5), 2739–2753. DOI: 10.3390/su6052739

Colvile, R. N., Hutchinson, E. J., Mindell, J. S., & Warren, R. F. (2001). The transport sector as a source of air pollution. *Atmospheric Environment*, *35*(9), 1537–1565. DOI: 10.1016/S1352-2310(00)00551-3

Cook, D. I., & Haverbeke, D. F. (1977). Suburban noise control with planting and solid barrier combinations [). University of Nebraska–Lincoln.]. *Research Bulletin (International Commission for the Northwest Atlantic Fisheries)*, •••, 100.

Cook, Q., Argenio, K., & Lovinsky-Desir, S. (2021). The impact of environmental injustice and social determinants of health on the role of air pollution in asthma and allergic disease in the United States. *The Journal of Allergy and Clinical Immunology*, *148*(5), 1089–1101. DOI: 10.1016/j.jaci.2021.09.018

Coulon, F., Jones, K., Li, H., Hu, Q., Gao, J., Li, F., Chen, M., Zhu, Y.-G., Liu, R., Liu, M., Canning, K., Harries, N., Bardos, P., Nathanail, P., Sweeney, R., Middleton, D., Charnley, M., Randall, J., Richell, M., & Pollard, S. (2016). China's soil and groundwater management challenges: Lessons from the UK's experience and opportunities for China. *Environment International*, *91*, 1–12. DOI: 10.1016/j.envint.2016.02.023

CPHEEO. (2022). *Central Public Health & Environmental Engineering Organisations*. Ministry of Housing and Urban Affairs.

Craglia, M., Granell, C., Ostermann, F., & Bröring, A. (2018). Digital Earth and smart cities: Governance, policy, and social innovation. *International Journal of Digital Earth*, *11*(3), 219–237.

Crawford, K. (2021). *Data poverty: The everyday politics of data-driven technology*. Data & Society Research Institute.

Cuthbert, M. O., Gleeson, T., Moosdorf, N., Befus, K. M., Schneider, A., Hartmann, J., & Lehner, B. (2019). Global patterns and dynamics of climate–groundwater interactions. *Nature Climate Change*, *9*(2), 137–141. DOI: 10.1038/s41558-018-0386-4

Dankan Gowda, V., Patil, S., Srinivas, V., Prasad, K., Jagtap, M. M., & Kaur, M. (2024). Smart urban ecosystems with IoT-based strategies for traffic optimization and pollution control. In Proceedings of the 2024 2nd International Conference on Advancement in Computation & Computer Technologies (InCACCT) (pp. 777–782). IEEE. DOI: 10.1109/InCACCT61598.2024.10551250

Dan, S., Pant, M., Kaur, T., & Pant, S. (2020). Toxic effect of formaldehyde. *Systematic Reviews*.

Dar, A.. (2024). Predicting risky driving behavior with classification algorithms. *European Transport Research Review*, *16*, 17.

Das, A., Kumar, R., Patel, S. S., Saha, M. C., & Guha, D. (2020). Source apportionment of potentially toxic elements in street dust of a coal mining area in Chhattisgarh, India, using multivariate and lead isotopic ratio analysis. *Environmental Monitoring and Assessment*, *192*(6), 396. DOI: 10.1007/s10661-020-08321-2

Das, D. (2020). Perspectives of smart cities in South Africa through applied systems analysis approach: A case of Bloemfontein. *Construction Economics and Building*, *20*(2), 65–88. DOI: 10.5130/AJCEB.v20i2.6657

Das, D. K. (2024). Exploring the symbiotic relationship between digital transformation, infrastructure, service delivery, and governance for smart sustainable cities. *Smart Cities*, *7*(2), 806–835. DOI: 10.3390/smartcities7020034

Das, D. K. (2025). Digital technology and AI for smart sustainable cities in the Global South: A critical review of literature and case studies. *Urban Science (Basel, Switzerland)*, *9*(3), 72. DOI: 10.3390/urbansci9030072

Das, D. K., & Chadchan, J. (2023). A proposed framework for an appropriate governance system to develop smart cities in India. *Territory, Politics, Governance*, •••, 1–22. DOI: 10.1080/21622671.2023.2229872

De Palma, A., & Lindsey, R. (2011). Traffic congestion pricing methodologies and technologies. *Transportation Research Part C, Emerging Technologies*, *19*(6), 1377–1399. DOI: 10.1016/j.trc.2011.02.010

Depsky, N. J., Cushing, L., & Morello-Frosch, R. (2022). High-resolution gridded estimates of population sociodemographics from the 2020 census in California. *PLoS One*, *17*(7), e0270746. DOI: 10.1371/journal.pone.0270746

Dey, S.. (2024). Factors associated with speeding: Literature review/meta-analysis. *Journal of Transport & Health*.

Dilawar, A., Chen, B., Ashraf, A., Alphonse, K., Hussain, Y., Ali, S., Jinghong, J., Shafeeque, M., Boyang, S., Sun, X., & Hussain, S. (2022). Development of a GIS based hazard, exposure, and vulnerability analyzing method for monitoring drought risk at Karachi, Pakistan. *Geomatics, Natural Hazards & Risk*, *13*(1), 1700–1720. DOI: 10.1080/19475705.2022.2090863

Di, Q., Dai, L., Wang, Y., Zanobetti, A., Choirat, C., Schwartz, J. D., & Dominici, F. (2017). Association of short-term exposure to air pollution with mortality in older adults. *Journal of the American Medical Association*, *318*(24), 2446–2456. DOI: 10.1001/jama.2017.17923

Dodel, M., & Hernandez, D. (2024). Smarter but more unequal transport? How socioeconomic and digital inequalities hinder adoption of mobility apps in the Global South. *Travel Behaviour & Society*, *38*, 100911. DOI: 10.1016/j.tbs.2024.100911

Dodman, D., Hayward, B., Pelling, M., Castán Broto, V., Chow, W., Chu, E., Dawson, R., Khirfan, L., McPhearson, T., Prakash, A., & Ziervogel, G. (2022). *Cities, settlements and key infrastructure*. In IPCC Sixth Assessment Report (AR6): Climate change 2022 – Impacts, adaptation and vulnerability. Intergovernmental Panel on Climate Change. https://www.ipcc.ch/report/ar6/wg2/

Dodson, J., & Sipe, N. (2008). *Shaping urban development: The role of planning and policy in managing urban sprawl*. University of Queensland Press.

Dowd, B. M., Press, D., & Huertos, M. L.DOWD. (2008). Agricultural nonpoint source water pollution policy: The case of California's Central Coast. *Agriculture, Ecosystems & Environment*, *128*(3), 151–161. DOI: 10.1016/j.agee.2008.05.014

Downs, A. (2004). *Still stuck in traffic: Coping with peak-hour traffic congestion*. Brookings Institution Press.

Druzhinin, P. V., Shkiperova, G. T., Potasheva, O. V., & Zimin, D. A. (2020). The Assessment of the Impact of the Economy's Development on Air Pollution. *Ekonomicheskie i Sotsialnye Peremeny*, *13*(2), 125–142. DOI: 10.15838/esc.2020.2.68.8

Du, G.. (2022). Sensation seeking → risky driving via emotion regulation. *Risk Analysis*, *42*(11), 2449–2466.

Duman, E.. (2023). Evaluating the effectiveness of machine learning techniques in predicting traffic accident severity. *PLoS One*, *18*(7), e0288298. PMID: 37418473

Duranton, G., & Turner, M. A. (2011). The fundamental law of road congestion: Evidence from US cities. *The American Economic Review*, *101*(6), 2616–2652. DOI: 10.1257/aer.101.6.2616

Dureja, A., Dureja, A., Kumar, V., & Sabharwal, S. (2024). Combining digital twin technology and intelligent transportation systems for smart mobility. In Springer (Ed.), [Title of the book] (pp. 281–296). Cham, Switzerland: *Springer*. ISBN 978-3-031-58522-7. DOI: 10.1007/978-3-031-58523-4_14

Emmert-Streib, F., Yang, Z., Feng, H., Tripathi, S., & Dehmer, M. (2020). An Introductory Review of Deep Learning for Prediction Models With Big Data. *Frontiers in Artificial Intelligence*, *3*(February), 1–23. DOI: 10.3389/frai.2020.00004

Erhardt, G., Mucci, R., Cooper, D., Sana, B., Chen, M., & Castiglione, J. (2021). Do transportation network companies increase or decrease transit ridership? Empirical evidence from San Francisco. *Transportation, 49*(2), 313–342. DOI: 10.1007/s11116-021-10178-4

Eubanks, V. (2018). *Automating inequality: How high-tech tools profile, police, and punish the poor*. St. Martin's Press.

Ewing, R., & Cervero, R. (2010). Travel and the built environment: A meta-analysis. *Journal of the American Planning Association, 76*(3), 265–294. DOI: 10.1080/01944361003766766

Fang, C. F., & Ling, D. L. (2003). Investigation of the noise reduction provided by tree belts. *Landscape and Urban Planning, 63*(4), 187–195. DOI: 10.1016/S0169-2046(02)00190-1

Farid, A. M., Liner, B., Alshareef, M., Singh Badhesha, P., Boccaletti, C., Alessandro, N., Cacho, A., Carlier, C.-I., Corriveau, A., & Khayal, I.. (2021). Smart city drivers & challenges in urban mobility, healthcare, and interdependent infrastructure systems. *IEEE Potentials, 40*(1), 11–16. DOI: 10.1109/MPOT.2020.3011399

Fayomi, G. U., Onyari, E. K., & Jaiyeola, A. T. (2024). Quantitative assessment of open drainage and storm water runoff capacity in Nigeria: Consequences and way forward. *Case Studies in Chemical and Environmental Engineering, 9*, 100766.

Fenger, J. (2009). Air pollution in the last 50 years–From local to global. *Atmospheric Environment, 43*(1), 13–22. DOI: 10.1016/j.atmosenv.2008.09.061

Ferrini, F., Fini, A., Mori, J., & Gori, A. (2020). Role of vegetation as a mitigating factor in the urban context. *Sustainability (Basel), 12*(10), 4247. DOI: 10.3390/su12104247

Floridi, L. (2018). Soft ethics and the governance of the digital. *Philosophy & Technology, 31*(1), 1–8. DOI: 10.1007/s13347-018-0303-9 PMID: 30322997

Franklin, J. P., & Karlstrom, A. (2009). Travel time reliability for Stockholm Roadways. *Transportation Research Record: Journal of the Transportation Research Board, 2134*(1), 106–113. DOI: 10.3141/2134-13

Friedrich, R., & Reis, S. (Eds.). (2004). *Emissions of air pollutants: measurements, calculations and uncertainties*. Springer Science & Business Media. DOI: 10.1007/978-3-662-07015-4

Fujii, S., & Taniguchi, A. (2005). Reducing solo car use through personalized travel feedback. *Transportation Research Part D, Transport and Environment, 10*(1), 1–19.

Ganguly, T., Selvaraj, K. L., & Guttikunda, S. K. (2020). National Clean Air Programme (NCAP) for Indian cities: Review and outlook of clean air action plans. *Atmospheric Environment: X, 8*, 100096. DOI: 10.1016/j.aeaoa.2020.100096

Gao, Z., Shi, Z., & Zhou, X. (2018). Integrating real-time IoT data into roadway network databases. *IEEE Transactions on Intelligent Transportation Systems*, *19*(4), 1100–1112.

Garau, C., & Pavan, V. M. (2018). Evaluating urban quality: Indicators and assessment tools for smart sustainable cities. *Sustainability (Basel)*, *10*(2), 575. DOI: 10.3390/su10030575

Gauderman, W. J., Urman, R., Avol, E., Berhane, K., McConnell, R., Rappaport, E., Chang, R., Lurmann, F., & Gilliland, F. (2015). Association of improved air quality with lung development in children. *The New England Journal of Medicine*, *372*(10), 905–913. DOI: 10.1056/NEJMoa1414123

Gehl, J. (2010). *Cities for people*. Island Press.

Genc, S., Zadeoglulari, Z., Fuss, S. H., & Genc, K. (2012). The adverse effects of air pollution on the nervous system. *Journal of Toxicology*, *2012*(1), 782462. DOI: 10.1155/2012/782462

Genders, W., & Razavi, S. (2016). Using a deep reinforcement learning agent for traffic signal control. *arXiv preprint arXiv:1611.01142*. /arXiv.1611.01142DOI: 10.48550

Genders, W. A. P., & Razavi, S. (2016). A survey of reinforcement learning methods for traffic signal control. *IEEE Transactions on Intelligent Transportation Systems*, *17*(7), 1946–1959.

Ghosh, B., & Arora, S. (2022). Smart as (un)democratic? The making of a smart city imaginary in Kolkata, India. *Environment and Planning C. Politics and Space*, *40*(1), 318–339. DOI: 10.1177/23996544211027583

Gidlöf-Gunnarsson, A., & Öhrström, E. (2007). Noise and well-being in urban residential environments: The potential role of perceived availability to nearby green areas. *Landscape and Urban Planning*, *83*(2–3), 115–126. DOI: 10.1016/j.landurbplan.2007.03.003

Gill, S. E., Handley, J. F., Ennos, A. R., & Pauleit, S. (2007). Adapting cities for climate change: The role of the green infrastructure. *Built Environment*, *33*(1), 115–133. DOI: 10.2148/benv.33.1.115

Gkartzonikas, C., & Gkritza, K. (2019). Towards real-time traffic flow and incident detection with IoT sensors. *Transport Reviews*, *39*(2), 181–198.

Golub, A., Satterfield, V., Serritella, M., Singh, J., & Phillips, S. (2019). Assessing the barriers to equity in smart mobility systems: A case study of Portland, Oregon. *Case Studies on Transport Policy*, *7*(4), 689–697. DOI: 10.1016/j.cstp.2019.10.002

Gómez-Hernández, J. J., & Xu, T. (2022). Contaminant Source Identification in Aquifers: A Critical View. *Mathematical Geosciences*, *54*(2), 437–458. DOI: 10.1007/s11004-021-09976-4

Goodchild, M. F. (2021). Responsible data governance in smart cities. *Computers, Environment and Urban Systems*, *88*, 101643.

Gössling, S., Cohen, S., & Haurie, A. (2016). *Sustainable transport and urban planning: Policies, practices and implications*. Springer.

Goyal, P., & Gandhi, G. (2016). Assessment of air quality during the 'odd-even scheme'of vehicles in Delhi. *Indian Journal of Science and Technology*, *9*(48), 1–7. DOI: 10.17485/ijst/2016/v9i30/97528

Goyal, S., Sahni, A., & Garg, T. (2018). Identification and prioritisation of barriers in building smart cities. *International Journal of Mechanical Engineering and Technology*, *9*, 819–827.

Graells-Garrido, E., Serra-Burriel, F., Rowe, F., Cucchietti, F. M., & Reyes, P. (2021). *A city of cities: Measuring how 15-minute urban accessibility shapes human mobility in Barcelona*. arXiv. https://arxiv.org/abs/2102.12315

Grambsch, A. (2001). *Climate change and air quality*. Department of Tansportation Centre for Climate Change and Environmental Forecasting.

Greater London Authority (GLA). (2020). ULEZ One Year Report. https://www.london.gov.uk

Griffin, R. D. (2016). *Principles of air quality management*. CRC Press. DOI: 10.1201/9781420006544

Grote, R., Samson, R., Alonso, R., Amorim, J. H., Cariñanos, P., Churkina, G., Fares, S., Le Thiec, D., Niinemets, Ü., Mikkelsen, T. N., & Paoletti, E. (2016). Functional traits of urban trees: Air pollution mitigation potential. *Frontiers in Ecology and the Environment*, *14*(10), 543–550. DOI: 10.1002/fee.1426

Gul, H., & Das, B. K. (2023). The impacts of air pollution on human health and well-being: A comprehensive review. *Journal of Environmental Impact and Management Policy*, *36*(36), 1–11. DOI: 10.55529/jeimp.36.1.11

Guo, A., He, T., Yue, W., Xiao, W., Yang, J., Zhang, M., & Li, M. (2023). Contribution of urban trees in reducing land surface temperature: Evidence from China's major cities. *International Journal of Applied Earth Observation and Geoinformation, 125*, 103570. DOI: 10.1016/j.jag.2023.103570

Gupta, U., Shukla, S., Bhardwaj, L. K., Rath, P., Upadhyay, D., Sharma, B., & Singh, H. (2024). Unraveling urban plant strategies for heavy metal uptake and detoxification. In *Urban Forests, Climate Change and Environmental Pollution: Physio-Biochemical and Molecular Perspectives to Enhance Urban Resilience* (pp. 93–119). Springer Nature Switzerland. DOI: 10.1007/978-3-031-67837-0_5

Gurjar, B. R., Butler, T. M., Lawrence, M. G., & Lelieveld, J. (2010). Evaluation of emissions and air quality in megacities. *Atmospheric Environment, 44*(35), 4156–4166. DOI: 10.1016/j.atmosenv.2010.07.028

Guttikunda, S. K., Dammalapati, S. K., Pradhan, G., Krishna, B., Jethva, H. T., & Jawahar, P. (2023). What is polluting Delhi's air? A review from 1990 to 2022. *Sustainability (Basel), 15*(5), 4209. DOI: 10.3390/su15054209

Guttikunda, S. K., Goel, R., & Pant, P. (2014). Nature of air pollution, emission sources, and management in the Indian cities. *Atmospheric Environment, 95*, 501–510. DOI: 10.1016/j.atmosenv.2014.07.006

Guttikunda, S. K., Nishadh, K. A., Gota, S., Singh, P., Chanda, A., Jawahar, P., & Asundi, J. (2019). Air quality, emissions, and source contributions analysis for the Greater Bengaluru region of India. *Atmospheric Pollution Research, 10*(5), 941–953. DOI: 10.1016/j.apr.2019.01.002

Gwilliam, K. (2011). Urban transport in developing countries. *Transport Reviews, 31*(2), 151–174. DOI: 10.1080/01441647.2010.519887

Haccuria, A., Michils, A., Michiels, S., & Van Muylem, A. (2014). Exhaled nitric oxide: A biomarker integrating both lung function and airway inflammation changes. *The Journal of Allergy and Clinical Immunology, 134*(3), 554–559. DOI: 10.1016/j.jaci.2013.12.1070

Hageman, K. J., Bogdal, C., & Scheringer, M. (2015). Long-range and regional atmospheric transport of POPs and implications for global cycling. [). Elsevier.]. *Comprehensive Analytical Chemistry, 67*, 363–387. DOI: 10.1016/B978-0-444-63299-9.00011-9

Haggerty, R., Sun, J., Yu, H., & Li, Y. (2023). Application of machine learning in groundwater quality modeling - A comprehensive review. *Water Research, 233*(402), 119745. DOI: 10.1016/j.watres.2023.119745

Han, L., Zhou, W., Li, W., & Qian, Y. (2022). Challenges in continuous air quality improvement: An insight from the contribution of the recent clean air actions in China. *Urban Climate, 46*, 101328. DOI: 10.1016/j.uclim.2022.101328

Hanna, E., & Comín, F. A. (2021). Urban green infrastructure and sustainable development: A review. *Sustainability (Basel), 13*(20), 11498. DOI: 10.3390/su132011498

Hannun, R. M., & Razzaq, A. H. A. (2022, March). Air pollution resulted from coal, oil and gas firing in thermal power plants and treatment: A review. []. IOP Publishing.]. *IOP Conference Series. Earth and Environmental Science, 1002*(1), 012008. DOI: 10.1088/1755-1315/1002/1/012008

Hao, Y., Li, Y., Han, X., Wang, T., Yang, R., Wang, P., Xiao, K., Li, W., Lu, H., Fu, J., Wang, Y., Shi, J., Zhang, Q., & Jiang, G. (2019). Air monitoring of polychlorinated biphenyls, polybrominated diphenyl ethers and organochlorine pesticides in West Antarctica during 2011–2017: Concentrations, temporal trends and potential sources. *Environmental Pollution, 249*, 381–389. DOI: 10.1016/j.envpol.2019.03.039

Harrison, R. M. (2006). Sources of air pollution. *Air Quality Guideliees. Air Quality Guidelines*, 9-31.

Harrison, C., & Donnelly, I. A. (2011). A theory of smart cities. In *Proceedings of the 55th Annual Meeting of the ISSS* (pp. 1–15).

Harris, P., & De Leeuw, E. (2022). Integrating health into urban infrastructure planning for equity. *Cities & Health, 6*(1), 1–4. DOI: 10.1080/23748834.2021.1938570

Hashim, W., Jamaludin, S. M., Mohamed, R. M. A. R., Gunasekaran, S. S., Thabit, R., & Gohel, H. (2024, July). Analysing type of plants for reducing noise pollution in climate change adaptation. []. IOP Publishing.]. *IOP Conference Series. Earth and Environmental Science, 1372*(1), 012056. DOI: 10.1088/1755-1315/1372/1/012056

Hassan, A. M., & Lee, H. (2015). The paradox of the sustainable city: Definitions and examples. *Environment, Development and Sustainability, 17*(6), 1267–1285. DOI: 10.1007/s10668-014-9604-z

Healey, P. (1997). *Collaborative planning: Shaping places in fragmented societies*. Macmillan. DOI: 10.1007/978-1-349-25538-2

Heliyon. (2021). https://www.cell.com/heliyon/issue?pii=S2405-8440(25)X0004-9

Hensher, D. A., Li, Z., & Mulley, C. (2020). Willingness to pay for Mobility-as-a-Service: Insights from Sydney. *Transportation, 47*(4), 1937–1956.

Hietanen, S. (2014). Mobility as a service: The new transport model? In *2014 World Transport Convention*.

Hoek, G., Krishnan, R. M., Beelen, R., Peters, A., Ostro, B., Brunekreef, B., & Kaufman, J. D. (2013). Long-term air pollution exposure and cardio-respiratory mortality: A review. *Environmental Health : A Global Access Science Source*, *12*(1), 43. DOI: 10.1186/1476-069X-12-43

Hörl, S., Cerqueira, R., & Axhausen, K. W. (2021). Gamified incentives for sustainable commuting. *Transportation Research Part A, Policy and Practice*, *145*, 246–261.

Hossain, M. A., & Huggins, R. (2021). The environmental and social impacts of unplanned and rapid industrialization in suburban areas: The case of the greater Dhaka region, Bangladesh. *Environment & Urbanization Asia*, *12*(1), 73–89. DOI: 10.1177/0975425321990319

Hossain, M. S., Hasan, M. M., & Islam, M. R. (2020). Urban transportation and environmental sustainability: Case study of Dhaka city. *Sustainable Cities and Society*, *56*, 102084. DOI: 10.1016/j.scs.2020.102084

Huang, Y. L., & Batterman, S. (2000). Selection and evaluation of air pollution exposure indicators based on geographic areas. *The Science of the Total Environment*, *253*(1-3), 127–144. DOI: 10.1016/S0048-9697(00)00425-3

Hung, Y. T., & Ashner, I. (2020). Sulfur dioxide emission and mitigation. In *Handbook of environment and waste management: acid rain and greenhouse gas pollution control* (pp. 627-658). DOI: 10.1142/9789811207136_0018

Hurtado, G.. (2023). Using contextual data to predict risky driving events. [REMOVED HYPERLINK FIELD]. *Accident; Analysis and Prevention*, *188*, 107028.

Iberdrola. (2023). *Adapting to climate change: What will the earth look like in 2030?* Nature.

Iqbal, M. M., Li, L., Hussain, S., Lee, J. L., Mumtaz, F., Elbeltagi, A., Waqas, M. S., & Dilawar, A. (2022). Analysis of seasonal variations in surface water quality over wet and dry regions. *Water (Basel)*, *14*(7), 1058. DOI: 10.3390/w14071058

Isiyaka, H. A., Mustapha, A., Juahir, H., & Phil-Eze, P. (2019). Water quality modelling using artificial neural network and multivariate statistical techniques. *Modeling Earth Systems and Environment*, *5*(2), 583–593. DOI: 10.1007/s40808-018-0551-9

Islam, S., & Hasan, M. (2021). Urban green space and social equity in Dhaka: Challenges and opportunities. *Urban Forestry & Urban Greening*, *61*, 127102. DOI: 10.1016/j.ufug.2021.127102

Iyer, L. S. (2021). AI-enabled applications towards intelligent transportation. *Transportation Engineering*, *5*, 100083. DOI: 10.1016/j.treng.2021.100083

Izah, S. C., Ogwu, M. C., Shahsavani, A., Kermani, M., & Hasanzadeh, V. (2024). Understanding the one health implications of air pollution. In *Air Pollutants in the Context of One Health: Fundamentals, Sources, and Impacts* (pp. 161–185). Springer Nature Switzerland. DOI: 10.1007/698_2024_1129

Jadeja, Y., Maheshwari, B., Packham, R., Bohra, H., Purohit, R., Thaker, B., Dillon, P., Oza, S., Dave, S., Soni, P., Dashora, Y., Dashora, R., Shah, T., Gorsiya, J., Katara, P., Ward, J., Kookana, R., Singh, P. K., Chinnasamy, P., & Chew, M. (2018). Managing aquifer recharge and sustaining groundwater use: Developing a capacity building program for creating local groundwater champions. *Sustainable Water Resources Management*, *4*(2), 317–329. DOI: 10.1007/s40899-018-0228-6

Jakeman, A. J., Barreteau, O., Hunt, R. J., Rinaudo, J.-D., & Ross, A. (2016). Erratum to: Integrated Groundwater Management. In *Integrated Groundwater Management* (pp. E1–E1). Springer International Publishing. DOI: 10.1007/978-3-319-23576-9_29

Jansen, M. (1989). Water supply and sewage disposal at Mohenjo-Daro. *World Archaeology*, *21*(2), 177–192. DOI: 10.1080/00438243.1989.9980100 PMID: 16470995

Jarosińska, D., Héroux, M. È., Wilkhu, P., Creswick, J., Verbeek, J., Wothge, J., & Paunović, E. (2018). Development of the WHO environmental noise guidelines for the European region: An introduction. *International Journal of Environmental Research and Public Health*, *15*(4), 813. DOI: 10.3390/ijerph15040813

Javaid, A. (2022). *Top 15 noisiest cities in the world: Check rankings of Moradabad, Kolkata, and Asansol*. Jagran Josh. https://www.jagranjosh.com/general-knowledge/top-15-noisiestcities-in-the-world-1648473437-1

Javidroozi, V., Carter, C., Grace, M., & Shah, H. (2023). Smart, sustainable, green cities: A state-of-the-art review. *Sustainability (Basel)*, *15*(6), 5353. DOI: 10.3390/su15065353

Jha, I. S., Sen, S., & Kumar, R. (2014). Smart grid development in India—A case study. In *Proceedings of the Eighteenth National Power Systems Conference (NPSC)* (pp. 1–6). IEEE. DOI: 10.1109/NPSC.2014.7103866

Jiang, Z., Yu, D., Luan, S., Zhou, H., & Meng, F. (2022). Integrating traffic signal optimization with vehicle microscopic control to reduce energy consumption in a connected and automated vehicles environment. *Journal of Cleaner Production*, *371*, 133694. DOI: 10.1016/j.jclepro.2022.133694

Jia, S., Bi, L., Zhu, W., & Fang, T. (2023). System dynamics modeling for improving the policy effect of traffic energy consumption and CO_2 emissions. *Sustainable Cities and Society*, *90*, 104398. DOI: 10.1016/j.scs.2023.104398

Joss, S. (2011). Eco-cities: The mainstreaming of urban sustainability; key characteristics and driving factors. *International Journal of Sustainable Development and Planning*, *6*(3), 268–285. DOI: 10.2495/SDP-V6-N3-268-285

Kallenborn, R., Hung, H., & Brorström-Lundén, E. (2015). Atmospheric long-range transport of persistent organic pollutants (POPs) into polar regions. [). Elsevier.]. *Comprehensive Analytical Chemistry*, *67*, 411–432. DOI: 10.1016/B978-0-444-63299-9.00013-2

Kanaroglou, P.. (2020). Real-time air quality monitoring with mobile sensors. *The Science of the Total Environment*, *738*, 139677.

Karim, M. R., & Hasan, M. (2013). Traffic congestion in Dhaka: Causes, impacts and policy implications. *Journal of the Bangladesh Institute of Planners*, *6*, 67–78. DOI: 10.3329/jbip.v6i0.17547

Karimpour, H.. (2024). Enhancing road safety with machine learning: Current advances and challenges. *Engineering Applications of Artificial Intelligence*, *137*, 108078.

Karim, S. A. A., Kadir, E. A., & Nasution, A. H. (2020). Optimization Based Model Using Fuzzy and Other Statistical Techniques Towards Environmental Sustainability. In Karim, S. A. A., Kadir, E. A., & Nasution, A. H. (Eds.), *Optimization Based Model Using Fuzzy and Other Statistical Techniques Towards Environmental Sustainability*. Springer Singapore., DOI: 10.1007/978-981-15-2655-8

Karri, V., Schuhmacher, M., & Kumar, V. (2016). Heavy metals (Pb, Cd, As and MeHg) as risk factors for cognitive dysfunction: A general review of metal mixture mechanism in brain. *Environmental Toxicology and Pharmacology*, *48*, 203–213. DOI: 10.1016/j.etap.2016.09.016

Karthika, I., & Pushpalatha, D., D, I. H., S, D., K, M., & S, H. (2025). Real time adaptive routing for efficient urban mobility. In 2025 International Conference on Visual Analytics and Data Visualization (ICVADV) (pp. 1052–1058). IEEE Xplore. DOI: 10.1109/ICVADV63329.2025.10961486

Katsanou, K., & Karapanagioti, H. K. (2017). Surface Water and Groundwater Sources for Drinking Water. In *Handbook of Environmental Chemistry* (Vol. 67, pp. 1–19). Issue December., DOI: 10.1007/698_2017_140

Kenda, K., Čerin, M., Bogataj, M., Senožetnik, M., Klemen, K., Pergar, P., Laspidou, C., & Mladenić, D. (2018). Groundwater Modeling with Machine Learning Techniques: Ljubljana polje Aquifer. *EWa*, *S3*, 697. DOI: 10.3390/proceedings2110697

Ketzler, R., Franco, D., & Trupp, M. (2020). Digital twin applications in urban traffic management. *Transportation Research Procedia*, *45*, 1–10.

Khadke, P. A., & Waghmare, P. B. (2018). Class and size wise distribution of urban centre and their determinants in Maharashtra. *Research Review International Journal of Multidisciplinary*, *3*, 722–730.

Khattak, A. J., & De Palma, A. (1997). The impact of weather on traffic flow. *Transportation Research Part A, Policy and Practice*, *31*(6), 475–491.

Khlie, K., & Benmamoun, Z. (2024). Towards smarter and greener cities: Harnessing AI and green technology for urban sustainability. *Journal of Infrastructure Policy Development*, *8*(8), 6300. DOI: 10.24294/jipd.v8i8.6300

Kingdon, C. (2023). Air pollution is the largest environmental risk to public health and children are especially vulnerable. *bmj, 381*.

Kitchin, R. (2016). The ethics of smart cities and urban science. *Philosophical Transactions. Series A, Mathematical, Physical, and Engineering Sciences*, *374*(2083), 20160115. DOI: 10.1098/rsta.2016.0115 PMID: 28336794

Köhler, E., & Strehler, M. (2012). Combining static and dynamic models for traffic signal optimization inherent load-dependent travel times in a cyclically time-expanded network model. *Procedia: Social and Behavioral Sciences*, *54*, 1125–1134. DOI: 10.1016/j.sbspro.2012.09.827

Kombo, O. H., Kumaran, S., Sheikh, Y. H., Bovim, A., & Jayavel, K. (2020). Long-term groundwater level prediction model based on hybrid KNN-RF technique. *Hydrology*, *7*(3), 1–24. DOI: 10.3390/hydrology7030059

Konarska, J., Uddling, J., Holmer, B., Lutz, M., Lindberg, F., Pleijel, H., & Thorsson, S. (2016). Transpiration of urban trees and its cooling effect in a high latitude city. *International Journal of Biometeorology*, *60*(1), 159–172. DOI: 10.1007/s00484-015-1014-x

Kopackova, H., & Libalova, P. (2017). Smart city concept as socio-technical system. *2017 International Conference on Information and Digital Technologies (IDT)*, 198–205. DOI: 10.1109/DT.2017.8024297

Kragh, J. (1979). Pilot study on railway noise attenuation by belts of trees. *Journal of Sound and Vibration*, *66*(3), 407–415. DOI: 10.1016/0022-460X(79)90859-9

Křeček, J., & Hořická, Z. (2006). Forests, air pollution and water quality: Influencing health in the headwaters of Central Europe's "Black Triangle". *Unasylva*, *57*(224), 46–49.

Krishnan, R.. (2024). Identification of the best machine learning model for predicting injury severities in RTAs. *International Journal of Injury Control and Safety Promotion*, *31*(3), 360–375. Advance online publication. DOI: 10.1080/17457300.2024.2335478

Kulshrestha, U., & Mishra, M. (2020). A review on long range transport of air pollution in south Asia. *Vayumandal*, *46*(2), 21–30.

Kumar, S. (2024, December 18). Singapore Smart Traffic System is Redefining Urban Mobility. IntelliStride. Retrieved June 23, 2025, from https://www.intellistride.com/blog/how-singapores-smart-traffic-system-is-redefining-urban-mobility/

Kumar, P., Gulia, S., Harrison, R. M., & Khare, M. (2017). The influence of odd–even car trial on fine and coarse particles in Delhi. *Environmental Pollution*, *225*, 20–30. DOI: 10.1016/j.envpol.2017.03.017

Kumar, P., Singh, A. B., Arora, T., Singh, S., & Singh, R. (2023). Critical review on emerging health effects associated with the indoor air quality and its sustainable management. *The Science of the Total Environment*, *872*, 162163. DOI: 10.1016/j.scitotenv.2023.162163

Kumar, S. R., Arumugam, T., Anandakumar, C., Balakrishnan, S., & Rajavel, D. (2013). Use of plant species in controlling environmental pollution. *Bulletin of Environment, Pharmacology and Life Sciences*, *2*(2), 52.

Kumar, V. T. M., & Dahiya, B. (2017). Smart economy in smart cities. In Kumar, V. T. M. (Ed.), *Smart economy in smart cities* (pp. 3–76). Springer., DOI: 10.1007/978-981-10-1610-3_1

Kundu, A. (2011). *Trends and Processes of Urbanisation in India. Human Settlements Group*. IIED.

Kyung, S. Y., & Jeong, S. H. (2020). Particulate-matter related respiratory diseases. *Tuberculosis and Respiratory Diseases*, *83*(2), 116. DOI: 10.4046/trd.2019.0025

Laa, B., Balac, M., & Schüssler, N. (2020). Integrating micro-mobility in multimodal routing. *Journal of Transport Geography*, *85*, 102710.

Land Transport Authority. (2021). *Singapore Electronic Road Pricing impact report.* Land Transport Authority.

Lebedev, A. T., Mazur, D. M., Polyakova, O. V., & Hänninen, O. (2014). Snow samples as markers of air pollution in mass spectrometry analysis. In *Environmental Indicators* (pp. 515–541). Springer Netherlands.

Lei, W., Qi, K., Hui, X., & Qidi, W. (2007). Traffic Intelligent Optimization and Local Traffic-Flow Control Inside Shanghai-EXPO-Area. In *2007 IEEE International Conference on Networking, Sensing and Control* (pp. 856–861). IEEE Institute of Electrical and Electronics Engineers. DOI: 10.1109/ICNSC.2007.372893

Lelieveld, J., Evans, J. S., Fnais, M., Giannadaki, D., & Pozzer, A. (2015). The contribution of outdoor air pollution sources to premature mortality on a global scale. *Nature*, *525*(7569), 367–371. DOI: 10.1038/nature15371

Leonard, R. J., McArthur, C., & Hochuli, D. F. (2016). Particulate matter deposition on roadside plants and the importance of leaf trait combinations. *Urban Forestry & Urban Greening*, *20*, 249–253. DOI: 10.1016/j.ufug.2016.09.008

Leong, W. C., Bahadori, A., Zhang, J., & Ahmad, Z. (2021). Prediction of water quality index (WQI) using support vector machine (SVM) and least square-support vector machine (LS-SVM). *International Journal of River Basin Management*, *19*(2), 149–156. DOI: 10.1080/15715124.2019.1628030

Liang, L., Wang, Z., & Li, J. (2019). The effect of urbanization on environmental pollution in rapidly developing urban agglomerations. *Journal of Cleaner Production*, *237*, 117649. DOI: 10.1016/j.jclepro.2019.117649

Li, B., Tao, S., & Dawson, R. W. (2002). Evaluation and analysis of traffic noise from the main urban roads in Beijing. *Applied Acoustics*, *63*(10), 1137–1142. DOI: 10.1016/S0003-682X(02)00024-5

Li, G., Wei, W., Shao, X., Nie, L., Wang, H., Yan, X., & Zhang, R. (2018). A comprehensive classification method for VOC emission sources to tackle air pollution based on VOC species reactivity and emission amounts. *Journal of Environmental Sciences (China)*, *67*, 78–88. DOI: 10.1016/j.jes.2017.08.003

Lin, B., & Zhu, J. (2018). Changes in urban air quality during urbanization in China. *Journal of Cleaner Production*, *188*, 312–321. DOI: 10.1016/j.jclepro.2018.03.293

Lioy, P. J. (2018). *Toxic Air Pollution: A Comprehensive Study Non-Criteria Air Pollutants*. CRC Press. DOI: 10.1201/9781351077347

Li, P., Wu, J., & Shukla, S. (2022). Achieving the One Health Goal: Highlighting Groundwater Quality and Public Health. *Water (Basel)*, *14*(21), 3540. DOI: 10.3390/w14213540

Li, R., Chen, W., Xiu, A., Zhao, H., Zhang, X., Zhang, S., & Tong, D. Q. (2019). A comprehensive inventory of agricultural atmospheric particulate matters (PM10 and PM2. 5) and gaseous pollutants (VOCs, SO2, NH3, CO, NOx and HC) emissions in China. *Ecological Indicators*, *107*, 105609. DOI: 10.1016/j.ecolind.2019.105609

Litman, T. (2021). *Evaluating transportation land use impacts*. Victoria Transport Policy Institute. https://www.vtpi.org/landuse.pdf

Litman, T. (2020). *Transportation and environmental policy in the 21st century*. Victoria Transport Policy Institute.

Liu, H., Tian, Z., & Huang, W. (2020). Emission impacts of adaptive traffic control. *Transportation Research Part D, Transport and Environment*, *82*, 102324.

Liu, X.. (2025). Traffic accident severity prediction based on MSCPO-XGBoost. *Scientific Reports*, *15*, 12345. PMID: 40670407

Liu, Y., Xu, J., Chen, D., Sun, P., & Ma, X. (2019). The association between air pollution and preterm birth and low birth weight in Guangdong, China. *BMC Public Health*, *19*(1), 3. DOI: 10.1186/s12889-018-6307-7

Liu, Y., Yang, C., Jiang, L., Xie, S., & Zhang, Y. (2019). Intelligent Edge Computing for IoT-Based Energy Management in Smart Cities. *IEEE Network*, *33*(2), 111–117. DOI: 10.1109/MNET.2019.1800254

Li, Y., Svenning, J.-C., Zhou, W., & Schell, C. J. (2023). Global inequality in cooling benefits from urban green spaces and its climate adaptation potential. *Nature Communications*, *14*, 6205. DOI: 10.1038/s41467-023-41846-5

Louf, R., & Barthélemy, M. (2013). Modeling the polycentric transition of cities. *Physical Review Letters*, *111*(19), 198702. DOI: 10.1103/PhysRevLett.111.198702 PMID: 24266493

Lu, C., Wang, F., Liu, Q., Deng, M., Yang, X., & Ma, P. (2023). Effect of NO2 exposure on airway inflammation and oxidative stress in asthmatic mice. *Journal of Hazardous Materials*, *457*, 131787. DOI: 10.1016/j.jhazmat.2023.131787

Lucas, K., Mattioli, G., Verlinghieri, E., & Guzman, A. (2016). Transport poverty and its adverse social consequences. *Proceedings of the Institution of Civil Engineers—Transport, 169*(6), 353–365. DOI: 10.1680/jtran.15.00073

Ma, J., Zhang, L., & Song, Y. (2021). Economic loss assessment of traffic congestion: A review and framework. *Transportation Research Part A, Policy and Practice*, *143*, 1–15. DOI: 10.1016/j.tra.2020.10.007

Mallik, C. (2019). Anthropogenic sources of air pollution. In *Air pollution: Sources, impacts and controls* (pp. 6–25). CAB International. DOI: 10.1079/9781786393890.0006

Manickathan, L., Defraeye, T., Allegrini, J., Derome, D., & Carmeliet, J. (2018). Parametric study of the influence of environmental factors and tree properties on the transpirative cooling effect of trees. *Agricultural and Forest Meteorology*, *248*, 259–274. DOI: 10.1016/j.agrformet.2017.10.014

Manisalidis, I., Stavropoulou, E., Stavropoulos, A., & Bezirtzoglou, E. (2020). Environmental and health impacts of air pollution: A review. *Frontiers in Public Health*, *8*, 14. DOI: 10.3389/fpubh.2020.00014

Marelle, L., Thomas, J. L., Raut, J. C., Law, K. S., Jalkanen, J. P., Johansson, L., Roiger, A., Schlager, H., Kim, J., Reiter, A., & Weinzierl, B. (2016). Air quality and radiative impacts of Arctic shipping emissions in the summertime in northern Norway: From the local to the regional scale. *Atmospheric Chemistry and Physics*, *16*(4), 2359–2379. DOI: 10.5194/acp-16-2359-2016

Marsden, G., Docherty, I., & Anable, J. (2021). Integrating cycling and micro-mobility into urban transport models. *Transport Policy*, *103*, 1–10.

Martinez, G. S., Spadaro, J. V., Chapizanis, D., Kendrovski, V., Kochubovski, M., & Mudu, P. (2018). Health impacts and economic costs of air pollution in the metropolitan area of Skopje. *International Journal of Environmental Research and Public Health*, *15*(4), 626. DOI: 10.3390/ijerph15040626

McDonald, B. C., Dallmann, T. R., Martin, E. W., & Harley, R. A. (2012). Long-term trends in nitrogen oxide emissions from motor vehicles at national, state, and air basin scales. *Journal of Geophysical Research*, *117*(D21), 2012JD018304. DOI: 10.1029/2012JD018304

McIlveen, N., McIlveen, P., & Lewis, I. (2024). Meta-analysis: TPB and risky driving (speeding, DUI, distraction, seatbelts). *Accident; Analysis and Prevention*.

Menon, A. (2000). Electronic road pricing in Singapore: Success factors. *Journal of Transport Economics and Policy*, *34*(3), 329–346.

Mikucioniene, D., Mínguez-García, D., Repon, M. R., Milašius, R., Priniotakis, G., Chronis, I., Kiskira, K., Hogeboom, R., Belda-Anaya, R., & Díaz-García, P. (2024). Understanding and addressing the water footprint in the textile sector: A review. *AUTEX Research Journal*, *24*(1), 1. DOI: 10.1515/aut-2024-0004

Ministry of Ecology and Environment. China (2018). China Air Quality Improvement Report. http://www.mee.gov.cn

Mishra, A. K., & Mohanty, P. K. (2018). Urban infrastructure financing in India: Applying the benefit and earmarking principles of taxation. *Journal of Social and Economic Development*, *20*(1), 110–128. DOI: 10.1007/s40847-018-0059-1

Mishra, P., Thakur, P., & Singh, G. (2022). Sustainable smart city to society 5.0: State-of-the-art and research challenges. *SAIEE African Research Journal, 113*(4), 152–164. DOI: 10.23919/SAIEE.2022.9945865

Mishra, R. K., Parida, M., & Rangnekar, S. (2010). Evaluation and analysis of traffic noise along bus rapid transit system corridor. *International Journal of Environmental Science and Technology, 7*(4), 737–750. DOI: 10.1007/BF03326183

Mohamed, A. M.. (2021). Machine learning for predicting traffic emissions. *The Science of the Total Environment, 781*, 146720.

Mohan, D.. (2024). Safety effectiveness of automated traffic enforcement systems (ATES). *Infrastructures, 9*(1), 25.

Moreno, C., Allam, Z., Chabaud, D., Gall, C., & Pratlong, F. (2021). Introducing the "15-Minute City": Sustainability, resilience and place identity in future post-pandemic cities. *Smart Cities, 4*(1), 93–111. DOI: 10.3390/smartcities4010006

Moreno, C., Allende, D., & Bosque, M. (2021). The 15-minute city concept in Paris. *Urban Planning, 6*(2), 42–60.

Moriarty, P. (2023). Advanced communication and computational technologies in a sustainable urban context: Smart grids, smart cities, and smart health. In Power Systems: Vol. Part F1423 (pp. 389–415). *Springer*. DOI: 10.1007/978-3-031-38506-3_15

MoRTH. (2023). *Road Accidents in India 2022*. Ministry of Road Transport and Highways, Government of India.

Musa, A. A., Malami, S. I., Alanazi, F., Ounaies, W., Alshammari, M., & Haruna, S. I. (2023). Sustainable traffic management for smart cities using Internet-of-Things-oriented intelligent transportation systems (ITS): Challenges and recommendations. *Sustainability (Basel), 15*(9859), 9859. Advance online publication. DOI: 10.3390/su15139859

Natarajan, N., & Sudheer, C. (2020). Groundwater level forecasting using soft computing techniques. *Neural Computing & Applications, 32*(12), 7691–7708. DOI: 10.1007/s00521-019-04234-5

Nava-Ruiz, C., Méndez-Armenta, M., & Ríos, C. (2012). Lead neurotoxicity: Effects on brain nitric oxide synthase. *Journal of Molecular Histology, 43*(5), 553–563. DOI: 10.1007/s10735-012-9414-2

Newman, P., & Kenworthy, J. (1991). Transport and urban form in thirty-two of the world's principal cities. *Transport Reviews, 11*(3), 249–272. DOI: 10.1080/01441649108716787

Newman, P., & Kenworthy, J. (1999). *Sustainability and cities: Overcoming automobile dependence*. Island Press.

Newman, P., & Kenworthy, J. (2015). *The end of automobile dependence: How cities are moving beyond car-based planning*. Island Press. DOI: 10.5822/978-1-61091-613-4

Nowak, D. J., Hirabayashi, S., Bodine, A., & Greenfield, E. (2014). Tree and forest effects on air quality and human health in the United States. *Environmental Pollution*, *193*, 119–129. DOI: 10.1016/j.envpol.2014.05.028 PMID: 25016465

OECD. (2020). *Developing an integrated approach to green infrastructure*. Organisation for Economic Co-operation and Development. https://www.oecd.org

OECD. (2020). *Nature-based solutions for cities*. Organisation for Economic Co-operation and Development.

OECD. (2021). *Regulating smart cities: Legal and institutional frameworks*. OECD Publishing.

Onuu, M. U. (2000). Road traffic noise in Nigeria: Measurements, analysis and evaluation of nuisance. *Journal of Sound and Vibration*, *233*(3), 391–405. DOI: 10.1006/jsvi.1999.2832

Ozdemir, B., Bayramoglu, E., & Demirel, O. (2014). Noise pollution and human health in Trabzon parks. *Studies on Ethno-Medicine*, *8*(2), 127–134. DOI: 10.1080/09735070.2014.11917627

Pal, A. K., Kumar, V., & Saxena, N. C. (2000). Noise attenuation by green belts. *Journal of Sound and Vibration*, *234*(1), 149–165. DOI: 10.1006/jsvi.2000.2863

Pangbourne, K., Mladenović, M. N., Stead, D., & Milakis, D. (2020). Questioning mobility as a service: Unanticipated implications for society and governance. *Transportation Research Part A, Policy and Practice*, *131*, 35–49. DOI: 10.1016/j.tra.2019.09.033

Papageorgiou, M., Diakaki, C., Dinopoulou, V., Kotsialos, A., & Wang, Y. (2003). Review of road traffic control strategies. *Proceedings of the IEEE*, *91*(12), 2043–2067. DOI: 10.1109/JPROC.2003.819610

Pasquale, F. (2015). *The black box society: The secret algorithms that control money and information*. Harvard University Press. DOI: 10.4159/harvard.9780674736061

Pathak, V., Tripathi, B. D., & Mishra, V. K. (2008). Dynamics of traffic noise in a tropical city Varanasi and its abatement through vegetation. *Environmental Monitoring and Assessment*, *146*(1-3), 67–75. DOI: 10.1007/s10661-007-0060-1

Paul, R. (2019). Urban metabolism of Bangalore City: A water mass balance analysis. *Journal of Industrial Ecology*, *22*, 1–23.

Peden, M.. (2023). Road traffic injuries and deaths: Global burden [review]. *Frontiers in Public Health*.

Peled, R. (2011). Air pollution exposure: Who is at high risk? *Atmospheric Environment*, *45*(10), 1781–1785. DOI: 10.1016/j.atmosenv.2011.01.001

Peng, T., Zhou, J., Zhang, C., & Fu, W. (2017). Streamflow Forecasting Using Empirical Wavelet Transform and Artificial Neural Networks. *Water (Basel)*, *9*(6), 406. DOI: 10.3390/w9060406

Pentland, A. (2012). *Building the new economy: Data as labor*. Harvard Business Review Press.

Peris-Ortiz, M., Bennett, D., & Yábar, D. P. B. (Eds.). (2016). *Sustainable smart cities: Creating spaces for technological, social, and business development*. Springer.

Pienkowski, M. (2021). Loud music and leisure noise is a common cause of chronic hearing loss, tinnitus and hyperacusis. *International Journal of Environmental Research and Public Health*, *18*(8), 4236. DOI: 10.3390/ijerph18084236

Pikridas, M., Vrekoussis, M., Sciare, J., Kleanthous, S., Vasiliadou, E., Kizas, C., Savvides, C., & Mihalopoulos, N. (2018). Spatial and temporal (short and long-term) variability of submicron, fine and sub-10 µm particulate matter (PM1, PM2. 5, PM10) in Cyprus. *Atmospheric Environment*, *191*, 79–93. DOI: 10.1016/j.atmosenv.2018.07.048

Pojani, D., & Stead, D. (2015). Sustainable urban transport in the developing world. *Transport Reviews*, *35*(3), 253–274.

Popescu, F., & Ionel, I. (2010). Anthropogenic air pollution sources. *Air quality*, 1-22.

Praharaj, S., & Han, H. (2019). Cutting through the clutter of smart city definitions: A reading into the smart city perceptions in India. *City, Culture and Society*, *18*, 100289. DOI: 10.1016/j.ccs.2019.05.005

Prakash, G. (2002). The urban turn. *Sarai reader.* Oxford University Press., *2*(7), 2-7.

Premsankar, G., Di Francesco, M., & Taleb, T. (2018). Edge Computing for the Internet of Things: A Case Study. *IEEE Internet of Things Journal*, *5*(2), 1275–1284. DOI: 10.1109/JIOT.2018.2805263

Qing, W. A. N. G. (2018). Urbanization and global health: The role of air pollution. *Iranian Journal of Public Health*, *47*(11), 1644.

Rahman, M. A., Stratopoulos, L. M., Moser-Reischl, A., Zölch, T., Häberle, K. H., Rötzer, T., Pretzsch, H., & Pauleit, S. (2020). Traits of trees for cooling urban heat islands: A meta-analysis. *Building and Environment*, *170*, 106606. DOI: 10.1016/j.buildenv.2019.106606

Rahman, M., & Hossain, M. (2020). Urban planning challenges and solutions for Dhaka: Towards sustainable mobility. *Journal of Urban Affairs*, *42*(4), 509–526. DOI: 10.1080/07352166.2019.1619802

Rajendran, R., Ragavan, R. P., Al-Sehemi, A. G., Uddin, M. S., Aleya, L., & Mathew, B. (2022). Current understandings and perspectives of petroleum hydrocarbons in Alzheimer's disease and Parkinson's disease: A global concern. *Environmental Science and Pollution Research International*, *29*(8), 10928–10949. DOI: 10.1007/s11356-021-17931-3

Raju, K. R. S. R., & Varma, G. H. K. (2017). Knowledge Based Real Time Monitoring System for Aquaculture Using IoT. *2017 IEEE 7th International Advance Computing Conference (IACC)*, 318–321. DOI: 10.1109/IACC.2017.0075

Raju, M. P., Bhardwaj, L. K., & Srivastava, A. K. (2026). Temporal trends in AQI and precursor pollutants: A long-term case study of Noida. *Journal of Environmental Science and Health. Part A, Toxic/Hazardous Substances & Environmental Engineering*, •••, 1–19.

Raju, M. P., Safai, P. D., Sonbawne, S. M., Buchunde, P. S., Pandithurai, G., & Dani, K. K. (2020). Black carbon aerosols over a high altitude station, Mahabaleshwar: Radiative forcing and source apportionment. *Atmospheric Pollution Research*, *11*(8), 1408–1417. DOI: 10.1016/j.apr.2020.05.024

Ramadan, M. N. A., Ali, M. A. H., Shin, Y. K., Alkhedher, M., & Alherbawi, M. (2024). Real-time IoT-powered AI system for monitoring and forecasting of air pollution in industrial environment. *Ecotoxicology and Environmental Safety*, *283*, 116856. DOI: 10.1016/j.ecoenv.2024.116856

Ramchunder, S. J., Brown, L. E., & Holden, J. (2009). Environmental effects of drainage, drain-blocking and prescribed vegetation burning in UK upland peatlands. *Progress in Physical Geography*, *33*(1), 49–79. DOI: 10.1177/0309133309105245

Rao, A. L., Santhikumar, R., & Venugopal, K. (2025). Visualizing Insight Enhancing Statistical Thinking in Higher Education. In Anastasiadou, S., & Seremeti, L. (Eds.), *Modes of Representation in Developing Statistical Thinking in Education* (pp. 181–214). IGI Global Scientific Publishing., DOI: 10.4018/979-8-3693-9934-7.ch010

Rao, P. S., Gavane, A. G., Ankam, S. S., Ansari, M. F., Pandit, V. I., & Nema, P. (2004). Performance evaluation of a green belt in a petroleum refinery: A case study. *Ecological Engineering*, *23*(2), 77–84. DOI: 10.1016/j.ecoleng.2004.06.013

Rautela, K. S., & Goyal, M. K. (2024). Transforming air pollution management in India with AI and machine learning technologies. *Scientific Reports*, *14*(1), 20412. DOI: 10.1038/s41598-024-71269-7

Reethof, G. (1973). Effect of plantings on radiation of highway noise. *Journal of the Air Pollution Control Association*, *23*(3), 185–189. DOI: 10.1080/00022470.1973.10469763

Reuther, A., Michaleas, P., Jones, M., Gadepally, V., Samsi, S., & Kepner, J. (2020). Survey of Machine Learning Accelerators. *2020 IEEE High Performance Extreme Computing Conference (HPEC), September*, 1–12. DOI: 10.1109/HPEC43674.2020.9286149

Revi, A., Satterthwaite, D., Aragón-Durand, F., Corfee-Morlot, J., Kiunsi, R., Pelling, M., Roberts, D., & Solecki, W. (2014). Urban areas. In C. B. Field et al. (Eds.), *Climate change 2014: Impacts, adaptation, and vulnerability (IPCC Fifth Assessment Report, Working Group II)*. Intergovernmental Panel on Climate Change. https://www.ipcc.ch/report/ar5/wg2/

Rocha Filho, G. P., Meneguette, R. I., Torres Neto, J. R., Valejo, A., Weigang, L., Ueyama, J., Pessin, G., & Villas, L. A. (2020). Enhancing intelligence in traffic management systems to aid in vehicle traffic congestion problems in smart cities. *Ad Hoc Networks*, *107*, 102265. DOI: 10.1016/j.adhoc.2020.102265

Ross, A. (2018). Speeding the transition towards integrated groundwater and surface water management in Australia. *Journal of Hydrology (Amsterdam)*, *567*, e1–e10. DOI: 10.1016/j.jhydrol.2017.01.037

Roy, P., & Srinivasan, K. K. (2025). Geographically weighted nonlinear regression for cost-effective policies to enhance bus ridership. *Sustainability (Basel)*, *17*(6), 2485. DOI: 10.3390/su17062485

Roy, R., & D'Angiulli, A. (2024). Air pollution and neurological diseases, current state highlights. *Frontiers in Neuroscience*, *18*, 1351721. DOI: 10.3389/fnins.2024.1351721

Roy, S., Byrne, J., & Pickering, C. (2012). A systematic quantitative review of urban tree benefits, costs, and assessment methods. *Urban Forestry & Urban Greening*, *11*(4), 351–363. DOI: 10.1016/j.ufug.2012.06.006

Sabrin, S., Karimi, M., Nazari, R., Pratt, J., & Bryk, J. (2021). Effects of different urban-vegetation morphology on the canopy-level thermal comfort and the cooling benefits of shade trees: Case-study in Philadelphia. *Sustainable Cities and Society*, *66*, 102684. DOI: 10.1016/j.scs.2020.102684

Sadowski, J. (2020). When data is power: Exclusion and discrimination in the smart city. *New Media & Society*, *22*(10), 1758–1776.

Sæbø, A., Popek, R., Nawrot, B., Hanslin, H. M., Gawronska, H., & Gawronski, S. W. (2012). Plant species differences in particulate matter accumulation on leaf surfaces. *The Science of the Total Environment*, *427*, 347–354. DOI: 10.1016/j.scitotenv.2012.03.084

Safaei, S.. (2024). Agent-based evaluation of speed enforcement cameras. *Scientific Reports*.

Saha, A. R., & Singh, N. (2017). Smart cities for a sustainable future: Can Singapore be a model for Delhi? *International Journal of Economic Research*, *14*, 367–379.

Sahoo, S. (2017). Machine learning algorithms for modeling groundwater level changes in agricultural regions of the US. *Water Resources Research*, *5*(3), 2–2. DOI: 10.1111/j.1752-1688.1969.tb04897.x

Sahour, S., Khanbeyki, M., Gholami, V., Sahour, H., Kahvazade, I., & Karimi, H. (2023). Evaluation of machine learning algorithms for groundwater quality modeling. *Environmental Science and Pollution Research International*, *30*(16), 46004–46021. DOI: 10.1007/s11356-023-25596-3

Salazar-Carrillo, J., Torres-Ruiz, M., Davis, C. A.Jr, Quintero, R., Moreno-Ibarra, M., & Guzmán, G. (2021). Traffic Congestion Analysis Based on a Web-GIS and Data Mining of Traffic Events from Twitter. *Sensors (Basel)*, *21*(9), 2964. DOI: 10.3390/s21092964 PMID: 33922627

Salmond, J. A., Tadaki, M., Vardoulakis, S., Arbuthnott, K., Coutts, A., Demuzere, M., Dirks, K. N., Heaviside, C., Lim, S., Macintyre, H., McInnes, R. N., & Wheeler, B. W. (2016). Health and climate related ecosystem services provided by street trees in the urban environment. *Environmental Health : A Global Access Science Source*, *15*(S1, Suppl. 1), 95. DOI: 10.1186/s12940-016-0103-6

Sánchez, E., Colmenarejo, M. F., Vicente, J., Rubio, A., García, M. G., Travieso, L., & Borja, R. (2007). Use of the water quality index and dissolved oxygen deficit as simple indicators of watersheds pollution. *Ecological Indicators*, *7*(2), 315–328. DOI: 10.1016/j.ecolind.2006.02.005

Santos, G., Behrendt, H., & Teytelboym, A. (2010). Part II: Policy instruments for sustainable road transport. *Research in Transportation Economics*, *28*(1), 46–91. DOI: 10.1016/j.retrec.2010.03.002

Sanyour, A., & Chaturvedi, S. (2025). Expert perspectives on climate change and public health in port cities: a case study of Mumbai. In *Urbanization, Climate Change, and Health: Integrating Strategies for Sustainable and Resilient Cities* (pp. 25–47). Springer Nature Switzerland. DOI: 10.1007/978-3-031-99132-5_2

Satterthwaite, D. (2016). *The urban crisis in the global South*. Routledge.

Savithiri, D., Kasthuri, R., Rajeswari, P., Vinothina, V., & Sivasubramanian, K. (2026). Self-help group model for financial sustainability of rural women in India: An explanatory study. In B. S. Awwad (Ed.), Artificial intelligence and technology: Systems management, decisions and control for sustainability in the digital age (Studies in Systems, Decision and Control, Vol. 645, pp. 969–976). Springer, Cham. DOI: 10.1007/978-3-032-10016-0_77

Saxena, P., & Sonwani, S. (2019). Criteria air pollutants and their impact on environmental health: Vol. 1. *No. 1*. Springer Singapore.

Scandella, M., Ghosh, A., Bin, M., & Parisini, T. (2022). Traffic-light control in urban environment exploiting drivers' reaction to the expected red lights duration. *Transportation Research Part C, Emerging Technologies*, *145*, 103910. DOI: 10.1016/j.trc.2022.103910

Segnestam, L. (2002). Indicators of environment and sustainable development: Theories and practical experience (Environmental Economics Series No. 89, pp. 1–4). *The World Bank*.

Sekhar, M., Tomer, S., Thiyaku, S., Giriraj, P., Murthy, S., & Mehta, V. (2017). Groundwater Level Dynamics in Bengaluru City, India. *Sustainability (Basel)*, *10*(1), 26. DOI: 10.3390/su10010026

Selvakumar, M., Reddy, M. A., Sathish, V., & Venkatesh, R. (2018). Potential influence of metro on bus: A case study. *Journal of the Institution of Engineers (India)*: Series A, 99, 379–384.

Sethi, D., & Kosmopoulos, P. G. (2025). Rooftop solar photovoltaic potential in polluted Indian cities: Atmospheric and urban impacts, climate trends, societal gains, and economic opportunities. *Remote Sensing (Basel)*, *17*(7), 1221. DOI: 10.3390/rs17071221

Shaheen, S., & Cohen, A. (2019). *Shared micro-mobility policy toolkit*. University of California Transportation Center.

Shah, N., & Patel, R. (2021). Spatial analysis of gastroenteritis prevalence in Mumbai's informal settlements. *Journal of Water, Sanitation, and Hygiene for Development : a Journal of the International Water Association, 11*(3), 456–468.

Shaikh, M., & Birajdar, F. (2024a). Advancements in Remote Sensing and GIS for Sustainable Groundwater Monitoring: Applications, Challenges, and Future Directions. *International Journal of Research in Engineering, Science and Management, 7*(3), 16–24. https://journal.ijresm.com/index.php/ijresm/article/view/2955

Shaikh, M. A. (2023). Groundwater management and sustainable farming practices: a socioeconomic analysis of their interplay in rural agriculture-a case study of Solapur, Maharashtra. *International Journal for Innovative Science Research Trends and Innovation,* 2023. DOI: 10.5281/zenodo.8402903

Shaikh, M., & Birajdar, F. (2024b). Groundwater and Public Health: Exploring the Connections and Challenges. *International Journal of Innovative Science and Research Technology, 9*(2), 1351–1361. DOI: 10.5281/zenodo.10730864

Sharma, N., Kumar, R., Vihari, N. S., Arora, M., & Saini, J. R. (2025). Prioritising smart city themes for multi-national enterprises and United Nations Sustainable Development Goals. *Sustainability (Basel), 17*(10), 4251. DOI: 10.3390/su17104251

Sharma, N., & Sahu, S. (2018). Fuzzy Logic based Traffic signal control. *Indian Journal of Science and Technology, 11*(23), 1–5. DOI: 10.17485/ijst/2018/v11i23/114380

Sharma, S. B., Jain, S., Khirwadkar, P., & Kulkarni, S. (2013). The effects of air pollution on the environment and human health. *Indian Journal of Research in Pharmacy and Biotechnology, 1*(3), 391.

Shehab, M. J., Kassem, I., Kutty, A. A., Kucukvar, M., Onat, N., & Khattab, T. (2021). 5G networks towards smart and sustainable cities: A review of recent developments, applications and future perspectives. *IEEE Access : Practical Innovations, Open Solutions, 10*, 2987–3006. DOI: 10.1109/ACCESS.2021.3139436

Sheik, A. G., Kumar, A., Sharanya, A. G., Amabati, S. R., Bux, F., & Kumari, S. (2024). Machine learning-based monitoring and design of managed aquifer rechargers for sustainable groundwater management: Scope and challenges. *Environmental Science and Pollution Research International, 0123456789*(59), 31572–31605. Advance online publication. DOI: 10.1007/s11356-024-35529-3

Shi, W., Kong, Q., & Liu, Y. (2008). A GPS/GIS Integrated System for Urban Traffic Flow Analysis. In 2008 11th International IEEE Conference on Intelligent Transportation Systems (pp. 844–849). IEEE Institute of Electrical and Electronics Engineers. DOI: 10.1109/ITSC.2008.4732569

Shin, D., Park, Y.-J., & Chung, C.-Y. (2020). Public perceptions of surveillance in smart cities. *Government Information Quarterly, 37*(2), 101444.

Shindell, D., & Faluvegi, G. (2009). Climate response to regional radiative forcing during the twentieth century. *Nature Geoscience, 2*(4), 294–300. DOI: 10.1038/ngeo473

Shortridge, J. E., Guikema, S. D., & Zaitchik, B. F. (2015). Empirical streamflow simulation for water resource management in data-scarce seasonal watersheds. In *Hydrology and Earth System Sciences Discussions* (Vol. 12, Issue 10, pp. 11083–11127). DOI: 10.5194/hessd-12-11083-2015

Shoushtarian, F., & Negahban-Azar, M. (2020). Worldwide Regulations and Guidelines for Agricultural Water Reuse: A Critical Review. *Water (Basel), 12*(4), 971. DOI: 10.3390/w12040971

Shrestha, A., & Mahmood, A. (2019). Review of Deep Learning Algorithms and Architectures. *IEEE Access : Practical Innovations, Open Solutions, 7*, 53040–53065. DOI: 10.1109/ACCESS.2019.2912200

Shruti, S., Singh, P. K., & Ohri, A. (2021). Evaluating the environmental sustainability of smart cities in India: The design and application of the Indian Smart City Environmental Sustainability Index. *Sustainability (Basel), 13*(1), 327. DOI: 10.3390/su13010327

Singh, S. K. Bader Eddin Alasali, Khuloud Nasser Al Balushi. (2025). Smart Transportation and Autonomous System Integration. In *Smart Cyber Physical Systems: Innovations and Real-World Implications* (Chapter 7). CRC Press. https://www.routledge.com/Smart-Cyber-Physical-Systems-Innovations-and-Real-World-Implications/Poonia-Upreti-Khan/p/book/9781032892931

Singh, S. K., & Jain, R. (2025). *Future of Mobility: Potential Challenges and Ethical Considerations*. In *AIoT for Smart Transportation: Transforming the Future of Mobility* (Chapter 12). IET Publishers. DOI: 10.1049/PBTR047E_ch12

Singh, A. K., Kumar, V. R. P., Dehdasht, G., Mohandes, S. R., Manu, P., & Pour Rahimian, F. (2023). Investigating the barriers to the adoption of blockchain technology in sustainable construction projects. *Journal of Cleaner Production, 403*, 136840. DOI: 10.1016/j.jclepro.2023.136840

Singh, P., & Yadav, D. (2021). Link between air pollution and global climate change. In *Global climate change* (pp. 79–108). Elsevier. DOI: 10.1016/B978-0-12-822928-6.00009-5

Singh, R. (2008). *3.3 Wastewater Related Risks and Social Vulnerability: A Case Study of Delhi. Resilience and Social Vulnerability, 121. Anand, N. (2017). Hydraulic city: Water and the infrastructures of citizenship in Mumbai*. Duke University Press.

Singh, S., & Saraswat, A. (2019). Design service volume, capacity, level of service calculation and forecasting for a semi-urban city. *Revue D'Intelligence Artificielle, 33*(2), 139–143. DOI: 10.18280/ria.330209

Singh, S., Singh, J., Goyal, S. B., Sehra, S. S., Ali, F., Alkhafaji, M. A., & Singh, R. (2023). A novel framework to avoid traffic congestion and air pollution for the sustainable development of smart cities. *Sustainable Energy Technologies and Assessments, 56*, 103125. DOI: 10.1016/j.seta.2023.103125

Sivarethinamohan, R., Sujatha, S., Priya, S., Gafoor, A., & Rahman, Z. (2021). Impact of air pollution in health and socio-economic aspects: Review on future approach. *Materials Today: Proceedings, 37*, 2725–2729. DOI: 10.1016/j.matpr.2020.08.540

Sivasubramanian, K. (2024). An analysis of green retailing for sustainable business through digital technology application in Indian retail market. In *Building climate neutral economies through digital business and green skills* (pp. 1–22). IGI Global., DOI: 10.4018/979-8-3693-5673-9.ch005

Sivasubramanian, K., Pushpa, A., Raju, V., & Kumar, M. D. (2020). Women and informal employment: An analysis of socio-economic and health conditions of women home-based workers in Chennai, India. *Journal of International Women's Studies, 21*(5), 97–106.

Smeets, J. (1982). Air quality limit and guide values for sulphur dioxide and suspended particulates—A European community directive. *Environmental Monitoring and Assessment, 1*(4), 373–382. DOI: 10.1007/BF00403837

Smith, S. F., Barlow, G. J., & Kohls, A. G. (2013). Distributed real-time traffic signal control. *Transportation Research Part C, Emerging Technologies, 24*, 168–184.

Smith, W. H. (2012). *Air pollution and forests: interactions between air contaminants and forest ecosystems*. Springer Science & Business Media.

Snow, A. (2017). *Automated Road Traffic Enforcement: Regulation, Governance & Public Perceptions*. RAC Foundation.

Snyder, F., & Ratti, C. (2023). Real-time urban emissions control through AI. *Environmental Research Letters, 18*(2), 024003.

Sochor, J., Strömberg, H., & Karlsson, I. C. M. (2015). Implementing Mobility as a Service: Challenges and opportunities. *Transportation Research Procedia*, *14*, 3264–3273.

Sokhi, R. S., Moussiopoulos, N., Baklanov, A., Bartzis, J., Coll, I., Finardi, S., Friedrich, R., Geels, C., Grönholm, T., Halenka, T., Ketzel, M., Maragkidou, A., Matthias, V., Moldanova, J., Ntziachristos, L., Schäfer, K., Suppan, P., Tsegas, G., Carmichael, G., & Kukkonen, J. (2021). Advances in air quality research—Current and emerging challenges. *Atmospheric Chemistry and Physics Discussion*, *2021*, 1–133. DOI: 10.5194/acp-2021-581

Solomatine, D. P. (1998). Genetic and other global optimization algorithms - comparison and use in calibration problems. *3rd International Conference on Hydroinformatics*, 1021–1028.

Son, H., Jang, J., Park, J., Balog, A., Ballantyne, P., Kwon, H. R., Singleton, A., & Hwang, J. (2025). Leveraging advanced technologies for (smart) transportation planning: A systematic review. *Sustainability (Basel)*, *17*(5), 2245. DOI: 10.3390/su17052245

Sotonye, N., & Rajabifard, A. (2021). Digital twin applications in emissions forecasting. *Environmental Modelling & Software*, *143*, 105110.

Sram, R. J., Binkova, B., Dostal, M., Merkerova-Dostalova, M., Libalova, H., Milcova, A., Rossner, P.Jr, Rossnerova, A., Schmuczerova, J., Svecova, V., Topinka, J., & Votavova, H. (2013). Health impact of air pollution to children. *International Journal of Hygiene and Environmental Health*, *216*(5), 533–540. DOI: 10.1016/j.ijheh.2012.12.001

Sram, R. J., Veleminsky, M., Veleminsky, M., & Stejskalová, J. (2017). The impact of air pollution to central nervous system in children and adults. *Neuroendocrinology Letters*, *38*(6), 389–396.

Stevanovic, A. (2010). Adaptive traffic control systems: Domestic and foreign state of practice. *Transportation Research Record: Journal of the Transportation Research Board*, *2173*(1), 46–54.

Stevens, M. (2023). Does Transit-Oriented Development (TOD) deliver on its transit promises? A meta-analysis. *Journal of the American Planning Association*, *89*(2), 214–228. DOI: 10.1080/01944363.2022.2086385

Stieb, D. M., Chen, L., Eshoul, M., & Judek, S. (2012). Ambient air pollution, birth weight and preterm birth: A systematic review and meta-analysis. *Environmental Research*, *117*, 100–111. DOI: 10.1016/j.envres.2012.05.007

Sultaana, M. A., & Saha, A. K. (2025). Unleashing the potential of 5G for smart cities: A focus on real-time digital twin integration. *Smart Cities*, *8*(2), 70. DOI: 10.3390/smartcities8020070

Sunio, G., & Schmöcker, J.-D. (2017). Real-time travel choice nudges in smartphone apps. *Transportation Research Part F: Traffic Psychology and Behaviour*, *49*, 35–47.

Sun, X.. (2024). Naturalistic driving studies applied to infrastructure: Systematic review. *Journal of Safety Research*.

Sun, Y.. (2025). AI-based prediction of traffic crash severity for improving road safety. *Scientific Reports*, *15*, 6789. PMID: 40721929

Su, Q., Dai, H., Xie, S., Yu, X., Lin, Y., Singh, V. P., & Karthikeyan, R. (2023). Water–energy–carbon nexus: Greenhouse gas emissions from integrated urban drainage systems in China. *Environmental Science & Technology*, *57*(5), 2093–2104. DOI: 10.1021/acs.est.2c08583 PMID: 36696288

Sureshchandra, M. S., Bhavsar, J. J., & Pitroda, R. J. (2016). Assessment of critical success factors for smart cities using the significance index method. *International Journal of Advanced Research in Innovative Ideas in Education*, *2*(2), 802–810.

Syafiuddin, A., Boopathy, R., & Hadibarata, T. (2020). Challenges and Solutions for Sustainable Groundwater Usage: Pollution Control and Integrated Management. *Current Pollution Reports*, *6*(4), 310–327. DOI: 10.1007/s40726-020-00167-z

Syal, S. (2024). The (unprivileged) polluter pays: Conflict of Rights in Delhi's stormwater drain-adjacent 'informal'settlements. *Planning Practice and Research*, *39*(6), 960–993. DOI: 10.1080/02697459.2024.2383824

Sze, V., Chen, Y.-H., Emer, J., Suleiman, A., & Zhang, Z. (2017). Hardware for machine learning: Challenges and opportunities. *2017 IEEE Custom Integrated Circuits Conference (CICC), 2017-April*, 1–8. DOI: 10.1109/CICC.2017.7993626

Talebpour, A., & Mahmassani, H. S. (2016). Influence of connected vehicles on traffic flow. *Transportation Research Part C, Emerging Technologies*, *71*, 143–163. DOI: 10.1016/j.trc.2016.07.007

Tallis, M., Taylor, G., Sinnett, D., & Freer-Smith, P. (2011). Estimating the removal of atmospheric particulate pollution by the urban tree canopy of London, under current and future environments. *Landscape and Urban Planning*, *103*(2), 129–138. DOI: 10.1016/j.landurbplan.2011.07.003

Tang, J., McNabola, A., & Misstear, B. (2020). The potential impacts of different traffic management strategies on air pollution and public health for a more sustainable city: A modelling case study from Dublin, Ireland. *Sustainable Cities and Society, 60*, 102229. DOI: 10.1016/j.scs.2020.102229

Tang, S. H., & Ong, P. P. (1988). A Monte Carlo technique to determine the effectiveness of roadside trees for containing traffic noise. *Applied Acoustics, 23*(4), 263–271. DOI: 10.1016/0003-682X(88)90040-0

Tavares, P., Ingi, D., Araújo, L., Pinho, P., & Bhusal, P. (2021). Reviewing the role of outdoor lighting in achieving Sustainable Development Goals. *Sustainability (Basel), 13*(22), 12657. DOI: 10.3390/su132212657

Tavella, R. A., da Silva Júnior, F. M. R., Santos, M. A., Miraglia, S. G. E. K., & Pereira Filho, R. D. (2025). A review of air pollution from petroleum refining and petrochemical industrial complexes: Sources, key pollutants, health impacts, and challenges. *ChemEngineering, 9*(1), 13. DOI: 10.3390/chemengineering9010013

Te Brake, W. H. (1975). Air pollution and fuel crises in preindustrial London, 1250–1650. *Technology and Culture, 16*(3), 337–359. DOI: 10.2307/3103030

Tesfaye, S., Hamba, N., Gerbi, A., & Neger, Z. (2020). Oxidative stress and carcinogenic effect of formaldehyde exposure: Systematic review & analysis. *Endocrinology & Metabolic Syndrome, 9*(6), 319–330.

Thakur, S., Bhardwaj, L. K., Rath, P., Sharma, Y., & Kumar, A. (2026). Relationship between Persistent Organic Pollutants (POPs) and Antarctic Environment. In *Xenobiotics in Environment* (p. 15). CRC Press. DOI: 10.1201/9781003521754-3

Thaler, R. H., & Sunstein, C. R. (2008). *Nudge: Improving decisions about health, wealth, and happiness.* Yale University Press.

Thompson, J. E. (2018). Airborne particulate matter: Human exposure and health effects. *Journal of Occupational and Environmental Medicine, 60*(5), 392–423. DOI: 10.1097/JOM.0000000000001277

Tiyasha, T., Tung, T. M., & Yaseen, Z. M. (2020). A survey on river water quality modelling using artificial intelligence models: 2000–2020. *Journal of Hydrology (Amsterdam), 585*(January), 124670. DOI: 10.1016/j.jhydrol.2020.124670

Tremper, A. H. (2015). *Impact of green screens on concentrations of particulate matter and oxides of nitrogen in near road environments* [Doctoral dissertation, King's College London].

Trigka, M., & Dritsas, E. (2025). Edge and cloud computing in smart cities. *Future Internet*, *17*(3), 118. DOI: 10.3390/fi17030118

Trindade, E. P., Hinnig, M. P. F., Da Costa, E. M., Marques, J. S., Bastos, R. C., & Yigitcanlar, T. (2017). Sustainable development of smart cities: A systematic review of the literature. *Journal of Open Innovation*, *3*(3), 11. DOI: 10.1186/s40852-017-0063-2

Tsocheva, I., Scales, J., Dove, R., Chavda, J., Kalsi, H., Wood, H. E., Colligan, G., Cross, L., Newby, C., Hall, A., Keating, M., Sartori, L., Moon, J., Thomson, A., Tomini, F., Murray, A., Hamad, W., Tijm, S., Hirst, A., & Griffiths, C. J. (2023). Investigating the impact of London's ultra low emission zone on children's health: Children's health in London and Luton (CHILL) protocol for a prospective parallel cohort study. *BMC Pediatrics*, *23*(1), 556. DOI: 10.1186/s12887-023-04384-5

Tu, G., & Zan, Z. (2025). The green paradox of new energy vehicles: A system dynamics analysis. *Sustainability (Basel)*, *17*(9), 3938. DOI: 10.3390/su17093938

Tyagi, V., Kumar, K., & Jain, V. K. (2006). A study of the spectral characteristics of traffic noise attenuation by vegetation belts in Delhi. *Applied Acoustics*, *67*(9), 926–935. DOI: 10.1016/j.apacoust.2005.09.002

UNGA. (2024). *Resolution A/RES/78/290: Improving Global Road Safety*. United Nations General Assembly. [REMOVED HYPERLINK FIELD]

UN-Habitat. (2022). *World cities report 2022: Envisioning resilient urban futures*. United Nations Human Settlements Programme.

Unsal, M. H. (2025). *Dust Pollution as an Indicator of Indoor Environment Quality* (Doctoral dissertation, Vilniaus universitetas.).

Van Kooten, R., Imhof, P., Brummelhuis, K., Van Pampus, M., Jamshidnejad, A., & De Schutter, B. (2017). ART-UTC: An adaptive real-time urban traffic control strategy. 2022 IEEE 25th International Conference on Intelligent Transportation Systems (ITSC), 1–6. DOI: 10.1109/ITSC.2017.8317813

Varet, F.. (2024). Validating a multidimensional model of traffic-rule legitimacy. *Psychology, Public Policy, and Law*. Advance online publication. DOI: 10.1037/law0000416

Varshini, S., Hungund, K., Kundu, S. K., Santhanam, H., & Dhyani, S. (2022). Assessing ecological risks of urban air and water environment to analyse the scenarios for mainstreaming nature-based solutions: a case study of Bengaluru City, India. In *Blue-Green Infrastructure Across Asian Countries: Improving Urban Resilience and Sustainability* (pp. 163–191). Springer Singapore. DOI: 10.1007/978-981-16-7128-9_8

Vedant, A. (2024). Urbanisation and Economic Development in India: A Case Study of Major Cities. *International Journal of Novel Research and Development*, *9*(8).

Veerappan, S. (2024). Edge-enabled smart storm water drainage systems: A real-time analytics framework for urban flood management. *Journal of Smart Infrastructure and Environmental Sustainability*, *1*(1), 52–59.

Venkatesh, G. (2014). A critique of the European Green City Index. *Journal of Environmental Planning and Management*, *57*(3), 317–328. DOI: 10.1080/09640568.2012.741520

Venugopal, K. (2025). iGenius: Exploring Gen Z's Reliance on AI Tools in Education. In Muhammad Anshari, Mohammad Nabil Almunawar, and Patricia Ordóñez de Pablos (Eds.), Impacts of Digital Technologies Across Generations (pp. 141-164). IGI Global Scientific Publishing. DOI: 10.4018/979-8-3693-6366-9.ch008

Venugopal, K. (2024a). Insights into Waste Management in Semi-Urban Areas: Understanding Community Perspectives. In Mehra, R., Kumar, P., & Hung, B. (Eds.), *Municipal Solid Waste Management and Recycling Technologies* (pp. 301–326). IGI Global Scientific Publishing., DOI: 10.4018/979-8-3693-4054-7.ch011

Venugopal, K. (2024b). Exploring Gen Z Perspectives toward ChatGPT in Academic Contexts. In Bansal, R., Chakir, A., Hafaz Ngah, A., Rabby, F., & Jain, A. (Eds.), *AI Algorithms and ChatGPT for Student Engagement in Online Learning* (pp. 51–69). IGI Global Scientific Publishing., DOI: 10.4018/979-8-3693-4268-8.ch004

Venugopal, K. (2024c). Reverse Logistics Analysis: Enhancing Small Business Performance Perspectives. In Mızrak, F. (Ed.), *Strategic Innovations for Dynamic Supply Chains* (pp. 247–274). IGI Global Scientific Publishing., DOI: 10.4018/979-8-3693-3575-8.ch011

Venugopal, K., Das, S., & Badawy, H. R. (2023). Prediction Analysis of Gen Zers' Attitudes on Ecological Consciousness. In Mishra, B. (Ed.), *Sustainable Science and Intelligent Technologies for Societal Development* (pp. 342–357). IGI Global Scientific Publishing., DOI: 10.4018/979-8-3693-1186-8.ch019

Venugopal, K., Das, S., Gopalakrishna, V., & Bansal, R. (2024). Customer Consciousness towards Food Wastage in the Indian Hospitality Industry. In Singh, A., Tyagi, P., & Garg, A. (Eds.), *Sustainable Disposal Methods of Food Wastes in Hospitality Operations* (pp. 38–56). IGI Global Scientific Publishing., DOI: 10.4018/979-8-3693-2181-2.ch003

Venugopal, K., Gopalakrishna, V., Ranganath, N. S., & Lakshmi, K. (2025). Reverse Mentoring and Knowledge Transfer: The Role of Gen Z in Upskilling Older Employees. In Chandan, H. (Ed.), *Enhancing Workplace Productivity through Gender Equity and Intergenerational Communication* (pp. 219–254). IGI Global Scientific Publishing., DOI: 10.4018/979-8-3373-2903-1.ch009

Venugopal, K., & Jagadeesh, B. (2025). Theoretical Insights into User Security and Privacy in Social Media Environments. In Kumar, R., Srivastava, S., & Elngar, A. (Eds.), *Human Impact on Security and Privacy: Network and Human Security, Social Media, and Devices* (pp. 289–312). IGI Global Scientific Publishing., DOI: 10.4018/979-8-3693-9235-5.ch015

Venugopal, K., & Nakkina, V. V. (2025). Technological Innovations and Their Effect on B2B Relationships in the Hospitality Industry. In Nadda, V., Sharma, A., Mediratta, H., & Dadwal, S. (Eds.), *Addressing Contemporary Challenges in the B2B Hospitality Sector* (pp. 195–232). IGI Global Scientific Publishing., DOI: 10.4018/979-8-3693-9760-2.ch007

Venugopal, K., Rajulu, B. G., Ravikumar, P., & Madhavi, K. (2026). Sustainable Learning Solutions: ChatGPT Integration in Academic Services. In Tariq, M. (Ed.), *Integrating ChatGPT Into System Applications and Services* (pp. 59–88). IGI Global Scientific Publishing., DOI: 10.4018/979-8-3693-9841-8.ch003

Venugopal, K., & Vakamullu, G. (2025). Assessment of Higher Education Students' Dependency on AI Tools in an Indian Indigenized Context. In Kgari-Masondo, M. (Ed.), *Indigenous Teaching Disciplines and Perspectives for Higher Education* (pp. 309–338). IGI Global Scientific Publishing., DOI: 10.4018/979-8-3693-9296-6.ch015

Voigt, H.-M. (1989). 1. Introduction. In *Evolution and Optimization* (pp. 11–16). De Gruyter., DOI: 10.1515/9783112568620-002

Voigt, P., & Von dem Bussche, A. (2017). *The EU General Data Protection Regulation (GDPR)*. Springer. DOI: 10.1007/978-3-319-57959-7

Wang, C., & Chen, X. (2020). Real-time emission modeling in urban traffic networks. *Transportation Research Part D, Transport and Environment, 84*, 102385.

Wang, J., Wu, Q., Liu, J., Yang, H., Yin, M., Chen, S., Guo, P., Ren, J., Luo, X., Linghu, W., & Huang, Q. (2019). Vehicle emission and atmospheric pollution in China: Problems, progress, and prospects. *PeerJ*, *7*, e6932. DOI: 10.7717/peerj.6932

Wang, S., Zhang, J., Zhang, Y., Wang, L., Sun, Z., & Wang, H. (2023). Review on source profiles of volatile organic compounds (VOCs) in typical industries in China. *Atmosphere (Basel)*, *14*(5), 878. DOI: 10.3390/atmos14050878

Wang, T., Xue, L., Feng, Z., Dai, J., Zhang, Y., & Tan, Y. (2022). Ground-level ozone pollution in China: A synthesis of recent findings on influencing factors and impacts. *Environmental Research Letters*, *17*(6), 063003. DOI: 10.1088/1748-9326/ac69fe

Wang, Z., Fan, M., & Fan, Y. (2024). The impact of digitally inclusive finance on environmental pollution: A case study of air pollution. *PLoS One*, *19*(7), e0305963. DOI: 10.1371/journal.pone.0305963

Weberg, O., Lund, J., Fors, V., & Resmini, A. (2025). Socially resilient mobility planning: Main challenges and design implications. *Transportation Research Interdisciplinary Perspectives*, *29*, 101334. DOI: 10.1016/j.trip.2025.101334

Weber, R., Watson, A., Forter, M., & Oliaei, F. (2011). Persistent organic pollutants and landfills-a review of past experiences and future challenges. *Waste Management & Research*, *29*(1), 107–121. DOI: 10.1177/0734242X10390730

Wei, H., Zheng, G., Liu, H. X., & Li, Z. (2019). Optimal traffic signal control with deep reinforcement learning. *Transportation Research Part C, Emerging Technologies*, *104*, 76–99.

Wei, X., Lyu, S., Yu, Y., Wang, Z., Liu, H., Pan, D., & Chen, J. (2017). Phylloremediation of air pollutants: Exploiting the potential of plant leaves and leaf-associated microbes. *Frontiers in Plant Science*, *8*, 1318. DOI: 10.3389/fpls.2017.01318

Wells, P. et al. (2020). Graduated Driver Licensing review of effectiveness. (Policy synthesis cited across OECD/ITF 2024).

WHO. (2021). *Global Plan for the Decade of Action for Road Safety 2021–2030*. World Health Organization. [REMOVED HYPERLINK FIELD]

WHO. (2023a). *Global Status Report on Road Safety 2023*. World Health Organization.

WHO. (2023b). *Global status report on road safety 2023 full report (PDF)*. World Health Organization. [REMOVED HYPERLINK FIELD]

Wiering, M. (2000). Multi-agent reinforcement learning for traffic light control. In *Proceedings of the Seventeenth International Conference on Machine Learning* (pp. 1151–1158).

Wikoff, D., Fitzgerald, L., & Birnbaum, L. (2012). Persistent organic pollutants: an overview. *Dioxins and health: including other persistent organic pollutants and endocrine disruptors*, 1-35.

Woolley, J., & Hutchinson, T. (2021). Why design can make safe behaviour hard; Safe System critique. *Journal of Road Safety*.

World Bank. (2020). *World development report 2020: Trading for development in the age of global value chains*. World Bank.

World Bank. (2023). *Bangladesh urbanization review: Managing cities for inclusive growth*. World Bank. https://www.worldbank.org/en/country/bangladesh/publication/bangladesh-urbanization-review

World Health Organization. (2011). *Health in the green economy: Co-benefits to health of climate change mitigation – transport sector*. WHO.

World Health Organization. (2017). *Urban green spaces and health: A review of evidence*. WHO Regional Office for Europe.

World Health Organization. (2021). *Compendium of WHO and other UN guidance on health and environment*.

World Health Organization. (2021). *WHO global air quality guidelines: particulate matter (PM2. 5 and PM10), ozone, nitrogen dioxide, sulfur dioxide and carbon monoxide*. World Health Organization.

Wu, J. Z., Ge, D. D., Zhou, L. F., Hou, L. Y., Zhou, Y., & Li, Q. Y. (2018). Effects of particulate matter on allergic respiratory diseases. *Chronic Diseases and Translational Medicine*, *4*(2), 95–102. DOI: 10.1016/j.cdtm.2018.04.001

Wunsch, A., Liesch, T., & Broda, S. (2018). Forecasting groundwater levels using nonlinear autoregressive networks with exogenous input (NARX). *Journal of Hydrology (Amsterdam)*, *567*, 743–758. DOI: 10.1016/j.jhydrol.2018.01.045

Yadava, R. N., & Bhatt, V. (2021). Carbon monoxide: Risk assessment, environmental, and health hazard. In *Hazardous Gases* (pp. 83–96). Academic Press. DOI: 10.1016/B978-0-323-89857-7.00030-X

Yadav, P., Dhupper, R., Singh, S. D., & Singh, B. (2019). Crop adaptation to air pollution I. Effect of particulate and SO2 pollution on growth, yield attributes and sulphur nutrition of wheat, barley and chickpea. *Indian Journal of Agricultural Research*, *53*(3), 303–308. DOI: 10.18805/IJARe.A-5273

Yang, C., & Barth, M. (2015). Eco-routing for reducing emissions: A review of methods. *Transportation Research Part D, Transport and Environment*, *34*, 36–52.

Yang, C., Liang, P., Fu, L., Cui, G., Huang, F., Teng, F., & Bangash, Y. A. (2022). Using 5G in smart cities: A systematic mapping study. *Intelligent Systems with Applications*, *14*, 200065. DOI: 10.1016/j.iswa.2022.200065

Yang, J., Chang, Y., & Yan, P. (2015). Ranking the suitability of common urban tree species for controlling PM2.5 pollution. *Atmospheric Pollution Research*, *6*(2), 267–277. DOI: 10.5094/APR.2015.031

Yang, X.-S. (2011). Optimization Algorithms. In *Studies in Computational Intelligence* (Vol. 356, pp. 13–31). Issue December., DOI: 10.1007/978-3-642-20859-1_2

Yang, Z., Peng, J., Wu, L., Ma, C., Zou, C., Wei, N., Zhang, Y., Liu, Y., Andre, M., Li, D., & Mao, H. (2020). Speed-guided intelligent transportation system helps achieve low-carbon and green traffic: Evidence from real-world measurements. *Journal of Cleaner Production*, *268*, 122230. DOI: 10.1016/j.jclepro.2020.122230

Yao, L.. (2012). Young drivers and speed selection guided by the TPB. *Accident; Analysis and Prevention*, *43*(5), 1349–1359.

Yao, R., Yan, Y., Wei, C., Luo, M., Xiao, Y., & Zhang, Y. (2022). Hydrochemical Characteristics and Groundwater Quality Assessment Using an Integrated Approach of the PCA, SOM, and Fuzzy c-Means Clustering: A Case Study in the Northern Sichuan Basin. *Frontiers in Environmental Science*, *10*(June), 1–12. DOI: 10.3389/fenvs.2022.907872

Yigitcanlar, T., Desouza, K., Butler, L., & Roozkhosh, F. (2020). Contributions and risks of artificial intelligence (AI) in building smarter cities: Insights from a systematic review of the literature. *Energies*, *13*(6), 1473. DOI: 10.3390/en13061473

Yin, J., & Tsai, F. T. C. (2018). Saltwater scavenging optimization under surrogate uncertainty for a multi-aquifer system. *Journal of Hydrology (Amsterdam)*, *565*(August), 698–710. DOI: 10.1016/j.jhydrol.2018.08.021

Yin, J., & Tsai, F. T.-C. (2019). Steady-State Approximate Freshwater–Saltwater Interface in a Two-Horizontal-Well Scavenging System. *Journal of Hydrologic Engineering*, *24*(10), 06019008. Advance online publication. DOI: 10.1061/(ASCE)HE.1943-5584.0001836

Yue, S., Wang, Y., Wang, J., & Chen, J. (2017). Relationships between lung cancer incidences and air pollutants. *Technology and Health Care*, *25*(1, suppl), 411–422. DOI: 10.3233/THC-171344

Zamil Al-Sudani, H. I. (2019). A Review on Groundwater Pollution. *International Journal of Recent Engineering Science*, *6*(5), 13–21. DOI: 10.14445/23497157/IJRES-V6I5P103

Zannin, P. H. T., Diniz, F. B., & Barbosa, W. A. (2002). Environmental noise pollution in the city of Curitiba, Brazil. *Applied Acoustics*, *63*(4), 351–358. DOI: 10.1016/S0003-682X(01)00052-4

Zhang, L., Liu, C., & Wang, Y. (2019). GPS-based urban congestion detection. *IEEE Transactions on Intelligent Transportation Systems*, *20*(8), 2934–2942.

Zhang, L., Tan, P. Y., Diehl, J. A., & Tan, H. T. W. (2022). Integrating biodiversity into urban green infrastructure planning: The case of Singapore. *Urban Forestry & Urban Greening*, *74*, 127650. DOI: 10.1016/j.ufug.2022.127650

Zhang, X., Han, L., Wei, H., Tan, X., Zhou, W., Li, W., & Qian, Y. (2022). Linking urbanization and air quality together: A review and a perspective on the future sustainable urban development. *Journal of Cleaner Production*, *346*, 130988. DOI: 10.1016/j.jclepro.2022.130988

Zhang, Y., Xie, Z., Hu, Q., & Hao, L. (2024). Groundwater Pollution Control and Groundwater Management. *Water (Basel)*, *16*(23), 3542. DOI: 10.3390/w16233542

Zhang, Z., Wang, H., & Zhao, J. (2011). Eco-routing through elevation-aware methods. In *2011 IEEE Intelligent Vehicles Symposium* (pp. 668–673).

Zhao, P., Park, H., & Ni, J. (2018). Urban spatial structure and residents' travel behavior: A case study in Beijing. *Transportation Research Part D, Transport and Environment*, *61*, 72–84. DOI: 10.1016/j.trd.2018.01.018

Zhou, Y.. (2021). Dynamic emissions modeling with IoT and AI. *Transportation Research Part D, Transport and Environment*, *95*, 102859.

Zuboff, S. (2019). *The age of surveillance capitalism: The fight for a human future at the new frontier of power*. PublicAffairs.

About the Contributors

Md. Shafiqul Islam has completed a Postdoctoral Fellowship at the London School of Economics and Political Science (LSE) in the United Kingdom. He earned his PhD in Disaster Management from the University of Dhaka. Currently, Dr. Islam is a faculty member in the Department of Environmental Science at the International University of Business Agriculture and Technology in Bangladesh. With over thirty years of experience, his work spans the education and development sectors, focusing on Sustainable Waste Management, Organic Farming, Climate Change Adaptation, Disaster Risk Reduction, Sustainable Development, and enhancing the social-ecological resilience of communities facing coastal disasters and environmental hazards. He has also facilitated corporate training and skill development, creating training modules in these areas. Dr. Islam has served as an editor, reviewer, and author for numerous books, articles, and book chapters, with his work published in international journals and by esteemed publishers such as IGI Global, Emerald, Springer Nature, Apple Academic Press, Taylor and Francis-Routledge, and UPL. He contributed to the UN Sustainable Development Report in 2016. His latest single-author book, titled "Cultivation and Drought Management in Agriculture," was published by Springer Nature in 2023. Waste Management case studies in developing countries: Sustainable practices and value creation in press, Springer nature, Protecting Biodiversity, culture, and communities through forest conservation approved by IGI Global, Think a Minute for Sustainability, Nova Science and Stay in Nature: Grow Organics, Nova Science publisher. Additionally, Dr. Islam has conducted social and environmental impact assessments for various projects and has collaborated with numerous universities, NGOs, and donor-funded initiatives, including DFID, USAID, the World Bank, the European Union, GWAB, and NETZ.

***

Roopa Adarsh is currently working as Associate Professor in the Department of Economics, Mount Carmel College, Bangalore. India

Amir Ahmad Dar is the Faculty in the Department of Mathematics at Lovely Professional University Punjab. The area of interest: Financial mathematics, financial economics, statistics and business mathematics.

Bashir Ahmad Dar teaches the courses of Political Science at Lovely Professional University, Punjab. He has published research papers in national and international journals and has presented his work at various national and international conferences

Nibedita Das is currently an Associate Professor with the Department of Information and Communication Technology, Tecnia Institute of Advanced Studies, Delhi, India. She holds PhD degree IRINS-IIT Dhanbad & certified from IIM Ahmadabad in the field of Renewable Energy and Leadership skill era. She has more than 14 years of academic and research experience. She is the author and co-author of approximately 22 scientific articles, book chapters and research papers in indexed international journals and prestigious Scopus indexed conference proceedings. She is an Editor of the IIP Series Futuristic Trends in Renewable & Sustainable Energy (USA); and also, fellow of IEEE Access (USA). She has received many awards (Young Achiever-2020, Women mobilizer-2024, State Level National Science Seminar (NSS): DST GUJCOST)-2024 honours, AIR-1 UR JRF-2012 and recognition for her academic and research contributions including the prestigious Rajya Puraskar Award- 1996 appreciated by the Hon'ble Governor, Odisha, India. She has India, UK, South Africa Patents and completed energy harvesting Solar project enacted AI & ML project worth 3 Lkah-2024 and working in various govt and non govt funded projects. Her filed of interests are Power Electronics, Control system, AI, ML, Deep Learning, NLP, renewable Energy and semiconductor hardware integrated software analysis for urban mobilization.

Babu Rao Gaddala is currently working as Associate Professor in the department of Chemical Engineering, Guru Ghasidas Vishwavidyalaya (A Central University),Bilaspur, Chhattisgarh, India.. I have consolidated experience of 17+ years in teaching in India & abroad (Oman). I have been awarded Ph.D in the month of January 2016 in the Department of Chemical Engineering from Andhra University, Visakhapatnam, India. I have obtained M.Tech degree first class with distinction in Chemical Engineering in 2007 & B.Tech in 2005 from Vignan's Engg College Guntur. I have published 33 research articles in peer reviewed journals (SCI/ SCOPUS indexed), 2 book chapters, Editor for 1 Edited book, 8 patents in national and international level. I was also participated and presented Papers in national (8) and international conferences (11) in India, Oman & Bahrain. I have attended 16 weeks of various FDP in premier organizations and online FDPs (10) and webinars (20). I have supervised 17 B.Tech Chemical Engineering & 1 M.Tech project.I have 20 national & international Professional Body Memberships which will give up to date knowledge of the field.

Sourabh Garg has been serving as an Assistant Professor in the Department of Mathematics at Manav Rachna International Institute of Research and Studies, Faridabad, India, since November 2024. With over 2.5 years of teaching experience, he holds a Ph.D. in Mathematics from IIT Mandi and an M.Phil. from the University of Delhi. His primary research interest lies in mathematical modelling in meteorology, with notable contributions to real-time monitoring of cloudburst events over the Northwest Himalayan region using the Weather Research and Forecasting (WRF) model. He has published extensively in SCI-indexed journals and presented his work at an international conference in Valencia, Spain, supported by an institutional academic travel grant. He has also secured commendable ranks in the CSIR-UGC examination three times.

Abby Balkis Gorayah is a Chartered Quantity Surveyor (MRICS) and Senior Lecturer and Program Manager at National University of Science and Technology, Muscat, Oman with over 30 years of industry and academic experience. Specialising in contract law, cost management, and project management, he has led major international projects, supervised research, and contributed to accreditation, consultancy, and professional development in construction education.

Jaishri Gothania is a PhD-qualified Data Scientist with over 9 years of experience in Artificial Intelligence, Machine Learning, and data analytics across academia and industry. She has authored 1 book, 17 book chapters, 9 research papers, and 3 conference papers, with publications indexed in Scopus and Web of Science. She holds 3 patents, including 2 granted innovations in AI and design. Her expertise spans data mining, intelligent systems, and applied AI solutions. She has also supervised postgraduate research, contributed to curriculum development, and actively participated in international conferences, editorial boards, and professional research communities.

Rituraj Jain is currently working in the Department of Information Technology, Marwadi University, Rajkot, Gujarat, India. He has published many papers in International Journals and Conferences. He has published 4 books. He is an Oracle Database SQL Certified Expert. He is also associated with professional bodies like the Indian Society of Technical Education (ISTE), Computer Science Teachers Association (CSTA), Academy & Industry Research Collaboration Center (AIRCC),

International Association of Computer Science and Information Technology (IACSIT), International Association of Engineers (IAENG). His research interests include Cloud Computing, Machine Learning, Deep Learning, Fuzzy Logic, IoT, Genetic Algorithms, Big Data and Software Engineering.

Babar Javed is an architect and spatial planner associated with the Manipal School of Architecture and Planning (MSAP), Manipal Academy of Higher Education (MAHE), Manipal, Karnataka. He has over 10 years of experience in consulting and academics. His key expertise includes spatial planning, land use, urban low-cost housing, climate-responsive design, urban informality, travel behavior, and sustainable mobility studies. He focuses on land use, urban studies, and transportation, and has completed a research project combining sustainable commuting, travel behavior, and employer-based commute incentives for the Government of Karnataka.

Kathaian R, is currently working as Associate Professor of Commerce, Vel Tech Rangarajan Dr.Sagunthala R & D Institute of Science and Technology, Avadi, Chennai, Tamilnadu, India. He is holding PhD and has vast years of experience in teaching and research at university level.

Hari Krishan Malhotra is currently working at Manav Rachna International institute of research and studies. He has done PhD in mathematics from university of Delhi. He has published many research papers in international SCI journals.He has more than 7 years experience in both teaching and research.

Sureshkumar Markanday Subramanian, native of Madurai, India, started his professional career with a B. Sc., in Chemistry (Madurai Kamaraj University, Madurai, India, 1994), then, he specialized in polymers with a post-graduate diploma in plastics processing technology (PGD-PPT, CIPET, Chennai, India). Later, he pursued M. Sc., and M. Phil., in Chemistry from Bharathidasan University, Tiruchirappalli, India, (1997 & 2005). He obtained his Ph. D., from the University of Pisa, Pisa, Italy in the year 2011 by winning the International Ph D Grant. As on date he is having 28 years of experience which includes technical, academic, research and academic administration. His experience includes inter-continental and international from the following countries; Italy, Canada, Tanzania, Zambia and Ethiopia. He is the recipient of national and international fellowships, awards for Best Teacher & Best Presentation. As on date he is having 160+ scholarly activities & achievements in the form Grants, MoU, Design Grants, Patents, Peer-reviewed research & review papers, book chapters, scientific articles, conference proceedings, faculty development program organization and participation, etc. Presently he is engaged in the R & D of bituminous materials, polymers, circular economy of plastics, AI in education and materials science, language and linguistics, etc. At present he is affiliated with ACE Engineering College, Ankushapur (Hyderabad), India as Associate Professor of Chemistry and Head of the Department – Chemistry.

Hafizur Rahman completed his MS in Environmental Science from the State University of Bangladesh and he is working with SGS in Bangladesh for environmental compliance issues and certification, environmental auditing and environmental management. His research interest in environmental pollution management including air quality and noise pollution.

Venkata Kiran Kumar Ravi is young individual who holds a doctoral degree in Mechanical Engineering and has significant academic and administrative experience in Engineering Colleges. He currently serves as academician in Department of Mechanical Engineering at Siddhartha Academy of Higher Education (Deemed to be University) Vijayawada Andhra Pradesh.His research areas include Finite Element Analysis, Composite Materials, Mechanical Vibrations, Fracture Mechanics, Data Science,Artificial Intelligence, Machine Learning, Deep Learning and Natural Language Processing.He is having certifications related to CAD/CAM, Data Science, Machine Learning, Deep Learning, NLP, Computer Vision, Generative AI from reputed institutes.He published papers and patents in Journals and presented articles in various Conferences.

Mahamadyaseen Shiggaon is an emerging researcher and academic from Karnataka, India, with a Bachelor's degree in Statistics and Computer Science from KLE PC Jabin Science College, Hubballi. His academic interests span data analysis, community-based research, and the application of technology for social development. He has undertaken field research and contributed to projects addressing urban infrastructure and innovation, including a research internship at the Institute for Inclusive Growth and Development (IIGD). Mahamadyaseen is committed to interdisciplinary learning and aims to contribute meaningfully to both scientific and social research through writing and collaborative inquiry.

Sitesh Kumar Singh is an Assistant Professor of Civil and Environmental Engineering at the National University of Science and Technology, Oman. He holds a PhD in Civil Engineering and has over nine years of international academic and research experience. His expertise includes sustainable transportation systems, project management, AI applications in civil engineering, and smart infrastructure. He is a Chartered Engineer (IEI), PMP® certified professional, and Fellow of Advance HE (UK). Dr. Singh holds multiple advanced degrees, including M.Tech, B.Tech, PGDIPR, PGDUPDL, and PGDEd. He has contributed extensively to teaching, curriculum development, and academic leadership, including involvement in international accreditation processes. He has supervised over 39 postgraduate and 2 doctoral students, contributing to academic leadership through workshops, conferences, and editorial roles, and has published over 5 books, 25 book chapters, 58 research articles & conference presentations, and 7 patents.

Varadharajan Bhoopathy is a Professor at Sree Rama Engineering College. He holds a Ph.D. in Information and Communication Engineering from Anna University and has over 16 years of academic experience. Specializing in wireless sensor networks and IoT, he has published numerous research papers and holds several international patents. Dr. Bhoopathy is a recipient of prestigious awards for his contributions to education and research. His commitment to academia continues to drive innovation and excellence in the field of computer science.

Koppala Venugopal, a professor in Management Studies at Aditya Institute of Technology and Management at Srikakulam, A.P. has had an adroit experience of academic, Research and community service. He has contributed 5 Projects, 2 patents, 10 books, 70 publications, 48 paper presentations, 56 resource talks and 12 memberships. He is guiding 4 scholars from various universities. His exceptional interest industry interface has made his efforts dexterous.

Index

U

Y

www.ingramcontent.com/pod-product-compliance
Lightning Source LLC
LaVergne TN
LVHW080905300926
849162LV00003B/560